Developments in Mathematics

Volume 60

The **Developments in Mathematics** (DEVM) book series is devoted to publishing well-written monographs within the broad spectrum of pure and applied mathematics.

Ideally, each book should be self-contained and fairly comprehensive in treating a particular subject. Topics in the forefront of mathematical research that present new results and/or a unique and engaging approach with a potential relationship to other fields are most welcome. High quality edited volumes conveying current state-of-the-art research will occasionally also be considered for publication. The DEVM series appeals to a variety of audiences including researchers, postdocs, and advanced graduate students.

More information about this series at http://www.springer.com/series/5834

Martin Bača • Mirka Miller • Joe Ryan •
Andrea Semaničová-Feňovčíková

Magic and Antimagic Graphs

Attributes, Observations, and Challenges in Graph Labelings

Springer

Martin Bača
Department of Applied Mathematics and Informatics
Technical University
Košice, Slovakia

Mirka Miller
School of Mathematical and Physical Sciences
University of Newcastle
Australia

Department of Mathematics
University of West Bohemia
Pilsen, Czech Republic

Joe Ryan
School of Electrical Engineering and Computing
University of Newcastle
Australia

Andrea Semaničová-Feňovčíková
Department of Applied Mathematics and Informatics
Technical University
Košice, Slovakia

ISSN 1389-2177 ISSN 2197-795X (electronic)
Developments in Mathematics
ISBN 978-3-030-24584-9 ISBN 978-3-030-24582-5 (eBook)
https://doi.org/10.1007/978-3-030-24582-5

Mathematics Subject Classification (2010): 05C78

This Springer imprint is published by the registered company Springer Nature Switzerland AG.
The registered company address is: Gewerbestrasse 11, 6330 Cham, Switzerland

This book is dedicated to Mirka Miller who devoted much time and effort to this work but sadly died before she could see it realized. We miss her dearly.

Preface

Various types of labelings of graphs have been intensively studied by combinatorialists for some time. The notion of magic labeling has its origin in very classical Chinese mathematics although at that time they were looked at as magic designs and patterns such as the magic square. Only recently have these labelings been associated with elements of a graph and investigated using notions and tools of modern graph theory. The effect of this association was the explosion of graph labeling schemes, of which magic and antimagic labelings are among the most abundant.

We have been inspired and motivated by Alison Marr and Wal Wallis who have published the second edition of the book on "Magic Graphs." Their book provides an introduction to the magic-type labelings, introduces the basic terminology, and gives a brief sketch of applications.

Our monograph focuses on variations of magic and antimagic types of labelings. Our primary aim is to present new results, new techniques, and new constructions on studied types of labelings. For several of them, we describe the interrelationship between these labeling schemes.

We have tried to make our explanation clear and relatively simple. However, the reader is assumed to have mathematical foundation to understand constructions and proofs of propositions. This book is especially relevant for senior undergraduate or postgraduate students with an interest in discrete mathematical structures or a major in graph labeling. However, we feel that understanding this work is within the grasp of the mathematically literate and interested layperson.

This book would not have been possible without the dedication, application, and inspiration of hundreds of researchers who have contributed to many of the results presented here. They are too numerous to list, but a skim through the bibliography will identify many of the scholars who have devoted their time and energies to pushing back the boundaries of knowledge in this particular field. To each of them we offer sincere and heartfelt thanks.

Throughout the text we mention Open Problems and Conjectures which arise from the ideas and results considered. Their solution can bring an extra impulse in development and open the door to other results. For convenience these challenges are collected in the final chapter. We feel that this is the chapter that readers will return to in years to come as they search for a new and interesting problem for themselves or to present to their students. Therefore, they are a challenge for everyone who is interested in this monograph.

Košice, Slovakia — Martin Bača
Newcastle, NSW, Australia — Mirka Miller
Pilsen, Czech Republic
Newcastle, NSW, Australia — Joe Ryan
Košice, Slovakia — Andrea Semaničová-Feňovčíková
July 2019

Contents

List of Figures

Chapter 1
Introduction

The area of graph theory has experienced fast development during the last 70 years, and among the huge diversity of concepts that appear while studying this subject, one that has gained a lot of popularity is the concept of labelings of graphs. In the intervening 50 years nearly 200 graph labeling techniques have been studied in over 2000 papers. A dynamic survey of graph labeling by Joseph Gallian [109] provides useful information that has been done for any particular type of labeling.

Graph labelings provide useful mathematical models for a wide range of applications, such as data security, cryptography (secret sharing schemes), astronomy, various coding theory problems, communication networks, mobile telecommunication systems, bioinformatics, and X-ray crystallography. More detailed discussions about applications of graph labelings can be found in Bloom and Golomb's papers [66] and [67].

Many studies in graph labeling refer to Rosa's research in 1967 [224]. Rosa introduced a function f from a set of vertices of a graph G to the set of integers $\{0, 1, \ldots, q\}$, where q is the number of edges in G, so that each edge xy is assigned the label $|f(x)-f(y)|$, with all labels distinct. Rosa called this labeling *β-valuation*. Independently, Golomb [114] studied the same type of labeling and called this labeling *graceful labeling*. The graceful labeling was broadly popularized in a paper by Gardner in 1972 [110], mainly for its connection to the Ringel's conjecture, which asserts that every tree of size q decomposes the complete graph K_{2q+1}. Ringel's conjecture can be derived by Kotzig's graceful conjecture, which asserts that every tree is graceful.

Although Erdős proved in an unpublished paper that almost all graphs are not graceful, many particular families of graphs have been proved to admit graceful labelings. Among the trees known to be graceful are caterpillars [224], trees with at most four end vertices [130], trees with diameter at most five [129], and trees with at most 27 vertices [10].

M. Bača et al., *Magic and Antimagic Graphs*, Developments in Mathematics 60,
https://doi.org/10.1007/978-3-030-24582-5_1

In 1963 Sedláček [230] published a paper about another kind of graph labeling. He called the labeling “magic.” His definition was motivated by the magic square notion in number theory. A *magic labeling* is a function from the set of edges of a graph G into the nonnegative real numbers, so that the sums of the edge labels around any vertex in G are all the same. Firstly, Sedláček established some sufficient conditions [231, 262] for the magicness of the graphs. In 1978 Doob [90] characterized regular magic graphs. The problem of characterizing all magic graphs was solved in 1980s when two different characterizations of all magic graphs were published; Jeurisen's and Jezný-Trenkler's. Jeurisen [152] used forbidden graphs and the cardinality of the neighborhood of independent set to characterize magic graphs. Jezný and Trenkler [153] characterized magic graphs using the separation of edges by a (1–2)-factor. The proofs of both characterizations are constructive, and they present the methods for the construction of magic labeling of the graph if such labeling exists.

Stewart [262] called a magic labeling *supermagic* if the set of edge labels consisted of consecutive integers. Up to now no characterization of all supermagic graphs is known. Only some special classes of the graphs are characterized. We know some necessary and some sufficient conditions for a graph to be supermagic.

Motivated by Sedláček's and Stewart's research, many new related definitions have been proposed and new results have been found. In general, a *graph labeling* is a mapping from elements of a graph (can be vertices, edges, or a combination) to a set of numbers (usually positive integers). If the domain of the mapping is the set of vertices or the set of edges, then the labeling is called *vertex labeling* or *edge labeling*, respectively. If the domain of the mapping is the set of vertices and edges, then the labeling is called *total labeling*. The mapping usually produces partial sums of the labeled elements of the graph. The partial sums will be either a set of *vertex-weights*, obtained for each vertex by adding all the labels of the vertex and its adjacent edges, or a set of *edge-weights*, obtained for each edge by adding the labels of an edge and its endpoints.

One of the situations that we are particularly interested in is when all the edge-weights or all the vertex-weights are the same. In such a case we call the labeled graph *edge-magic* or *vertex-magic*, respectively. Edge-magic and vertex-magic graphs are described in the book by Marr and Wallis [182].

Another situation that is of interest is when all the edge-weights or all the vertex-weights are different. In such a case we call the labeled graph *edge-antimagic* or *vertex-antimagic*, respectively. The study of these graphs was motivated by Hartsfield and Ringel [125], who considered labeling uniquely the edges of a graph containing q edges using the integers $1, 2, \dots, q$, and evaluating partial sums of labels at the vertices of the graph. If all the vertex-weights are different, then they call the graph *antimagic*.

Among the graphs known to be antimagic are paths, cycles, complete graphs, and wheels. It is easy to see that K_2 is not antimagic. In fact, Hartsfield and Ringel [125] put forth the following conjectures.

Conjecture 1.1 ([125]) Every connected graph other than K_2 is antimagic.

Conjecture 1.2 ([125]) Every tree other than K_2 is antimagic.

Alon et al. [15] used several probabilistic tools and some techniques from analytic number theory to show that this conjecture is true for all graphs having minimum degree $\Omega(\log |V(G)|)$.

The main aim of this monograph is to extend the knowledge of magic-type and antimagic-type of labelings. The second chapter summarizes known results in magic and supermagic graphs. At the beginning there is a historical survey of magic squares which are closely related to magic graphs. Then two different characterizations of all magic graphs and a characterization of regular magic graphs are presented. The properties of vertex-magic total and edge-magic total labelings are studied in Chaps. 3 and 4. These chapters are an extension of the book of Marr and Wallis [182] and the book of López and Muntaner-Batle [176]. The main topics of the monograph, vertex-antimagic total and edge-antimagic total labelings, are presented in Chaps. 5 and 6. Chapter 7 describes the construction of α-trees and also the connection between α-labeling and edge-antimagic labeling.

The monograph closes with an Index, in which the convention has been followed of italicizing the entries where a definition occurs.

We hope that the amount of figures in the monograph will help the reader to easily follow the text and they will contribute to the better understanding of the studied theme.

Almost every book contains errors, and this one will hardly be an exception. Please let us know about any errors and imperfections you find.

Acknowledgements We are indebted to the following friends and collaborators for many enjoyable and valuable discussions and help with this project: Ali Ahmad, Gohar Ali, Kashif Ali, S. Arumugam, Faraha Ashraf, Camino Balbuena, Christian Barrientos, Ewan Barker, Yasir Bashir, Edy Tri Baskoro, Francois Bertault, Gary Bloom (R.I.P.), Novi Herawati Bong, Ljiljana Brankovic, Yus M. Cholily, Dafik, Kinkar Chandra Das, Dalibor Fronček, Muhammad Irfan, Jaroslav Ivančo, Stanislav Jendroľ, Petr Kovář, Tereza Kovářová, Marcela Lascsáková, Anna S. Lladó, Yuqing Lin, Susana C. López, Jim A. MacDougall, Francesc A. Muntaner-Batle, Muthali Murugan, Muhammad F. Nadeem, Akito Oshima, Ali Ovais, Oudone Phanalasy, Zdeněk Ryjáček, Lienne Rylands, Muhammad K. Shafiq, Ayesha Shabbir, Muhammad K. Siddiqui, Denny R. Silaban, Anita A. Sillasen, Rinovia Simanjuntak, Slamin, Kiki A. Sugeng, Michal Tkáč, Marián Trenkler, Muhammad A. Umar, Tao-Ming Wang, Wal D. Wallis, and Maged Z. Youssef.

Finally, we are grateful for the constant support of Lynn Braddon and Thanikachalam Sabarigirinathan at Springer.

Chapter 2
Magic and Supermagic Graphs

2.1 Definition of Magic and Supermagic Labeling

In this chapter we will deal with magic and supermagic labeling. These labelings are special types of vertex-magic edge labelings.

Let a graph G and a mapping f from the edge set $E(G)$ of G into positive integers be given. The *index-mapping* of f is the mapping $f^{\star}$ from the vertex set $V(G)$ into positive integers defined by

$$f^{\star}(v) = \sum_{uv \in E(G)} f(uv), \quad \text{for every } v \in V(G). \tag{2.1}$$

An injective mapping f from $E(G)$ into the positive integers is called a *magic labeling* of G for an *index* λ if its index-mapping $f^{\star}$ satisfies

$$f^{\star}(v) = \lambda, \quad \text{for every } v \in V(G). \tag{2.2}$$

Jeurissen [152] called such a labeling *positive*.

A labeling is called *semi-magic* if it is a not injective mapping from $E(G)$ to the positive integers and satisfies Condition (2.2).

The value $f^{\star}(v)$ is also called the *vertex-weight* of the vertex v. According to this notation, the magic labeling of G is an injective mapping from $E(G)$ into positive integers such that the vertex-weight is the same at every vertex of G.

In other words, the labeling f is called a magic labeling of a graph G if it is an injection from the edge set $E(G)$ into positive integers such that the sum of the labels of all the edges incident to a given vertex is independent of this vertex. A graph that admits a magic labeling is called *magic*. The concept of magic labeling was introduced by Sedláček [230].

M. Bača et al., *Magic and Antimagic Graphs*, Developments in Mathematics 60,
https://doi.org/10.1007/978-3-030-24582-5_2

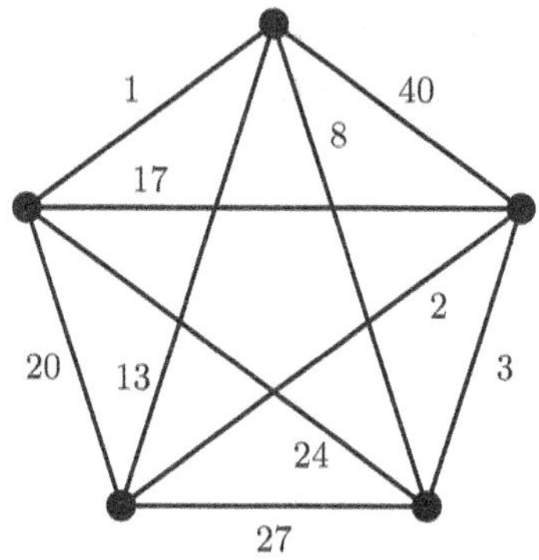

Fig. 2.1 Magic labeling of K_5 with magic index $\lambda = 62$

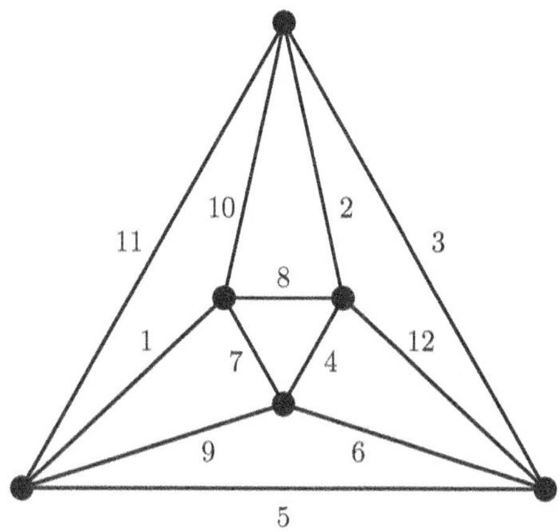

Fig. 2.2 Supermagic labeling of a graph with magic index $\lambda = 26$

Stewart [262] defined a supermagic labeling.

A magic labeling is called *supermagic* if the set of all the labels of the edges $\{f(e) : e \in E(G)\}$ consists of consecutive positive integers. We say that a graph G is *supermagic* if and only if there exists a supermagic labeling of G.

However, sometimes the same terminology has a different meaning. Some authors call a graph supermagic if the edges can be labeled with numbers $\{1, 2, \ldots, |E(G|\}$, see [251] and [125]. Note that for regular graphs these definitions are equivalent, see [136].

Figures 2.1 and 2.2 illustrate magic and supermagic graphs with corresponding labelings.

2.2 Magic Squares

A *magic square* $\mathcal{M}_n$ of order n is a $n \times n$ array of integers $1, 2, \ldots, n^2$ such that the sum of numbers along any row, column, and main diagonals is a fixed constant. It is easy to see that the constant is equal to

$$\frac{1+2+\cdots+n^2}{n} = \frac{n(n^2+1)}{2},$$

because the sum of all used numbers $1, 2, \ldots, n^2$ is equal to this constant multiplied by the numbers of columns (rows).

In 1963, the Czech mathematician Sedláček [230] pointed out the correspondence between a magic square $\mathcal{M}_n$ of order n and magic labeling of a complete bipartite graph $K_{n,n}$. He found that if we label every edge $u_i v_j$, $i, j = 1, 2, \ldots, n$ of $K_{n,n}$ with the number from ith row and jth column of the magic square $\mathcal{M}_n$, we obtain a supermagic labeling of $K_{n,n}$. As the magic square $\mathcal{M}_n$ exists for every positive integer n, $n \not= 2$, the graph $K_{n,n}$ is supermagic for every n, $n \not= 2$, see also [262].

Figure 2.3 illustrates a magic square $\mathcal{M}_3$ and the corresponding complete bipartite graph $K_{3,3}$ with its supermagic labeling.

Note [124] that every supermagic labeling of $K_{n,n}$ does not correspond to a magic square $\mathcal{M}_n$. Figure 2.4 depicts an array corresponding to the supermagic labeling of the complete bipartite graph $K_{3,3}$. This array is not a magic square, although the sums along every column and row are the same, equal to 15. However, the sums along the main diagonals are 12 and 24.

The origin of magic squares can be found in Chinese literature in 2800 BC in the legend Lo Shu – the legend of the giant tortoise, see [174]. On the shell of the tortoise that emerged from the flooding river Lo, a pattern with specially arranged dots forming the Lo Shu square was depicted – the magic square of order 3 with the constant of 15, see Fig. 2.5. The Chinese solar year consists of 24 cycles and each

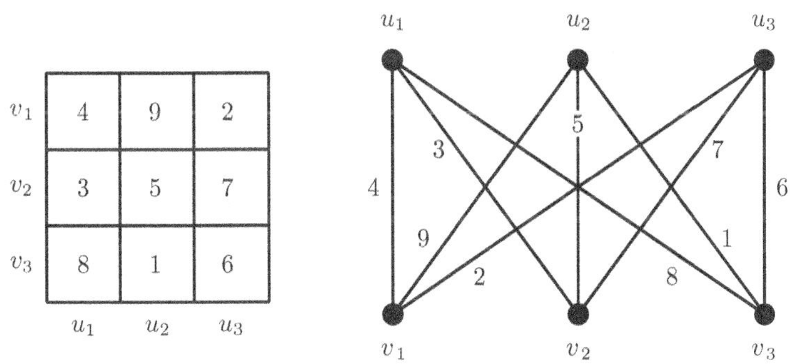

Fig. 2.3 Magic square $\mathcal{M}_3$ and the corresponding supermagic graph $K_{3,3}$

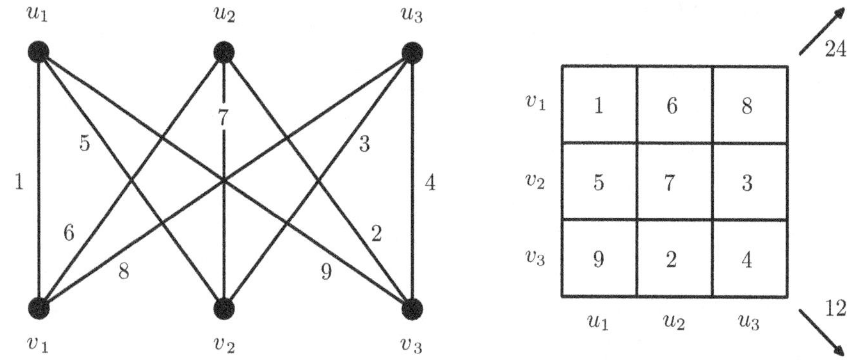

Fig. 2.4 Supermagic labeling of graph $K_{3,3}$ and the corresponding 3×3 array

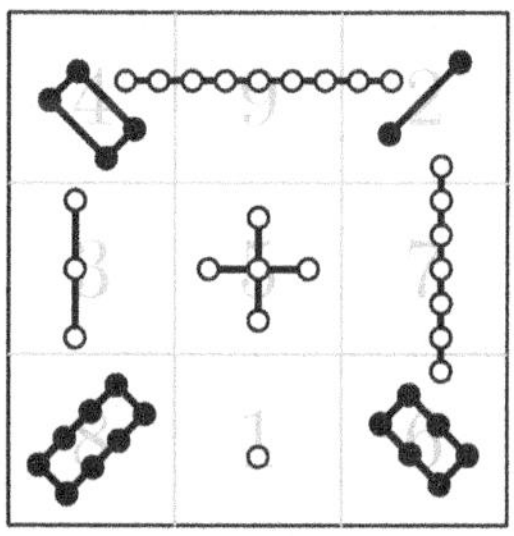

Fig. 2.5 Lo Shu square

of these cycles has 15 days. In the legend the mystic diagram on the tortoise shell was used to control the river and so helped people to fight the flood.

In the following period magic squares were often associated with mysticism and religion. They were used in astrology to produce horoscopes and talismans to provide people with health, long life, and happiness.

Magic squares spread into Ancient Greece, India, Egypt, and Arabia—to every developed civilization of the ancient world. They can be also found in Chinese mathematics of the thirteenth and Japanese mathematics of the seventeenth and eighteenth centuries.

The fascination of mankind with the magic squares is also evident from their occurrence in art. In 1514 German painter, engraver and mathematician Albrecht Dürer engraved a magic square of order 4 in his engraving *Melencolia I*, see Fig. 2.6, [92]. Very famous is also the "magic square" on the *Passion facade of the Sagrada Familia*, the church in Barcelona, Spain, see Fig. 2.7, [111]. Behind the sculpture *The Judas treason* from a Catalan sculptor Josep Maria Subirachs is the magic square of order 4. While the magic constant of the magic square of order 4 is 34, the pattern in this Passion is modified to have the magic constant 33—the age of Jesus

Fig. 2.6 Engraving Melencolia I from Albrecht Dürer and the detail of the magic square from the engraving

Fig. 2.7 Judas treason from Josep Maria Subirachs, Passion facade of Sagrada Familia, Barcelona, Spain

Christ at the crucifixion. Note that to achieve this the numbers 12 and 16 are not used; however, the numbers 10 and 14 are used twice.

A *semi-magic square* of order n is a $n \times n$ array of nonnegative integers such that the sum of numbers along any row and column is a fixed constant. The semi-magic square is also mentioned in Johann Wolfgang von Goethe's Faust, [113]. In the first part of the tragedy, part of the Witch's kitchen is hidden:

This you must ken!
From one make ten,
And two let be,
Make even three,
Then rich you'll be.
Skip o'er the four!
From five and six,
The Witch's tricks,
Make seven and eight,
'Tis finished straight;
And nine is one,
And ten is none,
That is the witch's one-time-one!

Figure 2.8 illustrates the semi-magic square mentioned in Goethe's Faust.

There exists a magic square for every positive integer n, $n \neq 2$. Nowadays we know several methods of constructing magic squares, but the standard way is to follow certain configurations which generate regular patterns. The book of Andrews [17] is probably the definitive work on magic squares. It shows how to construct normal magic squares as well as the many variations that exist. The book is highly technical and of more interest to the serious mathematician than the average magician.

It is an unsolved problem to determine the number of non-isomorphic normal magic squares of an arbitrary order. For $n = 3$, there is only one normal magic square. The 880 normal magic squares of order $n = 4$ were enumerated by Frénicle de Bessy (1693) and are illustrated in [57]. The number of normal magic squares of order $n = 5$ is 275 305 224; this was computed by Schroeppel in 1973, see [57]. The number of normal magic squares of order $n = 6$ is not known, but Pinn and Wieczerkowski [211] estimated it to be $(1.7745 \pm 0.0016) \times 10^{19}$, using Monte Carlo simulation and methods from statistical mechanics. Results of historical and computer enumeration of the number of non-isomorphic normal magic squares can

Fig. 2.8 Semi-magic square hidden in the fragment from Goethe's Faust

10	2	3
0	7	8
5	6	4

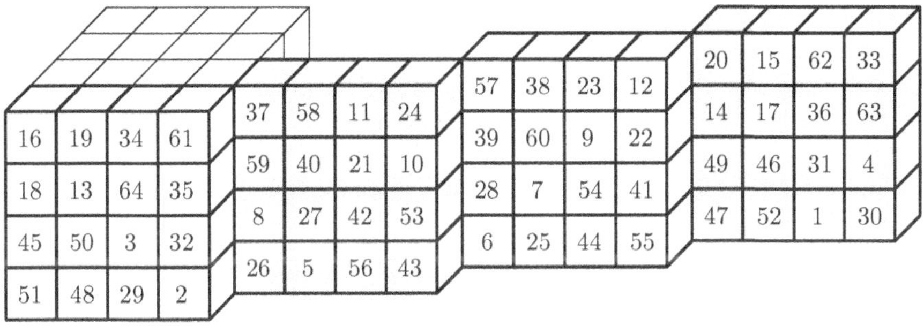

Fig. 2.9 Magic cube of order 4

be found in [283]. Abiyev [2] described a general method for generating magic squares of any order. By Abiyev's method magic squares of any order can be obtained for any type of numbers including complex numbers and magic squares generated by his method show some very interesting symmetrical properties, which are not possible to obtain via other techniques. These properties of Abiyev's magic squares facilitate their applications in cryptology, physics, mathematics, and genetics, for example see [3].

The natural generalization of magic squares is magic cubes and magic hypercubes. A *magic cube* of order n is a 3-dimensional matrix of order n, i.e., $n \times n \times n$ array, containing integers $1, 2, \ldots, n^3$ such that the sum of the numbers along any row, column, pillar, and the four main space diagonals is a fixed constant equal to $n(n^3 + 1)/2$. Trenkler [278] proved that a magic cube of order n exists for every positive integer $n \neq 2$.

Figure 2.9 illustrates a magic cube of order 4.

A *magic p-dimensional cube* of order n, denoted by M_n^p, is a p-dimensional matrix of order n

$$M_n^p = |m(i_1, i_2, \ldots, i_p) : 1 \leq i_1, i_2, \ldots, i_p \leq n|,$$

containing integers $1, 2, \ldots, n^p$ such that the sum of the numbers along every row and 2^{p-1} main space diagonals is equal to the number $n(n^p + 1)/2$. The row of M_n^p is an n-tuple of elements $m(i_1, i_2, \ldots, i_p)$ which have identical coordinates at $p - 1$ places. A diagonal of M_n^p is an n-tuple $\{m(x, i_2, \ldots, i_p) : x = 1, 2, \ldots, n, i_j = x$ or $i_j = 2^p + 1 - x$ for all $2 \leq j \leq p\}$. In [280] Trenkler proved that a magic p-dimensional cube of order n exists if and only if $p \geq 2$ and $n \neq 2$ or $p = 1$.

2.3 Characterization of Magic Graphs

It is easy to see that if G is a magic graph, then G can contain only one edge with end vertex of degree 1 and G cannot contain an edge uv with both end vertices of degree 2, i.e., $\deg(u) = \deg(v) = 2$. In both cases the magic constant λ constrains the labels of these edges. In the first case, the pending edge must be labeled by the number λ, and in the second case the edges adjacent with the edge in question must have the same label that is equal to $\lambda - f(uv)$. Thus, in particular, the 1-regular graph is magic if and only if it is isomorphic to K_2 and there exist no 2-regular magic graphs.

In 1975 Doob [90] published the following characterization of regular magic graphs. He proved

Theorem 2.3.1 ([90]) *Let G be a regular graph of degree $r \geq 3$. Then G is magic unless it has a connected component with one of the following properties.*

(i) There exist two edges whose deletion disconnects the component leaving a new component which is bipartite with two vertices of degree $r - 1$.
(ii) The vertices of the component can be partitioned into two sets V_1 and V_2 such that one edge has both endpoints in V_1, one has both endpoints in V_2, and all the other edges have one endpoint in V_1 and one in V_2.

While the previous theorem is a characterization of regular magic graphs, characterizing graph theoretic properties at first seems to be awkward. These properties, however, are often easy to apply and can be related to more familiar properties in many cases. One such property is the edge connectivity. Note that a graph has *edge connectivity* k if it is necessary to remove at least k edges to disconnect the graph. For bipartite regular graphs Doob [90] proved

Theorem 2.3.2 ([90]) *Let G be a regular bipartite graph. Then G is magic if and only if its edge connectivity is not* 2.

This characterization is based on the nonappearance of certain bipartite subgraphs. We say that a graph is *separated by an even cycle* if for any pairs of edges there is an even cycle that contains exactly one of them.

Theorem 2.3.3 ([90]) *Let G be a regular graph with degree $r > 4$. Then G is magic if and only if G is separable by even cycles.*

A graph H contained in a graph G is called a *subgraph* of G. A subgraph H of G is called a *spanning subgraph* of G if $V(H) = V(G)$. Alternatively, a spanning subgraph of G is also called a *factor* of G. A *k-factor* is a factor that is *k-regular*, that is, every vertex in the factor has degree k. A factor F is a (1-2)-*factor* of G if each of its components is a regular graph of degree one or two. By the symbol F^1, respectively F^2, we denote the subgraph of F which consists of all isolated edges, respectively, of all cycles of F, and the necessary vertices.

A (1-2)-factor *separates* the edges e_1 and e_2, if at least one of them belongs to F and neither F^1 nor F^2 contains both e_1 and e_2.

A characterization of all magic graphs using the notion of separating edges by a (1-2)-factor has been given by Jezný and Trenkler [153].

Theorem 2.3.4 ([153]) *A graph G is magic if and only if both the following statements hold.*

(i) Every edge of G belongs to a (1-2)-factor.
(ii) Every pair of edges e_1, e_2 is separated by a (1-2)-factor.

Independent of the results of Jezný and Trenkler [153], Jeurissen [151, 152] published a different characterization of all magic graphs.

If S is a set of vertices of a graph G, we denote by $\Gamma(S)$ the set of vertices of G adjacent to vertices of S.

Recall that a graph G is called *bipartite* if its vertex set can be partitioned into disjoint parts V_1 and V_2 such that every edge of G joins vertices of different parts; thus there are no edges between vertices in the same partite set. A bipartite graph is *balanced* if the two partite sets each contain the same number of vertices, i.e., $|V_1| = |V_2|$. Jeurissen [152] characterized connected magic bipartite graphs.

Theorem 2.3.5 ([152]) *A connected bipartite graph $G = V_1V_2$ is magic if and only if the following statements hold.*

(i) G is balanced.
(ii) $|\Gamma(S)| > |S|$ for all $S \subset V_1$, $\emptyset \neq S$.
(iii) G does not consist of two disjoint balanced bipartite graphs connected by a cross-bridge (see Fig. 2.10).

For connected magic non-bipartite graphs Jeurissen [152] proved

Theorem 2.3.6 ([152]) *A connected non-bipartite graph G with vertex set V is magic if and only if the following statements hold.*

(i) $|\Gamma(S)| > |S|$, for all $S \subset V$, $\emptyset \neq S$.
(ii) G is not a balanced graph V_1V_2 containing one handle at V_1 and one at V_2 (see Fig. 2.11).

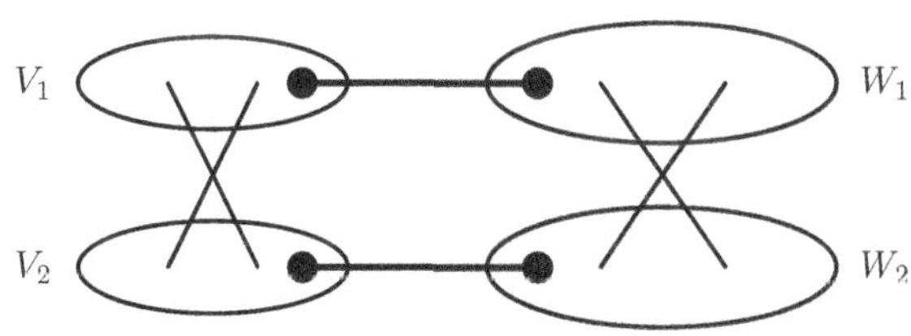

Fig. 2.10 Forbidden graph I

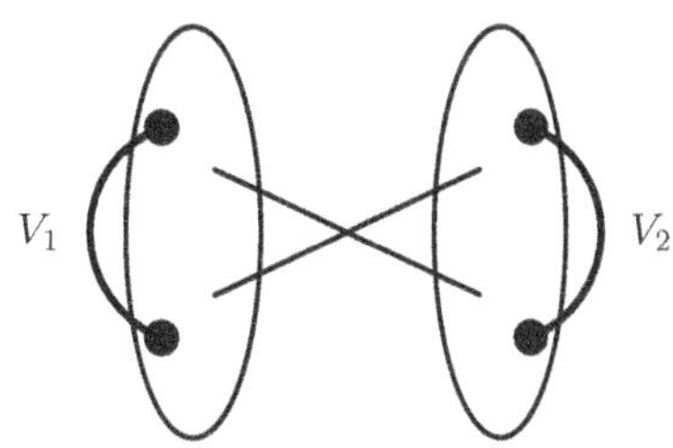

Fig. 2.11 Forbidden graph II of type $\mathcal{A}$

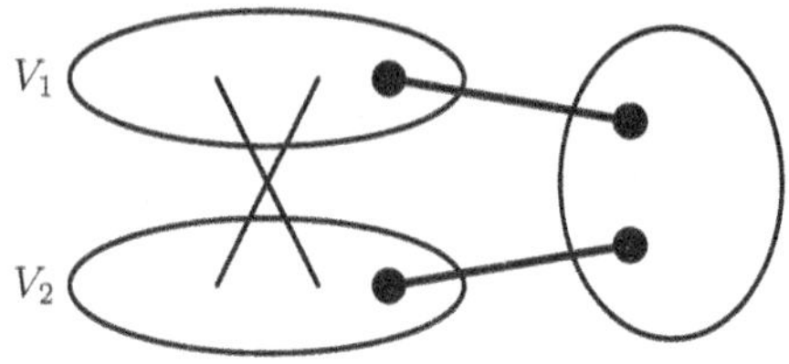

Fig. 2.12 Forbidden graph III of type $\mathcal{B}$

(iii) G is not a graph consisting of a balanced graph V_1V_2 connected by one edge from V_1 and one from V_2 to another (possibly disconnected) graph (see Fig. 2.12).

Only for the purpose of proving some results presented in this book we will call the graphs considered in Theorem 2.3.6 part (ii) as the graphs of type $\mathcal{A}$ and part (iii) as the graphs of type $\mathcal{B}$.

The proofs of the characterization by Jezný and Trenkler, as well as the one by Jeurissen, are constructive and they allow us to find a magic labeling of a graph G if such a labeling exists. We present an algorithm for finding a magic labeling of a graph G based on the proof given by Jezný and Trenkler, see [153].

In the following algorithm keep in mind that if G is a magic graph, then according to Theorem 2.3.4, every edge of G belongs to a (1-2)-factor, and every pair of edges e_1, e_2 is separated by a (1-2)-factor.

ALGORITHM for finding a magic labeling of a graph G
Input: A graph G.

Step 1. Let f be a labeling of edges of G with nonnegative integers such that the sum of the labels incident with each vertex is the same. Note that for every graph such a labeling exists, for example, $f(e) = 0$, for every $e \in E(G)$.

Step 2. If $f(e^*) = 0$, then let F be a (1-2)-factor that contains the edge e^*, i.e., $e^* \in E(F)$. We define a new labeling h, $h : E(G) \to \mathbb{Z}$ such that

$$h(e) = \begin{cases} f(e) + 2m & \text{if } e \in E(F^1) \\ f(e) + m & \text{if } e \in E(F^2) \\ f(e) & \text{if } e \notin E(F), \end{cases} \tag{2.3}$$

where $m = \max\{f(e) : e \in E(G)\} + 1$.
It is easy to see that the label of the edge e^* under the labeling h is positive. Furthermore, the vertex-weight of every vertex $v \in V(G)$ under the labeling h is

$$h^\star(v) = f^\star(v) + 2m,$$

as F is a spanning subgraph.
Thus, after a finite number of steps, every edge label will be positive and the vertex-weights will be the same.

Step 3. If there exist two edges e_1, e_2 with the same labels, then we find a (1-2)-factor F that separates them. Again using labeling h defined in (2.3) we obtain

$$h(e_1) \neq h(e_2)$$

and the vertex-weights are

$$h^\star(v) = f^\star(v) + 2m, \quad \text{for every } v \in V(G).$$

Thus, after a finite number of steps all edges will be labeled with different labels and the vertex-weights will be the same, which means that we obtain a magic labeling of G.

Output: A magic labeling of a graph G, if G is magic.

The previous algorithm allows us to find a magic labeling of a graph if such a labeling exists. However, the magic index of this labeling is very large. Thus we can state the following problem, see [152].

Open Problem 2.3.1 ([152]) Find the smallest magic index of a magic graph.

Jeurissen [152] proved that the smallest magic index of the Petersen graph is 26, that of $K_{3,3}$ is 15, that of K_5 is 20, and that of K_n, $n > 5$, $n \not\equiv 0 \pmod 4$ is $(n-2)(n-1)(n+1)/4$.

In 1983 Derings and Hünten [87] published another characterization of magic graphs.

Doob [90] proved the following result for disconnected regular graphs.

Theorem 2.3.7 ([90]) *Let G be a regular graph of degree $r \geq 3$ and $G_1, G_2, \ldots, G_n$ be the connected components of G. Then G is magic if and only if G_i is magic, $i = 1, 2, \ldots, n$.*

An analogous statement is not true for non-regular graph. Jeurissen [152] proved

Theorem 2.3.8 ([152]) *Let G be a component of a magic graph H. Let e be an edge of G such that if f_1 and f_2 are magic labelings of H for the same index, then $f_1(e) = f_2(e)$. Then one of the following statements must hold.*

(i) G is a one-edge graph.
(ii) $G - \{e\}$ is a bipartite graph V_1V_2 with $|V_1| = |V_2| + 1$, and e is a handle attached to V_1, (see Fig. 2.13).

This means that a magic graph can contain at most one component isomorphic to these graphs.

Theorem 2.3.9 ([152]) *A graph is magic if and only if each of its components is magic and at most one of them is a one-edge graph and at most one of them is a bipartite graph with a handle.*

Or, in the terminology of Trenkler,

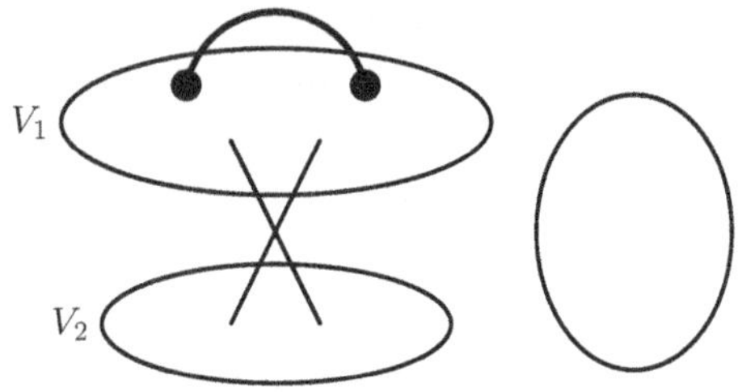

Fig. 2.13 A handle attached to V_1 in a bipartite graph $V_1 V_2$

Theorem 2.3.10 ([277]) *Let $G_1, G_2, \ldots, G_n$ be connected components of G. If G_i is magic, $i = 1, 2, \ldots, n$, and at most one of G_i has one edge which is contained in the cyclic parts of all its $(1$-$2)$-factors, then G is magic.*

2.4 Generalization of Magic Labeling

To each vertex v of a graph G let there be associated real number $\rho(v)$. If there exists a labeling f from the set of edges into positive real numbers such that

$$\sum_{uv \in E(G)} f(uv) = \rho(v), \quad \text{for every } v \in V(G), \tag{2.4}$$

then the labeling f is called a *ρ-positive labeling*. Moreover, if the labeling f is an injection, then it is called a *ρ-magic labeling*. A graph that admits a ρ-magic labeling (ρ-positive labeling) is called a *ρ-magic graph* (*ρ-positive graph*).

The motivation for the study of ρ-magic graphs was given by Doob [89].

By a *generalized even cycle D* we understand an even cycle C or two odd cycles C_1 and C_2 with one common vertex or two odd cycles C_1 and C_2 without common vertices joined by a path. A factor in which no component contains a generalized even cycle is called an *X-factor*. We say that a ρ-positive X-factor F of a ρ-positive graph G separates its edges e_1 and e_2 if at least one of them belongs to F and $f(e_1) \neq f(e_2)$, for some ρ-magic labeling f of G.

Šándorová and Trenkler [228] give a characterization of ρ-magic graphs.

Theorem 2.4.1 ([228]) *A graph G is ρ-magic if and only if every edge of G belongs to a ρ-positive X-factor and every two edges e_1, e_2 are separated by a ρ-positive X-factor.*

2.5 Conditions for a Graph to be Supermagic

Up to now, there is no known characterization of all supermagic graphs. Only some special classes of supermagic graphs have been characterized. Some necessary and some sufficient conditions are known for a graph to be supermagic. Because of the simpler structure of regular graphs, most of the published results are about supermagic regular graphs.

Let G be a supermagic graph of size m. Then G admits a supermagic labeling f,

$$f : E(G) \to \{a, a+1, \dots, a+m-1\}$$

for an index λ, where a is a positive integer.

According to the Conditions (2.1) and (2.2),

$$\begin{aligned} n\lambda = & \sum_{v \in V(G)} \sum_{uv \in E(G)} f(uv) = 2 \sum_{uv \in E(G)} f(uv) \\ =& 2\,(a + (a+1) \cdots + (a+m-1)) = (2a+m-1)m, \end{aligned} \tag{2.5}$$

and thus the magic index is

$$\lambda = \frac{(2a+m-1)m}{n}. \tag{2.6}$$

As the index λ is required to be a positive integer, using divisibility we can immediately exclude some graphs that are not supermagic. In [91] it was proved

Theorem 2.5.1 ([91]) *Let d be the greatest common divisor of integers n and m, and let $n_1 = n/d$. If n_1 and m are both even, then there exists no supermagic graph of order n and size m.*

Proof Let d denote the greatest common divisor of n and m and let $n_1 = n/d$, $m_1 = m/d$. Suppose that G is a supermagic graph of order n and size m. Then G admits a supermagic labeling $f : E(G) \to \{a, a+1, \dots, a+m-1\}$ for an index λ. By (2.6),

$$\lambda = \frac{(2a+m-1)m}{n} = \frac{(2a+m-1)m_1}{n_1}.$$

As m and n_1 are both even, the index $\lambda = (2a+m-1)m_1/n_1$ is not an integer, a contradiction. □

The Expression (2.6) for the magic index is simpler for regular supermagic graphs. There are two reasons for this. The first is that the number of edges of an r-regular graph of order n is $m = rn/2$. The second reason is that if G is an r-regular supermagic graph, then there exists a supermagic labeling of G which uses the labels $1, 2, \dots, rn/2$, i.e., the first $rn/2$ positive integers. Ivančo [136] proved that it is possible to label the edges of a supermagic regular graph with a supermagic labeling using consecutive positive integers with an arbitrary initial term a.

Thus according to this simplification we get for the magic index of the r-regular supermagic graphs

$$\lambda = \frac{r}{2}\left(1 + \frac{rn}{2}\right). \tag{2.7}$$

Using this formula and the divisibility, Ivančo [136] proved necessary conditions for the existence of a supermagic regular graph.

Theorem 2.5.2 ([136]) *Let G be an r-regular supermagic graph. Then the following statements hold.*

(i) If $r \equiv 1 \pmod 2$, then $|V(G)| \equiv 2 \pmod 4$.
(ii) If $r \equiv 2 \pmod 4$ and $|V(G)| \equiv 0 \pmod 2$, then G contains no component of an odd order.
(iii) If $|V(G)| > 2$, then $r > 2$.

Next we will present some sufficient conditions for a graph to be supermagic. They are based on the decomposition of graph G into factors, the spanning subgraphs of G, with some special properties.

Hartsfield and Ringel dealt with bipartite graphs decomposable into Hamilton cycles. They proved

Theorem 2.5.3 ([125]) *If a bipartite graph G is decomposable into two Hamilton cycles, then G is supermagic.*

It is easy to generalize this theorem.

Theorem 2.5.4 ([125]) *If a bipartite graph G is decomposable into even number of Hamilton cycles, then G is supermagic.*

Theorem 2.5.5 ([125]) *If a graph G is decomposable into two supermagic factors H_1 and H_2, with H_1 regular, then G is supermagic.*

Ivančo studied supermagic graphs decomposable into regular supermagic factors. He showed

Theorem 2.5.6 ([136]) *] Let $F_1, F_2, \ldots, F_k$ be mutually edge-disjoint regular supermagic factors of a graph G which form its decomposition. Then G is supermagic.*

Additionally, in this paper Ivančo also deals with the supermagicness of disjoint copies of supermagic graphs. The union of $m \geq 1$ disjoint copies of a graph G is denoted by mG.

Theorem 2.5.7 ([136]) *Let G be a supermagic graph decomposable into k pairwise edge-disjoint δ-regular factors. Then the following statements hold.*

(i) If k is even, then mG is supermagic, for every positive integer m.
(ii) If k is odd, then mG is supermagic, for every odd positive integer m.

A similar result was proved by Kovář [165]. Let us recall that in a proper edge coloring no two adjacent edges are assigned the same color.

Theorem 2.5.8 ([165]) *Let r be an integer, $r \geq 3$. Let G be an r-regular graph with a proper edge coloring, which has an supermagic labeling λ.*

(i) If r is odd, then mG has a supermagic labeling whenever m is an odd positive integer.
(ii) If r is even, then mG has a supermagic labeling for every positive integer m.

2.6 Number of Edges in Magic and Supermagic Graphs

Many graph properties can be described using the connections between the order, the size, the minimum degree, and the maximum degree of a graph. We can ask whether, using the connection between the size and the order, it is possible to decide whether some graph is or is not magic, or supermagic, respectively.

2.6.1 Magic Graphs

Doob [90] described regular graphs with large degree as being magic.

Theorem 2.6.1 ([90]) *Let G be a regular graph with degree $r > 4$ and n vertices. Then G is magic if $r > n/2$.*

A similar result holds for the number of edges.

Theorem 2.6.2 ([90]) *Let G be a regular graph with degree $r > 4$ and n vertices. Then G is magic if $|E(G)| > (n/2)^2$.*

In [279] Trenkler established a condition for the number of edges in a connected magic graph. He proved

Theorem 2.6.3 ([279]) *A connected magic graph with n vertices and m edges exists if and only if $n = 2$ and $m = 1$ or $n \geq 5$ and $5n/4 < m \leq n(n-1)/2$.*

Moreover, Trenkler [279] describes a construction of magic graphs of order n with a given number of edges.

For the number of edges in a magic graph we have

Theorem 2.6.4 ([91]) *A magic graph of order n and size m exists if and only if $n = 2$ and $m = 1$ or $n \in \{5, 6\}$ and $5n/4 < m \leq n(n-1)/2$ or $n \geq 7$ and $(5n-6)/4 < m \leq n(n-1)/2$. Moreover, any magic graph with at most $5n/4$ edges contains a component isomorphic to K_2.*

The previous assertions imply the following interpolation theorem.

Theorem 2.6.5 ([91]) *Let G_1 and G_2 be magic graphs of order n. Then there exists a magic graph of order n and size ε for each integer ε satisfying $|E(G_1)| \leq \varepsilon \leq |E(G_2)|$.*

Similar results are not valid for supermagic graphs. In [91] Drajnová, Ivančo, and Semaničová formulated a necessary condition for the existence of supermagic graphs, see Theorem 2.5.1. Using this theorem, we obtain, for example, that there exists no supermagic graph of order 8 and size $\varepsilon \equiv 2, 4, 6 \pmod 8$ (i.e., with 10, 12, 14, 18, 20, 22, 26, 28 edges). Theorem 2.5.1 suggests that the problem of characterizing the number of edges in supermagic graphs seems to be very difficult.

2.6.2 Supermagic Graphs

Let $M(n)$ ($m(n)$) denote the maximum (minimum) number of edges in a supermagic graph of order n. Evidently, $M(n)$ and $m(n)$ are not defined for $n = 1, 3, 4$ and $M(2) = m(2) = 1$. Stewart [263] characterized supermagic complete graphs K_n.

Theorem 2.6.6 ([263]) *A complete graph K_n of order n is supermagic if and only if $n = 2$ or $5 < n \not\equiv 0 \pmod 4$.*

According to Stewart's results, $M(n)$ for $n = 2$ and $5 < n \not\equiv 0 \pmod 4$ is equal to the number of edges in the complete graph K_n. In [91] Drajnová, Ivančo, and Semaničová proved that for $n \equiv 0 \pmod 4$ the value of $M(n)$ is equal to the number of edges of the complete graph with one edge deleted, $K_n - \{e\}$. Moreover, they proved the general result that by deleting an edge from the complete graph K_n, $n \geq 6$, we obtain a supermagic graph.

Theorem 2.6.7 ([91]) *For every positive integer $n \geq 6$, the complete graph K_n without an edge is supermagic.*

Proof We will consider the following cases.

Case A: $6 \leq n \not\equiv 0 \pmod 4$ By Theorem 2.6.6, the complete graph K_n is supermagic, thus there exists a supermagic labeling $f : E(K_n) \to \{1, 2, \dots, n(n-1)/2\}$ for an index λ. Let $\hat{e}$ be an edge of K_n such that $f(\hat{e}) = 1$. We define a labeling $g : E(K_n - \hat{e}) \to \{1, 2, \dots, n(n-1)/2 - 1\}$ by

$$g(e) = f(e) - 1, \quad \text{for every } e \in E(K_n - \hat{e}).$$

Since K_n is an $(n-1)$-regular graph, we have

$$g^\star(v) = f^\star(v) - (n-1), \quad \text{for every } v \in V(K_n).$$

Therefore, g is a supermagic labeling of $K_n - \hat{e}$.

Case B: $8 \leq n \equiv 0 \pmod 4$ Let $\hat{e}$ be an arbitrary edge of the complete graph K_n. Denote the vertices of K_n by $v_1, v_2, \dots, v_n$ in such a way that $\hat{e} = v_{n-1}v_n$.

Let G be a subgraph of K_n induced by the set $\{v_1, v_2, \dots, v_{n-2}\}$. The graph G is isomorphic to K_{n-2}, and by Theorem 2.6.6, there exists a supermagic labeling f

from $E(G)$ into $\{1, 2, \ldots, (n-2)(n-3)/2\}$. Clearly, $f^{\star}(v_i) = ((n-2)(n-3)/2 + 1)(n-3)/2$, for all i, $1 \le i \le n-2$.

Let $a = (n^3 - 6n^2 + 7n + 4)/4$ be a positive integer and define a mapping $g : E(K_n - \hat{e}) \to \{a, \ldots, a + n(n-1)/2 - 2\}$ by

$$g(v_i v_j) = \begin{cases} a - 1 + f(v_i v_j) & \text{for } 1 \le j \le n-2 \text{ and } 1 \le i \le n-2 \\ a + \frac{(n-2)(n-3)}{2} - 1 + i & \text{for } j = n-1 \text{ and } n-8 \ge i \equiv 0, 3 \pmod 4 \\ & \text{or } j = n-1 \text{ and } i = n-7, n-5 \\ & \text{or } j = n \text{ and } n-8 \ge i \equiv 1, 2 \pmod 4 \\ & \text{or } j = n \text{ and } i = n-6, n-4, n-3, n-2 \\ a + \frac{(n-2)(n+1)}{2} - i & \text{otherwise.} \end{cases}$$

It is easy to see that the mapping g is a bijection and for its index-mapping we get

$$g^{\star}(v_i) = \frac{1}{4}(n^4 - 6n^3 + 9n^2 + 4n - 12), \quad \text{for } 1 \le i \le n.$$

Thus, g is a supermagic labeling and $K_n - \hat{e}$ is a supermagic graph. □

Trenkler [279] proved

Theorem 2.6.8 ([279]) *In a connected magic graph of order at least* 5 *the minimum degree is greater than* 2 *and the number of vertices of degree* 2 *is less than the number of vertices of degree at least* 3.

Theorem 2.6.9 ([91]) *There exist only two non-isomorphic supermagic graphs of order* 5.

Proof Stewart proved that K_5 is not supermagic. According to Theorem 2.6.8, if G is a supermagic graph of order 5, then $7 \le m \le 9$. Figure 2.14 illustrates all non-isomorphic connected graphs of order 5.

Consider the following cases.

Case A: $m = 9$ Let G be a graph of order 5 and size 9. Up to isomorphism there exists exactly one such graph, (in Fig. 2.14 denoted by G_{20}). Suppose G is supermagic, thus there exists a supermagic labeling $f : E(G) \to \{a, a+1, \ldots, a+8\}$ for an index λ. According to the Conditions (2.1) and (2.2)

$$\lambda = \frac{(2a+8)9}{5}. \tag{2.8}$$

As λ is a positive integer,

$$a \equiv 1 \pmod 5. \tag{2.9}$$

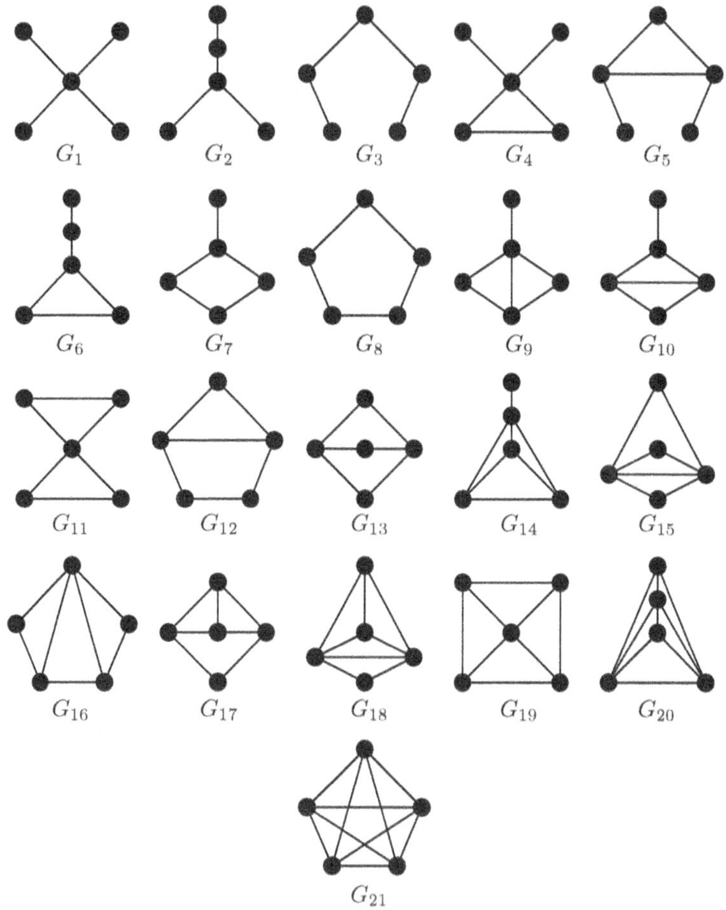

Fig. 2.14 List of all non-isomorphic connected graphs of order 5

As the edges incident with two vertices u, v of degree 3 are mutually distinct, we have

$$2\lambda = f^{\star}(u) + f^{\star}(v) \leq (a+8) + (a+7) + \cdots + (a+3).$$

According to (2.8), and after some manipulations, we get

$$a \leq \frac{7}{2} \tag{2.10}$$

Combining (2.9) and (2.10), we get $a = 1$. Thus $\lambda = 18$. There are three possibilities how to label the edges incident with the vertices u and v: 1. $\{9, 8, 1\}$, $\{7, 6, 5\}$; 2. $\{9, 7, 2\}$, $\{8, 6, 4\}$ and 3. $\{9, 5, 4\}$, $\{8, 7, 3\}$. The remaining edges, forming a triangle, can be labeled with 1. $\{2, 3, 4\}$; 2. $\{1, 3, 5\}$; 3. $\{1, 2, 6\}$.

In the first case consider the vertex w incident with the edges labeled with the numbers 2 and 4. As $\lambda = 18$, the sum of the labels of the other two edges adjacent with w must be 12. But this is not possible using the labels from the sets $\{9, 8, 1\}$ and $\{7, 6, 5\}$, a contradiction.

Analogously we get a contradiction in the second and in the third case.

Case B: $m = 8$ Let G be a graph of order 5 and size 8. Up to isomorphism there exist exactly two such graphs (in Fig. 2.14 denoted by G_{18} and G_{19}). Suppose G is supermagic. Then there exists a supermagic labeling $f : E(G) \to \{a, a+1, \ldots, a+7\}$ for an index λ,

$$\lambda = \frac{(2a+7)8}{5}. \tag{2.11}$$

Since λ is a positive integer, it follows that

$$a \equiv 4 \pmod{5}. \tag{2.12}$$

The graph G_{18} contains a vertex v of degree 2 and thus

$$\lambda = f^\star(v) \leq (a+7) + (a+6).$$

Substituting (2.11) for λ, we have

$$a \leq \frac{3}{2}. \tag{2.13}$$

But according to (2.12) this is not possible, and thus the graph G_{18} is not supermagic.

Figure 2.15 depicts a supermagic labeling of the graph G_{19}.

Case C: $m = 7$ Let G be a graph of order 5 and size 7. Up to isomorphism there exist exactly four such graphs (in Fig. 2.14 denoted by G_{14}, G_{15}, G_{16}, and G_{17}). Note that graphs G_{15} and G_{14} are not supermagic.

Suppose G is supermagic. Then there exists a supermagic labeling $f : E(G) \to \{a, a+1, \ldots, a+6\}$ for an index λ,

$$\lambda = \frac{(2a+6)7}{5}. \tag{2.14}$$

As λ is a positive integer, then

$$a \equiv 2 \pmod{5}. \tag{2.15}$$

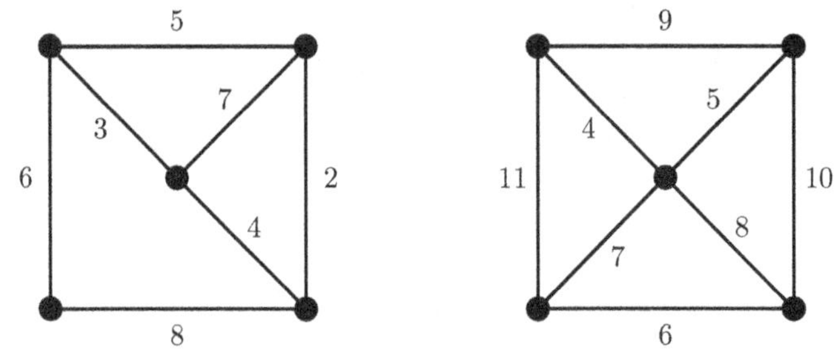

Fig. 2.15 Two non-isomorphic supermagic graphs of order 5

In the graph G_{16} the edges incident with two vertices u, v of degree 2 are mutually distinct, thus

$$2\lambda = f^{\star}(u) + f^{\star}(v) \leq (a+6) + (a+5) + (a+4) + (a+3).$$

Substituting (2.14) for λ, we get

$$a \leq \frac{3}{4}. \tag{2.16}$$

This is a contradiction, thus the graph G_{16} is not supermagic.

A supermagic labeling of the graph G_{17} is depicted in Fig. 2.15. □

Summarizing previous results we get the following theorem for the maximum number $M(n)$ of edges in a supermagic graph.

Theorem 2.6.10 ([91]) *Let $n \geq 5$ be a positive integer. Then*

$$M(n) = \begin{cases} 8 & \text{for } n = 5 \\ \frac{n(n-1)}{2} & \text{for } 6 \leq n \not\equiv 0 \pmod 4 \\ \frac{n(n-1)}{2} - 1 & \text{for } 8 \leq n \equiv 0 \pmod 4. \end{cases}$$

For the minimum number $m(n)$ of edges in supermagic graphs, we have

Theorem 2.6.11 ([91]) *Let $n \geq 5$ be a positive integer. Then*

$$m(n) \geq 3n - \frac{1}{2} - \sqrt{3n^2 - 2n + \frac{1}{4}}.$$

Proof Suppose that G is a supermagic graph of order n with $m = m(n)$ edges. G admits a supermagic labeling $f : E(G) \to \{a, a+1, \ldots, a+m-1\}$ for the index

$$\lambda = \frac{(2a+m-1)m}{n}. \tag{2.17}$$

Let V_3 denote the set of vertices of degree at least 3, the cardinality of this set is denoted by n_3. By n_2 denote the number of 2-vertices (i.e., vertices of degree 2). As every vertex of a supermagic graph G has degree at least 2, $n = n_2 + n_3$. For the number of edges we have

$$2m = \sum_{v \in V(G)} \deg(v) = 2n_2 + \sum_{v \in V_3} \deg(v) \geq 2n_2 + 3n_3 = 3n - n_2,$$

thus

$$m \geq \frac{3n}{2} - \frac{n_2}{2}. \tag{2.18}$$

If G contains no 2-vertex then $m \geq 3n/2$ and the assertion is satisfied. So we can assume that $n_2 \geq 1$.

In any supermagic graph there does not exist an edge joining vertices of degree 2, i.e., every vertex of degree 2 is adjacent to two distinct vertices of degrees at least 3. This means all edges incident with the n_2 vertices of degree 2 are mutually distinct and their number is $2n_2$. The sum of the labels of the edges incident with 2-vertices has to be less than or equal to the sum of the largest values which can be assigned to any $2n_2$ edges in the supermagic labeling f,

$$n_2\lambda \leq (a+m-1)+(a+m-2)+\cdots+(a+m-2n_2) = (2a+2m-2n_2-1)n_2.$$

As $n_2 \neq 0$, by (2.6) we get

$$\frac{(2a+m-1)m}{n} = \lambda \leq 2a + 2m - 2n_2 - 1.$$

This inequality yields

$$2n_2 \leq 2a\left(1 - \frac{m}{n}\right) - \frac{m^2}{n} + \frac{m}{n} + 2m - 1. \tag{2.19}$$

Any supermagic graph of order $n > 2$ has more edges than vertices and so $1 - m/n < 0$. Since $a \geq 1$,

$$2a\left(1 - \frac{m}{n}\right) \leq 2\left(1 - \frac{m}{n}\right).$$

Using this in (2.19), we obtain

$$2n_2 \leq 1 - \frac{m^2}{n} - \frac{m}{n} + 2m.$$

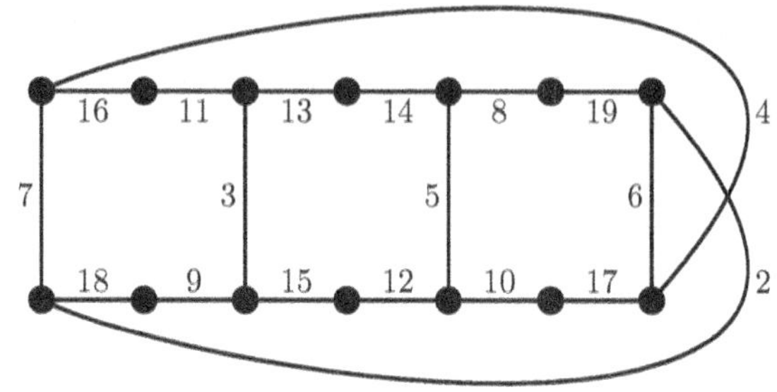

Fig. 2.16 Supermagic graph of size $9n/7$ for $n = 14$

Combining this with $n_2 \geq 3n - 2m$ (from (2.18)), we get

$$0 \geq m^2 + m(1 - 6n) + 6n^2 - n.$$

$$m \geq 3n - \frac{1}{2} - \sqrt{3n^2 - 2n + \frac{1}{4}},$$

which is the desired lower bound for $m(n)$. □

According to the previous theorem, $m(n) \geq \lceil (3 - \sqrt{3})n \rceil$. However, it seems that it is not possible to reach this bound. In [91] it is proved that for $n = 14, 42, 70$ we have $m(n) \geq 9n/7$. The corresponding supermagic graph for $n = 14$ is depicted in Fig. 2.16.

To establish the lower bound for minimum number of edges in a supermagic graph of order n, Drajnová et al. [91] presented two classes of non-regular supermagic graphs. Let d and p be nonnegative integers such that $k = d + p \geq 2$. By $M_{d,p}$ denote the graph with the vertex set $\{u_1, u_2, \ldots, u_{2k}, v_1, v_2, \ldots, v_p\}$ and the edge set consisting of edges

$$e_1 = u_1u_2, e_2 = u_2u_3, \ldots, e_{2k-1} = u_{2k-1}u_{2k}, e_{2k} = u_{2k}u_1$$

$$f_1 = u_1u_{k+1}, f_2 = u_2u_{k+2}, \ldots, f_d = u_du_{k+d}$$

$$l_1 = v_1u_{d+1}, l_2 = v_2u_{d+2}, \ldots, l_p = v_pu_{d+p}$$

$$r_1 = v_1u_{2k}, r_2 = v_2u_{2k-1}, \ldots, r_p = v_pu_{2k-p+1}.$$

Theorem 2.6.12 ([91]) *The graph $M_{d,p}$ is supermagic for every odd positive integer d.*

Proof For every odd positive integer d there exists a positive integer s such that $d = 2s - 1$. Put $a = p + s$. By g we denote the mapping from the edge set of $M_{d,p}$

to the set $\{a, a+1, \ldots, a+|E(M_{d,p})|-1\}$ defined by

$$g(e) = \begin{cases} a + \frac{i-1}{2} & \text{for } e = e_i \text{ and } i = 1, 3, \ldots, 2k-1 \\ a + k + \frac{i-1-d}{2} & \text{for } e = e_i \text{ and } i = d+1, d+3, \ldots, 2k \\ a + k + \frac{2k+i-1-d}{2} & \text{for } e = e_i \text{ and } i = 2, 4, \ldots, d-1 \\ a + 2k - 1 + i & \text{for } e = r_i \text{ and } i = 1, 2, \ldots, p \\ a + 3k - i & \text{for } e = f_i \text{ and } i = 1, 2, \ldots, d \\ a + 3k + p - i & \text{for } e = l_i \text{ and } i = 1, 2, \ldots, p. \end{cases}$$

It is easy to see that g is a bijection and its index-mapping $g^\star$ satisfies

$$g^\star(v) = 8p + 12s - 6, \quad \text{for every } v \in V(M_{d,p}).$$

Thus g is a supermagic labeling and $M_{d,p}$ is a supermagic graph. □

Theorem 2.6.13 ([91]) *For every positive integer $k \geq 2$ there exists a supermagic graph of order $3k$ and size $4k$.*

Proof Consider a cycle C_{2k} with vertices $u_1, u_2, \ldots, u_{2k}$ and edges $e_1 = u_1u_2$, $e_2 = u_2u_3, \ldots, e_{2k-1} = u_{2k-1}u_{2k}$, $e_{2k} = u_{2k}u_1$. Let f be a mapping from $E(C_{2k})$ to the set of positive integers defined by

$$f(e_i) = \begin{cases} k - 1 + \frac{i-1}{2} & \text{for } i = 1, 3, \ldots, k, k+4, k+6, \ldots, 2k-1 \\ 4k - 2 & \text{for } i = 2 \\ k - 1 + \frac{i}{2} & \text{for } i = k+1 \\ 2k - 2 + \frac{i-3}{2} & \text{for } i = k+2 \\ 2k - 2 + \frac{i-2}{2} & \text{for } i = 4, 6, \ldots, k-1, k+3, k+5, \ldots, 2k, \end{cases}$$

for k odd, and

$$f(e_i) = \begin{cases} k - 1 + \frac{i-1}{2} & \text{for } i = 1, 3, k+1 \\ 4k - 2 & \text{for } i = 2 \\ 2k - 2 + \frac{i-2}{2} & \text{for } i = k+2, 2k \\ 2k - \frac{i}{2} & \text{for } i = 4, 6, \ldots, k, k+4, k+6, \ldots, 2k-2 \\ 3k - 3 - \frac{i-3}{2} & \text{for } i = 5, 7 \ldots, k-1, k+3, k+5, \ldots, 2k-1, \end{cases}$$

for k even.

Let S_k be a graph with vertex set $V(C_{2k}) \cup \{v_1, v_2, \ldots, v_k\}$ and edge set $E(C_{2k}) \cup \bigcup_{i=1}^{k} \{v_i u_{i1}, v_i u_{i2}\}$, where u_{i1}, u_{i2} are vertices of C_{2k} such that

$$f^\star(u_{i1}) = 3k - 3 + i \quad \text{and} \quad f^\star(u_{i2}) = 5k - 1 - i.$$

Consider a mapping $g : E(S_k) \to \{k-1, k, \ldots, 5k-2\}$ defined by

$$g(e) = \begin{cases} f(e) & \text{for } e \in E(C_{2k}) \\ 5k-1-i & \text{for } e = v_i u_{i1} \\ 3k-3+i & \text{for } e = v_i u_{i2}. \end{cases}$$

It is easy to check that g is a bijection. Moreover its index-mapping $g^\star$ satisfies

$$g^\star(v) = 8k-4, \quad \text{for every } v \in V(S_k).$$

Thus g is a supermagic labeling and S_k is a desired supermagic graph. □

Figure 2.17 depicts the graphs $M_{3,0}$, $M_{1,3}$ and S_4 and their supermagic labelings. Combining Theorems 2.6.12 and 2.6.13, we immediately obtain

Theorem 2.6.14 ([91]) *Let $n \geq 5$ be a positive integer. Then*

$$m(n) \leq \begin{cases} \frac{4n}{3} & \text{for } n \equiv 0 \pmod 3 \\ \frac{4n}{3} + \frac{5}{3} & \text{for } n \equiv 1 \pmod 3 \\ \frac{4n}{3} + \frac{1}{3} & \text{for } n \equiv 2 \pmod 3). \end{cases}$$

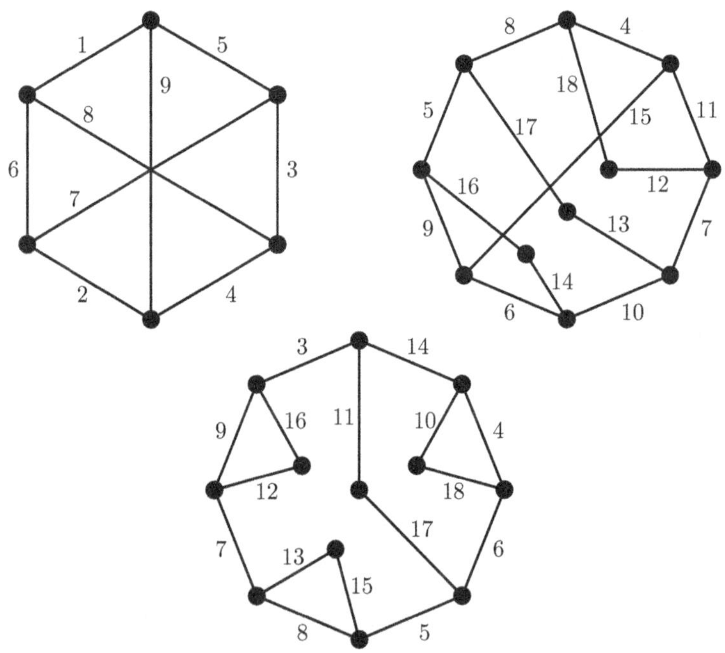

Fig. 2.17 Supermagic labelings of graphs $M_{3,0}$, $M_{1,3}$ and S_4

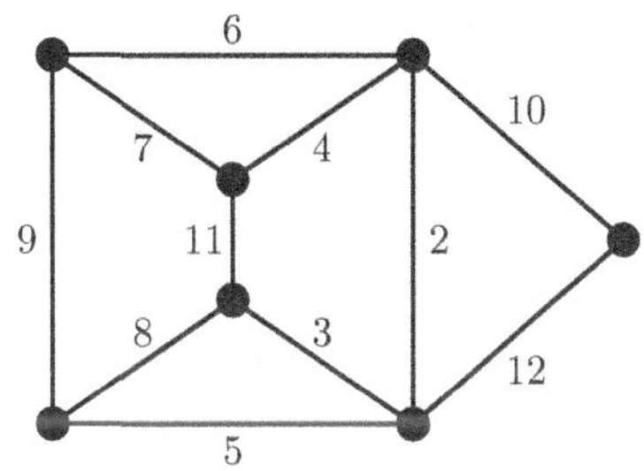

Fig. 2.18 Supermagic labeling of the graph of order 7 with the minimum number of edges

Proof It is obvious that the graphs $S_{n/3}$ (for $n \equiv 0 \pmod 3$), the graph in Fig. 2.18, and the graphs $M_{5,(n-10)/3}$ (for $n \equiv 1 \pmod 3$) and $M_{1,(n-2)/3}$ (for $n \equiv 2 \pmod 3$) are supermagic graphs of order n with the required number of edges. □

If the order of a supermagic graph is a prime number, the following result was proved.

Theorem 2.6.15 ([91]) *Let G be a supermagic graph of order $n \geq 5$ and size m. If the greatest common divisor of the numbers n and m is 1, then $m > 4n/3$. Moreover, if m is an even integer, then $m > (4n+2)/3$.*

Proof Consider a supermagic labeling $f : E(G) \to \{a, a+1, \ldots, a+m-1\}$ for an index $\lambda = (2a+m-1)m/n$. As n and m are coprime and λ is a positive integer, then $\gamma = (2a+m-1)/n$ is also a positive integer. From this we can obtain

$$\lambda = \gamma m \tag{2.20}$$

$$a = \frac{1}{2}(\gamma n - m + 1). \tag{2.21}$$

Let n_2 denote the number of 2-vertices in G. The values of the edges (mutually distinct) incident with the 2-vertices are at most $a+m-1, a+m-2, \ldots, a+m-2n_2$. Thus

$$\lambda \leq (a+m-1) + (a+m-2n_2) = 2a + 2m - 2n_2 - 1.$$

Substituting (2.20) and (2.21) in this inequality we get

$$n_2 \leq \frac{1}{2}((1-\gamma)m + \gamma n). \tag{2.22}$$

As in the proof of Theorem 2.6.11, we get $n_2 \geq 3n - 2m$ (from (2.18)), and combining this with (2.22), we have

$$(5-\gamma)m \geq (6-\gamma)n. \tag{2.23}$$

Since γ is a positive integer it is sufficient to consider the following cases.

Case A: $\gamma \geq 5$ According to Theorem 2.6.4 we have

$$m > \tfrac{5n}{4} = (1 + \tfrac{1}{4})n \geq (1 + \tfrac{1}{\gamma-1})n = \tfrac{\gamma}{\gamma-1}n.$$

Therefore, $m(\gamma-1) > \gamma n > \gamma n-2$. Hence $m-2 < \gamma(m-n) = (2a + m - 1)(m-n)/n$. After some manipulation we obtain

$$(a + m - 1) + (a + m - 2) < \frac{(2a + m - 1)m}{n} = \lambda.$$

This means $n_2 = 0$, and then $m \geq \frac{3n}{2} > \frac{4n+2}{3}$.

Case B: $\gamma \in \{3, 4\}$ By (2.23), we get $m \geq n(6 - \gamma)/(5 - \gamma) \geq 3n/2 > (4n + 2)/3$.

Case C: $\gamma = 2$ According to (2.23), we have $m \geq 4n/3$. Moreover, $m \neq 4n/3$. In the opposite case we get $m = 4k$ and $n = 3k$ for some integer $k > 1$. This means the greatest common divisor of n and m is k, a contradiction. Note also that (2.21) implies $m = 2(n - a) + 1$, therefore m is an odd integer in this case.

Case D: $\gamma = 1$ From (2.21), we get $m = n-2a+1 < n$, contrary to Theorem 2.6.4. Thus, this case is impossible.

□

Combining Theorems 2.6.14 and 2.6.15 gives the following result.

Theorem 2.6.16 ([91]) *Let $n \geq 5$ be a prime number. Then*

$$m(n) = \begin{cases} \frac{4n}{3} + \frac{5}{3} & \text{for } n \equiv 1 \pmod 3 \\ \frac{4n}{3} + \frac{1}{3} & \text{for } n \equiv 2 \pmod 3). \end{cases}$$

For dense bipartite graph Ivančo [137] proved the following connection between magicness and the minimum degree of a graph.

Theorem 2.6.17 ([137]) *Let G be a balanced bipartite graph with minimum degree $\delta(G)$. If $\delta(G) > |V(G)|/4 \geq 2$, then G is a magic graph.*

In [137] it is shown that the bound $\delta(G) > |V(G)|/4$ in Theorem 2.6.17 can be replaced by the condition $\deg(u) + \deg(v) > |V(G)|/2$ for all nonadjacent vertices $u \in V_1(G)$ and $v \in V_2(G)$ of a balanced bipartite graph G with the partite sets V_1 and V_2. Ivančo [137] also proved that these bounds are the best possible.

2.7 Magic and Supermagic Line Graphs

The *line graph* $L(G)$ of a graph G is the graph with vertex set $V(L(G)) = E(G)$, where two edges $e, e' \in E(G)$ are adjacent in $L(G)$ whenever they have a common end vertex in G.

Denote by $\mathcal{F}_1$ the family of connected graphs which contain an edge uv such that $\deg(u)+\deg(v) = 3$. By $\mathcal{F}_2$ we denote the family of all connected unicyclic graphs with a 1-factor. $\mathcal{F}_3$ denotes the family of connected graphs which contain edges vu and uw such that $\deg(v) + \deg(u) = \deg(u) + \deg(w) = 4$. $\mathcal{F}_4$ is the family of six graphs illustrated in Fig. 2.19. Finally, let $\mathcal{F} = \mathcal{F}_1 \cup \mathcal{F}_2 \cup \mathcal{F}_3 \cup \mathcal{F}_4$.

2.7.1 *Magic Line Graphs*

In [140] Ivančo, Lastivková, and Semaničová characterized magic line graphs of connected graphs.

Theorem 2.7.1 ([140]) *Let G be a connected graph of size at least 5. The line graph $L(G)$ is magic if and only if $G \notin \mathcal{F}$.*

Proof Assume that the line graph of G is not magic. If each vertex of G has degree at most 2, then G is either a path or a cycle, i.e., $G \in \mathcal{F}_1 \cup \mathcal{F}_3$. Next, we suppose that the maximum degree of G is at least 3. So, $L(G)$ is non-bipartite. According to Theorem 2.3.6, we consider the following cases.

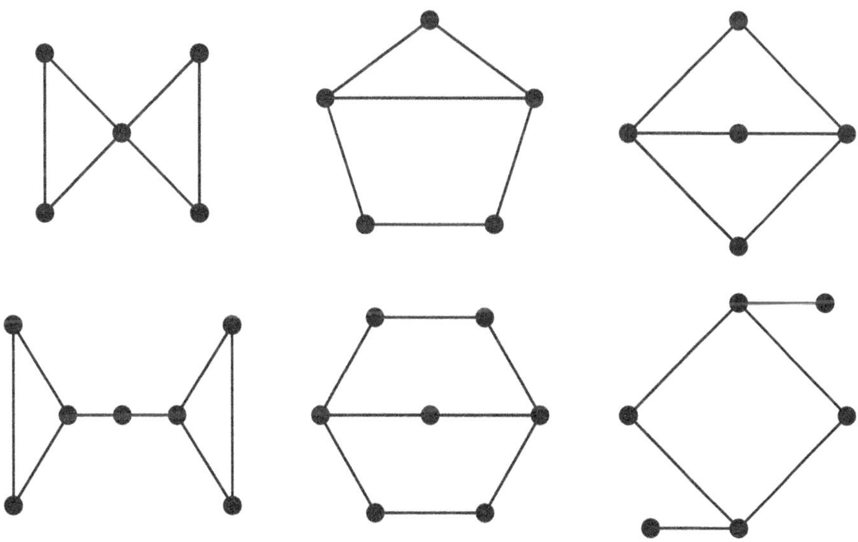

Fig. 2.19 The family of graphs denoted by $\mathcal{F}_4$

Case A There is an independent set $S \subset V(L(G))$ such that $|\Gamma(S)| \leq |S|$. Suppose that $S = \{e_1, e_2, \ldots, e_k\}$ has the smallest possible cardinality. If $|S| = 1$, then $|\Gamma(\{e_1\})| = 1$, i.e., e_1 is a terminal edge of G with end vertices of degree 1 and 2. Thus $G \in \mathcal{F}_1$.

If $|S| > 1$, then any edge of G is adjacent to at least two others. The edges $e_1, e_2, \ldots, e_k$ are independent, thus any edge of G is adjacent to at most two of them. Therefore,

$$\begin{aligned}|S| \geq |\Gamma(S)| &= |\Gamma(\{e_1\}) \cup \Gamma(\{e_2\}) \cup \cdots \cup \Gamma(\{e_k\})| \\ &\geq \tfrac{1}{2}(|\Gamma(\{e_1\})| + |\Gamma(\{e_2\})| + \cdots + |\Gamma(\{e_k\})|) \geq \tfrac{1}{2}2k = |S|.\end{aligned}$$

This means $|\Gamma(S)| = |S|$ and any edge of $\Gamma(S)$ is adjacent to exactly two edges of S. As G is a connected graph, $|E(G)| = |S \cup \Gamma(S)| = 2|S| = |V(G)|$. So, G is unicyclic and S is its 1-factor, i.e., $G \in \mathcal{F}_2$.

Case B Suppose that $L(G)$ is of type $\mathcal{B}$. Then there is a set $E' \subset E(G)$ such that the subgraph L' of $L(G)$ induced by E' is a balanced bipartite graph connected by a pair of edges to another subgraph. Since L' is bipartite, every vertex of the subgraph G' of G induced by E' is of degree at most two, i.e., every component of G' is either a path or an even cycle. Moreover, the set $E(G) - E'$ contains either one edge incident with a 2-vertex (i.e., vertex of degree 2) of G', or a pair of edges incident with two 1-vertices of G'. Consider the following subcases.

Case B1 G' contains an even cycle. Then only one edge of $E(G) - E'$ is incident with its vertex. Thus, some two adjacent edges of this cycle have both end vertices of degree 2 in G, i.e., $G \in \mathcal{F}_3$.

Case B2 G' consists of two paths. Then a pair of edges of $E(G) - E'$ is incident with its terminal vertices. The other terminal vertices of G' are terminal in G, too. Evidently, in this case $G \in \mathcal{F}_1$.

Case B3 G' is a path connected by one edge to another subgraph. Then either $|E'| > 2$ and $G \in \mathcal{F}_1$, or $|E'| = 2$ and $G \in \mathcal{F}_3$, because both edges of E' have end vertices of degree 1 and 3 in G.

Case B4 G' is a path connected by a pair of edges to another subgraph. Then any two adjacent edges of this path have both end vertices of degree 2 in G, i.e., $G \in \mathcal{F}_3$.

Case C Suppose that $L(G)$ is of type $\mathcal{A}$. Moreover, assume that $G \notin \mathcal{F}_1 \cup \mathcal{F}_2 \cup \mathcal{F}_3$. For $d \leq 2$, every d-vertex of G is adjacent to some vertex of degree at least 3, because $G \notin \mathcal{F}_1 \cup \mathcal{F}_3$. As $L(G)$ is a balanced bipartite graph with two added edges, $6 \leq |E(G)| \equiv 0 \pmod 2$ and G contains either one 4-vertex or two 3-vertices. One can easily see that $G \in \mathcal{F}_4$ in this case.

The converse implication is obvious. □

The complexity of deciding whether a graph G belongs to the family $\mathcal{F}_i$ $(i = 1, 2, 3, 4)$ is polynomial. Using the Even–Kariv algorithm for finding 1-factor in G we get that testing whether the line graph of a given graph is magic has computational complexity $O(n^{5/2})$. Moreover, each graph of the family $\mathcal{F}$ contains a vertex of degree at most two, thus

Corollary 2.7.1 ([140]) *Let G be a connected graph with minimum degree at least 3. Then $L(G)$ is a magic graph.*

2.7.2 Supermagic Line Graphs

The problem of characterizing supermagic line graphs of general graphs seems to be difficult. In [136] Ivančo solved it for regular graphs and Ivančo et al. [140] dealt with regular bipartite graphs.

Theorem 2.7.2 ([136]) *Let G be an r-regular graph, where $r \geq 5$. If either $r \equiv 2 \pmod 4$ and $|V(G)| \equiv 1 \pmod 2$ or $r \equiv 1 \pmod 4$, then $L(G)$ is supermagic.*

Theorem 2.7.3 ([136]) *Let G be a 3-regular graph containing a 1-factor. Then $L(G)$ is supermagic.*

Note that all the edges of a graph G incident with a vertex v induce a subgraph $K(v)$ of $L(G)$, which is isomorphic to the complete graph of order $\deg(v)$. Subgraphs $K(v)$, for all $v \in V(G)$, are edge-disjoint and form a decomposition of $L(G)$. If vertices u and v of G are not adjacent, then $K(u)$ and $K(v)$ are vertex-disjoint subgraphs of $L(G)$. So, for a bipartite graph G with parts V_1 and V_2, the subgraph $R_1(G) = \bigcup_{v\in V_1} K(v)$ $(R_2(G) = \bigcup_{v\in V_2} K(v))$ consists of mutually disjoint complete subgraphs of $L(G)$. Moreover, $R_1(G)$ and $R_2(G)$ are spanning subgraphs of $L(G)$ which form its decomposition.

Let r_1, r_2, q be positive integers and let $\mathcal{G}(q; r_1, r_2)$ be the family of all bipartite graphs of size q whose every edge joins an r_1-vertex to an r_2-vertex. Clearly, there is a vertex partition $\{V_1, V_2\}$ of $G \in \mathcal{G}(q; r_1, r_2)$, where V_i consists of r_i-vertices of G $(i = 1, 2)$. Then $|V_i|r_i = q$ and $R_i(G) = \frac{q}{r_i} K_{r_i}$ is a factor of $L(G)$ for $i \in \{1, 2\}$. So, combining Theorems 2.5.6 and 2.8.1, we immediately obtain

Corollary 2.7.2 ([140]) *Let $r_1 \geq 5$, $r_2 \geq 5$, and q be positive integers such that one of the following conditions is satisfied.*

(i) $r_1 \equiv 1 \pmod 4$, $r_2 \equiv 1 \pmod 4$.
(ii) $r_1 \equiv 1 \pmod 4$, $r_2 \equiv 2 \pmod 4$, $q \equiv 2 \pmod 4$.
(iii) $r_1 \equiv 1 \pmod 4$, $r_2 \equiv 3 \pmod 4$, $q \equiv 1 \pmod 2$.
(iv) $r_1 \equiv 2 \pmod 4$, $r_2 \equiv 2 \pmod 4$, $q \equiv 2 \pmod 4$.
(v) $r_1 \equiv 3 \pmod 4$, $r_2 \equiv 3 \pmod 4$, $q \equiv 1 \pmod 2$.

If $G \in \mathcal{G}(q; r_1, r_2)$, then $L(G)$ is a supermagic graph.

It is possible to extend this result to regular bipartite graphs.

Lemma 2.7.1 ([140]) *Let m and $r \geq 3$ be positive integers. Suppose $v_{i,1}, v_{i,2}, \ldots, v_{i,r}$ are vertices of the ith component of mK_r for $i \in \{1, 2, \ldots, m\}$. Then there is a bijective mapping $f : E(mK_r) \to \{1, 2, \ldots, m\binom{r}{2}\}$ such that*

$$f^*(v_{1,j}) = f^*(v_{2,j}) = \cdots = f^*(v_{m,j}), \quad \textit{for all } j \in \{2, 3, \ldots, r\}.$$

Proof Evidently, it is sufficient to consider $m \geq 2$. If mK_r is supermagic, then its supermagic labeling has the desired properties. So, according to Theorem 2.8.1, it remains to consider the following cases.

Case A: $r = 3$ Define a mapping $f : E(mK_3) \to \{1, 2, \ldots, 3m\}$ by

$$f(v_{i,j}v_{i,k}) = \begin{cases} i & \text{if } \{j,k\} = \{1,2\} \\ 1 + 2m - i & \text{if } \{j,k\} = \{2,3\} \\ 2m + i & \text{if } \{j,k\} = \{1,3\}. \end{cases}$$

Clearly, f is the desired mapping because

$$f^\star(v_{i,j}) = \begin{cases} 2m + 2i & \text{if } j = 1 \\ 1 + 2m & \text{if } j = 2 \\ 1 + 4m & \text{if } j = 3. \end{cases}$$

Case B: $r = 4$ In this case we define a bijection $f : E(mK_4) \to \{1, 2, \ldots, 6m\}$ by

$$f(v_{i,j}v_{i,k}) = \begin{cases} i & \text{if } \{j,k\} = \{1,2\} \\ m + i & \text{if } \{j,k\} = \{3,4\} \\ 1 + 4m - 2i & \text{if } \{j,k\} = \{2,3\} \\ 2 + 4m - 2i & \text{if } \{j,k\} = \{1,4\} \\ 4m + i & \text{if } \{j,k\} = \{1,3\} \\ 5m + i & \text{if } \{j,k\} = \{2,4\}. \end{cases}$$

For its index-mapping we get

$$f^\star(v_{i,j}) = \begin{cases} 2 + 8m & \text{if } j = 1 \\ 1 + 9m & \text{if } j = 2 \\ 1 + 9m & \text{if } j = 3 \\ 2 + 10m & \text{if } j = 4. \end{cases}$$

Case C: $4 < r \equiv 0 \pmod 4$ Then there is an integer $p \geq 2$ such that $r = 4p$. The subgraph $H_{i,s}$ of mK_d induced by $\{v_{i,4s-3}, v_{i,4s-2}, v_{i,4s-1}, v_{i,4s}\}$ is a complete graph for all $i \in \{1, 2, \ldots, m\}$ and $s \in \{1, 2, \ldots, p\}$. Therefore, the spanning subgraph $H := \bigcup_{i=1}^{m} \bigcup_{s=1}^{p} H_{i,s}$ of mK_r is isomorphic to mpK_4. As in Case B, there is a bijection $h : E(H) \to \{1, 2, \ldots, 6mp\}$ such that $h^{\star}(v_{1,j}) = h^{\star}(v_{2,j}) = \cdots = h^{\star}(v_{m,j})$, for all $j \in \{1, 2, \ldots, r\}$. Similarly, the spanning subgraph $B := mK_r - E(H)$ of mK_r is isomorphic to $mK_{p[4]}$. By Theorem 2.8.1, $mK_{p[4]}$ is a supermagic graph. Thus, there exists a supermagic labeling $g : E(B) \to \{1, 2, \ldots, |E(B)|\}$ of B for an index λ, i.e., $g^{\star}(v_{i,j}) = \lambda$, for all $i \in \{1, 2, \ldots, m\}$ and $j \in \{1, 2, \ldots, r\}$. Since H and B form a decomposition of mK_r, we can define a mapping $f : E(mK_r) \to \{1, 2, \ldots, m\binom{r}{2}\}$ by

$$f(e) = \begin{cases} h(e) & \text{if } e \in E(H) \\ 6mp + g(e) & \text{if } e \in E(B). \end{cases}$$

As $f^{\star}(v_{i,j}) = h^{\star}(v_{i,j}) + 6mp(r-4) + \lambda$, we have $f^{\star}(v_{1,j}) = f^{\star}(v_{2,j}) = \cdots = f^{\star}(v_{m,j})$, for all $j \in \{1, 2, \ldots, r\}$.

Case D: $6 \leq r \equiv 2 \pmod 4$ and $m \equiv 0 \pmod 2$ Then there is a positive integer p such that $r = 4p + 2$. The subgraph G of mK_r induced by $\bigcup_{i=1}^{m} \bigcup_{j=3}^{r} \{v_{i,j}\}$ is isomorphic to mK_{4p}. As in Case C (B if $p = 1$), there is a bijection $t : E(G) \to \{1, 2, \ldots, m\binom{4p}{2}\}$ such that $t^{\star}(v_{1,j}) = t^{\star}(v_{2,j}) = \cdots = t^{\star}(v_{m,j})$, for all $j \in \{3, 4, \ldots, r\}$. Consider a mapping $f : E(mK_r) \to \{1, 2, \ldots, m\binom{r}{2}\}$ given by

$$f(v_{i,j}v_{i,k}) = \begin{cases} (k-3)m + i & \text{if } j = 2, 3 \leq k, k \equiv 1 \pmod 2 \\ 1 + (k-2)m - i & \text{if } j = 2, 4 \leq k < r, k \equiv 0 \pmod 2 \\ 1 + (k-1)m - 2i & \text{if } j = 2, k = r \\ (k-3)m + 2i & \text{if } j = 1, k = r \\ (2r - k - 2)m + i & \text{if } j = 1, 4 \leq k < r, k \equiv 0 \pmod 2 \\ 1 + (2r - k - 1)m - i & \text{if } j = 1, 3 \leq k, k \equiv 1 \pmod 2 \\ 2(r-2)m + i & \text{if } j = 1, k = 2 \\ (2r-3)m + t(v_{i,j}v_{i,k}) & \text{if } 2 < j < k \leq r. \end{cases}$$

It is not difficult to check that f is a bijection and for its index-mapping we have

$$f^{\star}(v_{i,j}) = \begin{cases} 2p + (8p(3p+1) - 1)m + 2i & \text{if } j = 1 \\ 2p + (8p(p+1) + 1)m & \text{if } j = 2 \\ 1 + 2(r-2)m + (2r-3)m(r-3) + t^{\star}(v_{i,j}) & \text{if } 3 \leq j \leq r. \end{cases}$$

Case E: $7 \leq r \equiv 3 \pmod 4$ and $m \equiv 0 \pmod 2$ Then the subgraph G of mK_r induced by $\bigcup_{i=1}^{m} \bigcup_{j=3}^{r} \{v_{i,j}\}$ is isomorphic to mK_{r-2}. By Theorem 2.8.1, the graph G is supermagic and so there is a supermagic labeling $t : E(G) \to \{1, 2, \ldots, m\binom{r-2}{2}\}$ of G for an index λ. Consider a mapping $f : E(mK_r) \to$

$\{1, 2, \ldots, m\binom{r}{2}\}$ given by

$$f(v_{i,j}v_{i,k}) = \begin{cases} (k-3)m+i & \text{if } j=2, 3 \le k \equiv 1 \pmod 2 \\ 1+(k-2)m-i & \text{if } j=2, 4 \le k \equiv 0 \pmod 2 \\ 1+(2r-k-1)m-i & \text{if } j=1, 3 \le k \equiv 1 \pmod 2 \\ (2r-k-2)m+i & \text{if } j=1, 4 \le k \equiv 0 \pmod 2 \\ 1+(2r-3)m-i & \text{if } j=1, k=2 \\ (2r-3)m+t(v_{i,j}v_{i,k}) & \text{if } 2 < j < k \le r. \end{cases}$$

It is easy to verify that f is a bijection. Moreover, for its index-mapping we get

$$f^\star(v_{i,j}) = \begin{cases} \frac{1}{2}(r+1)+(\frac{1}{2}(r-3)(3r+1)+5)m-2i & \text{if } j=1 \\ \frac{1}{2}(r-1)+(\frac{1}{2}(r-1)(r+1)-1)m & \text{if } j=2 \\ 1+2(r-2)m+(r-3)(2r-3)m+\lambda & \text{if } 3 \le j \le r, \end{cases}$$

which completes the proof. □

Theorem 2.7.4 ([140]) *Let G be a bipartite regular graph of degree $r \ge 3$. Then the line graph $L(G)$ is supermagic.*

Proof Suppose that V_1, V_2 are parts of G. As G is a bipartite r-regular graph, there exist mutually edge-disjoint 1-factors $F_1, F_2, \ldots, F_r$ of G which form its decomposition. Put $m = |V_1|$ (clearly, $|V_1| = |V_2|$) and denote the vertices of G by $u_1, u_2, \ldots, u_m, v_1, v_2, \ldots, v_m$ in such a way that $E(F_1) = \{u_1v_1, u_2v_2, \ldots, u_mv_m\}$, $V_1 = \{u_1, u_2, \ldots, u_m\}$ and $V_2 = \{v_1, v_2, \ldots, v_m\}$.

The subgraphs $R_1(G)$, $R_2(G)$ of the line graph $L(G)$ consist of complete graphs with r vertices. Therefore, they are isomorphic to mK_r. Denote by $a_{i,j}$ ($b_{i,j}$), $i \in \{1, 2, \ldots, m\}$, $j \in \{1, 2, \ldots, r\}$, the vertex of $R_1(G)$ ($R_2(G)$) which corresponds to the edge of G incident with u_i (v_i) and which belongs to F_j, i.e., the vertex of $L(G)$ corresponding to $u_rv_s \in E(F_j)$ is denoted by $a_{r,j}$ in $R_1(G)$ and by $b_{s,j}$ in $R_2(G)$.

By Lemma 2.7.1, there is a bijective mapping $g_1 : E(R_1(G)) \to \{1, 2, \ldots, m\binom{r}{2}\}$ such that $g_1^\star(a_{1,j}) = g_1^\star(a_{2,j}) = \cdots = g_1^\star(a_{m,j})$, for all $j \in \{2, 3, \ldots, r\}$. Then a mapping $g_2 : E(R_2(G)) \to \{1+m\binom{r}{2}, 2+m\binom{r}{2}, \ldots, 2m\binom{r}{2}\}$ given by

$$g_2(b_{i,j}b_{i,k}) = 1 + 2m\binom{r}{2} - g_1(a_{i,j}a_{i,k})$$

is bijective, too. Moreover, $g_2^\star(b_{i,j}) = (r-1)(1+2m\binom{r}{2}) - g_1^\star(a_{i,j})$. Consider the mapping $f : E(L(G)) \to \{1, 2, \ldots, 2m\binom{r}{2}\}$ defined by

$$f(e) = \begin{cases} g_1(e) & \text{if } e \in E(R_1(G)) \\ g_2(e) & \text{if } e \in E(R_2(G)). \end{cases}$$

Evidently, f is a bijection. Let x be an edge of G which belongs to F_1. Then there exists $i \in \{1, 2, \ldots, m\}$ such that $x = u_i v_i$, i.e., the vertex of $L(G)$ corresponding to x is denoted by $a_{i,1}$ in $R_1(G)$ and by $b_{i,1}$ in $R_2(G)$. Thus

$$f^\star(x) = g_1^\star(a_{i,1}) + g_2^\star(b_{i,1}) = (r-1)\left(1 + 2m\tbinom{r}{2}\right).$$

Similarly, for an edge $y \in E(F_j)$, $j \in \{2, 3, \ldots, r\}$, there exist $r, s \in \{1, 2, \ldots, m\}$, $r \neq s$, such that $y = u_r v_s$. Then

$$f^\star(y) = g_1^\star(a_{r,j}) + g_2^\star(b_{s,j}) = g_1^\star(a_{s,j}) + g_2^\star(b_{s,j}) = (r-1)\left(1 + 2m\tbinom{r}{2}\right).$$

Therefore, f is a supermagic labeling of $L(G)$ for index $(r-1)(1 + 2m\binom{r}{2})$. □

Corollary 2.7.3 ([140]) *Let k_1, k_2, q, and $r \geq 3$ be positive integers such that one of the following conditions is satisfied.*

(i) $r \equiv 0 \pmod 2$.
(ii) $r \equiv 1 \pmod 2$, $k_1 \equiv 1 \pmod 4$, $k_2 \equiv 1 \pmod 4$.
(iii) $r \equiv 1 \pmod 2$, $k_1 \equiv 1 \pmod 4$, $k_2 \equiv 2 \pmod 4$, $q \equiv 2 \pmod 4$.
(iv) $r \equiv 1 \pmod 2$, $k_1 \equiv 1 \pmod 4$, $k_2 \equiv 3 \pmod 4$, $q \equiv 1 \pmod 2$.
(v) $r \equiv 1 \pmod 2$, $k_1 \equiv 3 \pmod 4$, $k_2 \equiv 3 \pmod 4$, $q \equiv 1 \pmod 2$.

If $G \in \mathcal{G}(q; k_1 r, k_2 r)$, then $L(G)$ is a supermagic graph.

Proof Suppose that u_i, $i \in \{1, 2, \ldots, m\}$, where $m = q/(k_1 r)$, $(v_j$, $j \in \{1, 2, \ldots, n\}$, where $n = q/(k_2 r))$ denotes a $(k_1 r)$-vertex ($(k_2 r)$-vertex) of a graph G belonging to $\mathcal{G}(q; k_1 r, k_2 r)$. Then there is a graph $G' \in \mathcal{G}(q; r, r)$ with vertex set $V(G') = (\bigcup_{i=1}^{m} \bigcup_{r=1}^{k_1} \{u_i^r\}) \cup (\bigcup_{j=1}^{n} \bigcup_{s=1}^{k_2} \{v_j^s\})$ such that for any edge $u_i v_j \in E(G)$ there exists an edge $u_i^r v_j^s \in E(G')$, where $r \in \{1, 2, \ldots, k_1\}$ and $s \in \{1, 2, \ldots, k_2\}$ (i.e., G' is obtained from G by distributing every vertex into vertices of degree r).

The subgraph $K(u_i)$ $(K(v_j))$ of $L(G)$ is decomposable into $k_1 K_r$ and $K_{k_1[r]}$ ($k_2 K_r$ and $K_{k_2[r]}$). Thus, it is not difficult to see that $L(G)$ is decomposable into factors F_1, F_2, F_3, where F_1 is isomorphic to $L(G')$, F_2 is isomorphic to $mK_{k_1[r]}$ (if $k_1 > 1$) and F_3 is isomorphic to $nK_{k_2[r]}$ (if $k_2 > 1$). Combining Theorems 2.7.4, 2.5.6, and 2.8.1, we obtain the assertion. □

Ivančo, Lastivková, and Semaničová also proved the following negative statement.

Theorem 2.7.5 ([140]) *Let q, r_1, r_2 be positive integers such that either $r_1 + r_2 \leq 4$ and $q > 2$, or $4 < r_1 + r_2 \equiv 1 \pmod 2$ and $q \equiv 0 \pmod 4$. If $G \in \mathcal{G}(q; r_1, r_2)$, then the line graph $L(G)$ is not supermagic.*

Proof The line graph $L(G)$ of a graph $G \in \mathcal{G}(q; r_1, r_2)$ is $(r_1 + r_2 - 2)$-regular of order q. Evidently, $L(G)$ is not magic when $r_1 + r_2 \leq 4$ and $q > 2$. The other case immediately follows from the fact, see [136], that a supermagic regular graph H of odd degree satisfies $|V(H)| \equiv 2 \pmod 4$. □

2.8 Regular Magic and Supermagic Graphs

2.8.1 Complete Graphs and Complete Multipartite Graphs

Stewart [262] proved that the complete graph K_n is magic for $n = 2$ and all $n \geq 5$. In [263] Stewart showed that a complete graph K_n is supermagic if and only if $n = 2$ or $5 < n \not\equiv 0 \pmod 4$, see Theorem 2.6.6.

Doob [90] proved that the complete bipartite graph $K_{m,n}$ is magic if and only if $m = n \neq 2$. Stewart [262] showed that in this case $K_{n,n}$ is also supermagic.

The characterization of supermagic regular complete multipartite graphs is given in [136]. Let $K_{k[n]}$ denote a complete k-partite graph whose every part has n vertices. Note that $K_{k[1]}$ is a complete graph on k vertices.

Theorem 2.8.1 ([136]) *The graph $mK_{k[n]}$ is supermagic if and only if one of the following conditions is satisfied.*

(i) $n = 1,\ k = 2,\ m = 1.$
(ii) $n = 1,\ k = 5,\ m \geq 2.$
(iii) $n = 1,\ 5 < k \equiv 1 \pmod 4,\ m \geq 1.$
(iv) $n = 1,\ 6 \leq k \equiv 2 \pmod 4,\ m \equiv 1 \pmod 2.$
(v) $n = 1,\ 7 \leq k \equiv 3 \pmod 4,\ m \equiv 1 \pmod 2.$
(vi) $n = 2,\ k \geq 3,\ m \geq 1.$
(vii) $3 \leq n \equiv 1 \pmod 2,\ 2 \leq k \equiv 1 \pmod 4,\ m \geq 1.$
(viii) $3 \leq n \equiv 1 \pmod 2,\ 2 \leq k \equiv 2 \pmod 4,\ m \equiv 1 \pmod 2.$
(ix) $3 \leq n \equiv 1 \pmod 2,\ 2 \leq k \equiv 3 \pmod 4,\ m \equiv 1 \pmod 2.$
(x) $4 \leq n \equiv 0 \pmod 2,\ k \geq 2,\ m \geq 1.$

Note that Shiu et al. [250] proved that, for $n > 2$, $sK_{n,n}$ is supermagic if and only if n is even or both s and n are odd.

According to Theorem 2.3.5, a bipartite magic graph must be balanced. Thus, for supermagic bipartite graph G we have $|V(G| \equiv 0 \pmod 2$. In [136] Ivančo proved that every supermagic graph of odd degree has order $|V(G| \equiv 2 \pmod 4$. Thus no odd regular graph of order $4k$ is supermagic. In [137] Ivančo proposed the following conjecture.

Conjecture 2.8.1 ([137]) Let G be an r-regular bipartite graph of order $2n$. If $r > n/2$, then G is supermagic except for $n \equiv 0 \pmod 2$ and $r \equiv 1 \pmod 2$.

Ivančo [137] proved

Theorem 2.8.2 ([137]) *Let G be an r-regular bipartite graph of order $2n$ such that one of the following conditions is satisfied.*

(i) $r \equiv 0 \pmod 4$ *and* $r - 2 > \frac{n}{2}$.
(ii) $r \equiv 1 \pmod 4$, $n \equiv 1 \pmod 2$, $r - 11 > \frac{n}{2}$ *and* $r \geq \frac{3n+2}{4}$.
(iii) $r \equiv 2 \pmod 4$, $n \equiv 1 \pmod 2$ *and* $r - 8 > \frac{n}{2}$.

(iv) $r \equiv 2 \pmod 4$, $n \equiv 0 \pmod 2$, $r - 8 > \frac{n}{2}$ *and* $r \geq \frac{3n+2}{4}$.
(v) $r \equiv 3 \pmod 4$, $n \equiv 1 \pmod 2$, $r - 5 > \frac{n}{2}$ *and* $r \geq \frac{3n+2}{4}$.

Then G is a supermagic graph.

Furthermore, Ivančo [136] showed

Theorem 2.8.3 ([136]) *Let G be a bipartite* 4*-regular which can be decomposed into pairwise edge-disjoint subgraphs isomorphic to* C_4*. Then G is a supermagic graph.*

2.8.2 Cartesian Product of Graphs

The *Cartesian product* $G_1 \Box G_2$ of graphs G_1, G_2 is a graph whose vertices are all ordered pairs $[v_1, v_2]$, where $v_1 \in V(G_1)$, $v_2 \in V(G_2)$ and two vertices $[v_1, v_2]$, $[u_1, u_2]$ are joined by an edge in $G_1 \Box G_2$ if and only if either $v_1 = u_1$ and v_2, u_2 are adjacent in G_2, or v_1, u_1 are adjacent in G_1 and $v_2 = u_2$.

For Cartesian product of two graphs Trenkler [277] proved

Theorem 2.8.4 ([277]) *If G is a semi-magic graph none of whose components is* K_2 *and for each edge* $e \in E(G)$ *there exists a* $(1\text{-}2)$*-factor F such that e does not belong to the cycle part of F, then* $G \Box K_2$ *is magic.*

Theorem 2.8.5 ([277]) *If G is a semi-magic graph none of whose components is* K_2 *and H is a graph every one of whose connected components has at least 3 vertices, then* $G \Box H$ *is magic.*

Theorem 2.8.6 ([277]) *The Cartesian product of a magic graph of order at least 5 and* K_2 *is a magic graph.*

Sedláček [231] proved that, for n even, $n \geq 4$, the prism $C_n \Box K_2$ is magic but not supermagic. For n odd, $n \geq 3$, the prism $C_n \Box K_2$ is not magic, see also [277]. In [277] Trenkler also proved that $P_n \Box K_2$ is not magic, for all $n \geq 1$ and a graph $C_n \Box P_m$ is magic if and only if $4 \leq n \equiv 0 \pmod 2$ and $m = 2$ or $n \geq 3$ and $m \geq 3$. Moreover, $P_n \Box P_m$ is magic if and only if $3 \leq n \leq m$ and $n, m \equiv 0 \pmod 2$.

Ivančo proved the following theorem.

Theorem 2.8.7 ([136]) *Let* G_1, G_2 *be regular graphs satisfying* $|V(G_1)|G_2$ *and* $|V(G_2)|G_1$ *are supermagic graphs. Then* $G_1 \Box G_2$ *is a supermagic graph.*

Theorem 2.8.8 ([136]) *The Cartesian product of two cycles of order n,* $C_n \Box C_n$ *is a supermagic graph, for any integer* $n \geq 3$.

For the Cartesian product of two even cycles Ivančo obtained

Theorem 2.8.9 ([136]) *Let* $n \geq 2, k \geq 2$ *be integers. Then* $C_{2n} \Box C_{2k}$ *is supermagic.*

Ivančo [136] gave a characterization of supermagic cubes Q_n. The graph Q_n of the n-dimensional cube can be defined such that $Q_1 = K_1$ and $Q_{k+1} = Q_k \square K_2$ for any positive integer k.

Theorem 2.8.10 ([136]) *The n-dimensional cube Q_n is supermagic if and only if either $n = 1$ or $4 \leq n \equiv 0 \pmod 2$.*

Hartsfield and Ringel [124] proved

Theorem 2.8.11 ([124]) *If $n \equiv 1 \pmod 4$, then the graph $K_n \square K_2$ is supermagic.*

For the Cartesian product of complete graphs Ivančo proved

Theorem 2.8.12 ([136])

(i) Let $k \geq 5$ and $p \geq 5$ be odd integers. Then $K_k \square K_p$ is supermagic.
(ii) Let $n \geq 4$ and $q \geq 4$ be even integers. Then $K_{k[n]} \square K_{p[q]}$ is supermagic.

Kovář [165] proved the following result.

Theorem 2.8.13 ([165]) *Let G be a supermagic r-regular graph on m vertices with a proper edge r coloring and let H be a supermagic s-regular graph on n vertices with a proper edge s coloring. Suppose r is even or n is odd and s is even or m is odd. Then $G \square H$ is supermagic.*

2.8.3 Lexicographic Product (Composition) of Graphs

Lexicographic product (or a *composition*) $G_1 \circ G_2$ of graphs G_1, G_2 is a graph whose vertices are all ordered pairs $[v_1, v_2]$, where $v_1 \in V(G_1)$, $v_2 \in V(G_2)$ and two vertices $[v_1, v_2]$, $[u_1, u_2]$ are joined by an edge in $G_1 \circ G_2$ if and only if either v_1, u_1 are adjacent in G_1 or $v_1 = u_1$ and v_2, u_2 are adjacent in G_2. Note that G_2 can contain isolated vertices.

Shiu et al. [251] studied the composition of graphs. They proved that the composition of a cycle C_m and a totally disconnected graph D_n is supermagic if $3 \leq m \equiv 1 \pmod 2$ and $n = 2$ or $n = 6$, or if $m \geq 3$, $n \geq 3$, and $n \neq 6$, or if $4 \leq m \equiv 0 \pmod 2$. The composition of K_2 and a totally disconnected graph D_n, $K_2 \circ D_n \cong K_{n,n}$, is supermagic if $n = 1$ or $n \geq 3$. They showed that the composition of an r-regular supermagic graph and D_n is also supermagic. Ivančo [136] proved a more general assertion.

Theorem 2.8.14 ([136]) *Let G_1, G_2 be regular graphs satisfying the following conditions.*

(i) $|V(G_2)| \geq 3$.
(ii) $|V(G_1)|G_2$ is supermagic or G_2 is totally disconnected.
(iii) $|V(G_2)| \equiv 0 \pmod 2$ or $|V(G_2)||E(G_1)| \equiv 1 \pmod 2$.

Then $G_1 \circ G_2$ is a supermagic graph.

Ho and Lee [128] proved that the composition of a complete graph K_k and a graph D_n, i.e., $K_{k[n]}$, is supermagic for $k = 3$ or 5, and $n = 2$ or n odd.

2.8.4 Zykovian Product of Graphs

The *Zykovian product* $G_1[G_2]$ of graphs G_1, G_2 is a graph $G_1 \otimes G_2$ with the vertex set $V(G_1 \otimes G_2) = V(G_1) \cup V(G_2)$ and the edge set $E(G_1 \otimes G_2) = E(G_1) \cup E(G_2) \cup \{uv,$ for all $u \in V(G_1), v \in V(G_2)\}$.

Trenkler [277] proved

Theorem 2.8.15 ([277]) *Let G be a semi-magic graph none of whose components are isomorphic to K_2 or K_3. Then $G \otimes K_1$ is magic.*

In [279] Trenkler proved that the Zykovian product of a totally disconnected graph on n vertices and a complete graph K_n is not a magic graph.

2.8.5 Circulant Graphs

Let n, m and $a_1, a_2, \ldots, a_m$ be positive integers, $1 \leq a_i \leq \lfloor n/2 \rfloor$ and $a_i \neq a_j$, for all $1 \leq i, j \leq m$. An undirected graph with the set of vertices $V = \{v_1, v_2, \ldots, v_n\}$ and the set of edges $E = \{v_i v_{i+a_j} : 1 \leq i \leq n, 1 \leq j \leq m\}$, the indices being taken modulo n, is called a *circulant graph* and denoted by $C_n(a_1, a_2, \ldots, a_m)$. The numbers $a_1, a_2, \ldots, a_m$ are called the *generators* and we say that the edge $v_i v_{i+a_j}$ is *of type* a_j.

It is easy to see that the circulant graph $C_n(a_1, a_2, \ldots, a_m)$ is a regular graph of degree r, where

$$r = \begin{cases} 2m - 1 & \text{if } \frac{n}{2} \in \{a_1, a_2, \ldots, a_m\} \\ 2m & \text{otherwise.} \end{cases}$$

The circulant graph $C_n(a_1, a_2, \ldots, a_m)$ is connected, see [73], if and only if for the greatest common divisor of the numbers $a_1, a_2, \ldots, a_m, n$

$$\gcd(a_1, a_2, \ldots, a_m, n) = 1. \tag{2.24}$$

More precisely, $C_n(a_1, a_2, \ldots, a_m)$ has $d = \gcd(a_1, a_2, \ldots, a_m, n)$ connected components which are isomorphic to $C_{n/d}(a_1/d, a_2/d, \ldots, a_m/d)$.

Heuberger [127] characterized bipartite circulant graphs.

Theorem 2.8.16 ([127]) *Let $G = C_n(a_1, a_2, \ldots, a_m)$ be a connected circulant graph. Then G is bipartite if and only if $a_1, a_2, \ldots, a_m$ are odd and n is even.*

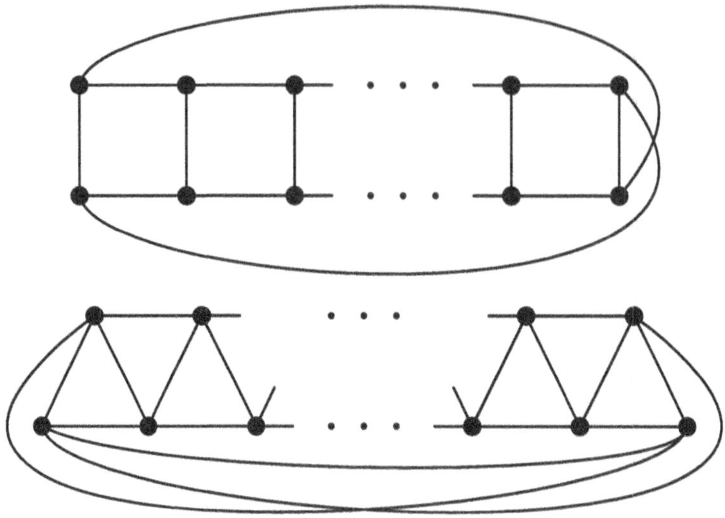

Fig. 2.20 Möbius ladders M_n for n even and n odd

The *Möbius ladder* M_n, where $6 \leq n \equiv 0 \pmod 2$, is the 3-regular graph consisting of the cycle C_n of length n, in which all pairs of the opposite vertices are connected. For $5 \leq n \equiv 1 \pmod 2$, the Möbius ladder M_n is the 4-regular graph consisting of the cycle C_n of length n together with two chords at each vertex v joining v to the two most opposite vertices of C_n. Figure 2.20 illustrates Möbius ladders M_n, for n even and n odd.

For n even, the Möbius ladder M_n is isomorphic to the 3-regular circulant graph $C_n(1, n/2)$. Sedláček [231] proved the following result.

Theorem 2.8.17 ([231]) *Let $n \geq 6$ be an even integer. The Möbius ladder M_n is supermagic if and only if $n \equiv 2 \pmod 4$.*

Note that Sedláček [231] proved that the Möbius ladder M_{2m+1} is magic. It is known, see [262], that M_5 is not supermagic. Let us recall the following open problem

Open Problem 2.8.1 ([231]) Decide whether the Möbius ladder M_{2m+1} is supermagic for some m, $m \neq 2$.

Notice that the complete graph K_n is the $(n-1)$-regular circulant graph $C_n(1, 2, \dots, \lfloor n/2 \rfloor)$.

Lemma 2.8.1 ([234]) *Let $G = C_n(a_1, n/2)$ be a 3-regular circulant graph and let $d = \gcd(a_1, n/2)$. Then G is magic if and only if $n/d \equiv 2 \pmod 4$ and $a_1/d \equiv 1 \pmod 2$.*

Proof According to Theorem 2.3.7, G is magic if and only if its components are magic graphs. Since $d = \gcd(a_1, n/2) = \gcd(a_1, n/2, n)$, G consists of d

connected components isomorphic to $C_{n/d}(a_1/d, n/(2d))$. Since $\gcd(a_1/d, n/(2d)) = \gcd(a_1/d, n/(2d), n/d) = 1$, then it is easy to see that $k = \gcd(a_1/d, n/d)$ is equal to either 2 or 1.

Suppose $k = 2$. Then $C_{n/d}(a_1/d, n/(2d))$ is isomorphic to the Cartesian product $C_{n/(2d)} \Box K_2$, where $n/(2d) \equiv 1 \pmod 2$. In [277] Trenkler proved that $C_{2l+1} \Box K_2$ is not magic for all positive integers l. So in this case the graph $C_{n/d}(a_1/d, n/(2d))$ is not magic.

Suppose $k = 1$. Then $n/d \equiv 0 \pmod 2$ and $a_1/d \equiv 1 \pmod 2$. It is easy to see that $C_{n/d}(a_1/d, n/(2d))$ is isomorphic to $C_{n/d}(1, n/(2d))$, and this graph is isomorphic to Möbius ladder $M_{n/d}$. According to Theorem 2.8.17 we have that $M_{n/d}$ is magic for $n/d \equiv 2 \pmod 4$ and for $n/d \equiv 0 \pmod 4$ $M_{n/d}$ is not magic. □

Lemma 2.8.2 ([234]) *Every 4-regular circulant graph is magic.*

Proof Let $G = C_n(a_1, a_2)$ be a 4-regular circulant graph with the minimum number of vertices that is not magic. According to Theorem 2.3.7, G is connected, i.e., $\gcd(a_1, a_2, n) = 1$.

Since G is not magic, it is of type $\mathcal{A}$. Thus $n \equiv 0 \pmod 2$ and G is a connected non-bipartite graph. So, one of the generators is odd and the other one is even. Without loss of generality, let

$$a_1 \equiv 1 \pmod 2, \quad a_2 \equiv 0 \pmod 2.$$

Put $d_1 = \gcd(a_1, n)$ and $d_2 = \gcd(a_2, n)$. Then

$$1 \le d_1 \equiv 1 \pmod 2, \quad \text{and}$$
$$2 \le d_2 \equiv 0 \pmod 2.$$

The edges of type a_1 form d_1 disjoint cycles of length n/d_1, which we denote by $C^1_{a_1}, C^2_{a_1}, \ldots, C^{d_1}_{a_1}$. Because $n/d_1 \equiv 0 \pmod 2$, they are of even length. The edges of type a_2 form d_2 disjoint cycles of length n/d_2. They are denoted by $C^1_{a_2}, C^2_{a_2}, \ldots, C^{d_2}_{a_2}$.

We will consider the following cases.

Let $d_1 = 1$. In [127] it is shown that if $\gcd(a_1, n) = 1$, then there exists a positive integer $2 \le k < n/2$ such that G is isomorphic to the graph $C_n(1, k)$. Since $a_2 \equiv 0 \pmod 2$, $k \equiv 0 \pmod 2$. For every edge e of G it is easy to find two edge-disjoint odd cycles, see Fig. 2.21, such that e does not lie on this cycle. (In Fig. 2.21 the edge e is denoted by e_1 if it is of type a_1 and e_2 if it is of type a_2.) This contradicts the fact that G is of type $\mathcal{A}$.

Let $d_1 \ge 3$. First we show that there exists an odd cycle in G consisting of the edges of both types. Note that there exists an odd cycle in G if there exist $x, y \in \mathbb{Z}$

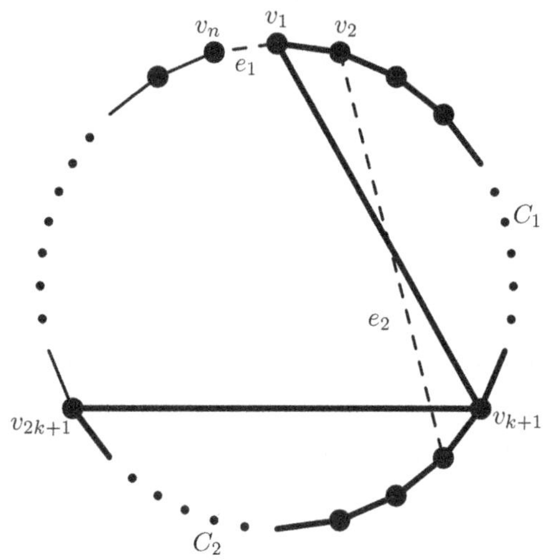

Fig. 2.21 Two edge-disjoint cycles not containing edge e

such that

$$a_1x + a_2y \equiv 0 \pmod{n} \tag{2.25}$$

$$x + y \equiv 1 \pmod{2} \tag{2.26}$$

$$x + y \leq n. \tag{2.27}$$

Since n/d_1 and a_1/d_1 are coprime, there exist integers r, s such that $1 = rn/d_1 + sa_1/d_1$, i.e., s is a modular inverse of the element a_1/d_1 (modulo n/d_1).

Set $y = d_1$ and x to the smallest positive integer such that $x \equiv -sa_2 \pmod{n/d_1}$. (Evidently $x < n/d_1$.) It is not difficult to show that the couple (x, y) is the solution of (2.25), (2.26) and (2.27).

Let v_i be an arbitrary vertex of the graph G. Let $\mathfrak{C}_{v_i}$ denote the cycle $\mathfrak{C}_{v_i} = v_i v_{i+a_1} \dots v_{i+xa_1} v_{i+xa_1+a_2} \dots v_{i+xa_1+(y-1)a_2} v_i$. According to (2.25) and (2.27), $\mathfrak{C}_{v_i}$ is a cycle in G. According to (2.26), it is of odd length.

Note that the edges $v_i v_{i+a_1}, v_{i+a_1} v_{i+2a_1}, \dots, v_{i+(x-1)a_1} v_{i+xa_1}$ are of type a_1 and there exists $C^j_{a_1}$, $j \in \{1, 2, \dots, d_1\}$ such that all of them lie on this cycle. The edges $v_{i+xa_1} v_{i+xa_1+a_2}, v_{i+xa_1+a_2} v_{i+xa_1+2a_2}, \dots, v_{i+xa_1+(y-1)a_2} v_i$ are of type a_2 and there exists $C^k_{a_2}$, $k \in \{1, 2, \dots, d_2\}$ such that all of them lie on this cycle.

Since G is of type $\mathcal{A}$, there exists a set of edges $\{e_1, e_2\}$ such that $G - \{e_1, e_2\}$ is a bipartite graph.

Consider the following cases.

Case A Let the edges e_1, e_2 be of type a_2.

If e_1, e_2 lie on the same cycle, say $C^1_{a_2}$, then consider the vertex $v_i \notin V(C^1_{a_2})$. Then $\mathfrak{C}_{v_i}$ is an odd cycle containing neither e_1 nor e_2, a contradiction.

Let e_1, e_2 lie on different cycles, say $e_1 \in C^1_{a_2}$, $e_2 \in C^2_{a_2}$. Moreover, suppose that $e_1 = v_j v_{j+a_2}$, the indices being taken modulo n. Then consider the cycle $\mathfrak{C}_{v_j}$. Evidently $e_1 \notin \mathfrak{C}_{v_j}$. If $e_2 \in C^2_{a_2}$ is some edge of $\mathfrak{C}_{v_j}$, then also $v_{j+xa_1+(y-1)a_2}v_j \in C^2_{a_2}$. So v_j is incident with the edge $e_1 \in C^1_{a_2}$ and also with the edge $v_{j+xa_1+(y-1)a_2}v_j \in C^2_{a_2}$. This contradicts the fact that $C^1_{a_2}$ and $C^2_{a_2}$ are disjoint.

Case B Let the edges e_1, e_2 be of type a_1. Analogously to Case A, we get a contradiction.

Case C Let e_1 be of type a_1 and e_2 be of type a_2. Without loss of generality, let $e_1 \in C^1_{a_1}$ and $e_2 \in C^1_{a_2}$. Then consider the vertex v_i, such that $v_i \in V(C^2_{a_2})$ and $v_i \notin V(C^1_{a_1})$. The cycle $\mathfrak{C}_{v_i}$ is an odd cycle in $G - \{e_1, e_2\}$, a contradiction.

□

Lemma 2.8.3 ([234]) *Every 5-regular circulant graph is magic.*

Proof Let $G = C_n(a_1, a_2, n/2)$ be a 5-regular circulant graph with the smallest number of vertices that is not a magic graph. Thus G is connected, i.e.,

$$1 = \gcd(a_1, a_2, \frac{n}{2}, n) = \gcd(a_1, a_2, \frac{n}{2}).$$

Since G is not a magic graph, it is of type $\mathcal{A}$. So it is a non-bipartite graph. Thus at least one of the generators $a_1, a_2, n/2$ is odd, and at least one of them is even. Consider the 4-regular graph $H = C_n(a_1, a_2)$. According to Lemma 2.8.2, H is a magic graph.

Suppose that H is connected, i.e., $\gcd(a_1, a_2, n) = 1$. Since G is not magic, by repeated application of Theorem 2.9.4, H is bipartite. Then a_1, a_2 are odd and $n/2 \equiv 0 \pmod 2$. It is not difficult to check that the edges of type $n/2$ are joining vertices in the same partition classes of the 4-edge connected graph H. Since their number is $n/2 \geq 4$, the graph G is not of type $\mathcal{A}$, a contradiction.

Suppose H is disconnected. As $p = \gcd(a_1, a_2, n) \neq 1$ and $\gcd(a_1, a_2, n/2) = 1$, we get $p = 2$. This means that H consists of two magic components isomorphic to $C_{n/2}(a_1/2, a_2/2)$. Thus $C_n(a_1, a_2, n/2)$ is isomorphic to $C_{n/2}(a_1/2, a_2/2) \square K_2$. According to Theorem 2.8.6, G is a magic graph, a contradiction. □

Lemma 2.8.4 ([234]) *Let $G = C_n(a_1, a_2, \dots, a_m)$ be an r-regular circulant graph, where $r \geq 6$. Then G is a magic graph.*

Proof Let $G = C_n(a_1, a_2, \dots, a_m)$ be an r-regular circulant graph, $r \geq 6$, with the smallest possible number of vertices and the smallest possible number of edges, that is not a magic graph. According to Theorems 2.3.7 and 2.3.3, G is connected and of type $\mathcal{A}$. So $n \equiv 0 \pmod 2$ and G is a non-bipartite graph. Hence at least one of the generators, say a_1, is even (as G is non-bipartite) and at least one of the generators, say a_2, is odd (as G is connected). Since $r \geq 6$, there exists a generator $a_3 \notin \{a_1, a_2\}$. Consider the graph

$$H = C_n(a_1, a_2, a_4, \dots, a_m).$$

Then H is a non-bipartite circulant graph of degree $r_H \geq 4$ and with fewer edges than G. According to Lemma 2.8.2 for $r_H = 4$, or to Lemma 2.8.3 for $r_H = 5$, or according to the choice of G for $r_H \geq 6$, we have that H is a magic graph.

Since G is not magic, then by repeated applications of Theorem 2.9.4, H is disconnected. Since a_2 is odd, we have $\gcd(a_1, a_2, a_4, \ldots, a_m, n) \neq 2$, thus H consists of at least three isomorphic, non-bipartite components. So there exist three edge-disjoint odd cycles in H, and so in G. Hence, G is not of type $\mathcal{A}$, a contradiction. □

The following theorem provides a characterization of magic circulant graphs. The proof follows from Lemmas 2.8.1, 2.8.2, 2.8.3, and 2.8.4.

Theorem 2.8.18 ([234]) *Let $G = C_n(a_1, a_2, \ldots, a_m)$ be a circulant graph of degree $r \geq 3$. Then G is a magic graph if and only if*

$$r = 3 \quad \text{and} \quad \frac{n}{d} \equiv 2 \pmod 4, \quad \frac{a_1}{d} \equiv 1 \pmod 2, \quad \text{where } d = \gcd\left(a_1, \frac{n}{2}\right),$$

or

$$r \geq 4.$$

Theorem 2.5.2 gives necessary conditions for a circulant graph to be supermagic. Hence we have

Corollary 2.8.1 ([234]) *Let $G = C_n(a_1, a_2, \ldots, a_m)$ be an r-regular circulant graph and let $d = \gcd(a_1, a_2, \ldots, a_m, n)$. If*

$$r \equiv 1 \pmod 2 \quad \text{and} \quad n \equiv 0 \pmod 4,$$

or

$$r \equiv 2 \pmod 4, \; n \equiv 0 \pmod 2 \quad \text{and} \quad \frac{n}{d} \equiv 1 \pmod 2,$$

then G is not a supermagic graph.

In [234] Semaničová gave a characterization of 3-regular circulant supermagic graphs.

Theorem 2.8.19 ([234]) *Let $G = C_n(a_1, n/2)$ be a 3-regular circulant graph and let $d = \gcd(a_1, n/2)$. Then G is supermagic if and only if $n/d \equiv 2 \pmod 4$, $a_1/d \equiv 1 \pmod 2$ and $d \equiv 1 \pmod 2$.*

Proof Let $G = C_n(a_1, n/2)$ be a 3-regular circulant graph and $d = \gcd(a_1, n/2)$.

Suppose that G is supermagic. Thus G is magic and according to Lemma 2.8.1, $n/d \equiv 2 \pmod 4$ and $a_1/d \equiv 1 \pmod 2$. If $d \equiv 0 \pmod 2$, then $n \equiv 0 \pmod 4$ and so, by Corollary 2.8.1, $d \equiv 1 \pmod 2$.

On the other hand, suppose $n/d \equiv 2 \pmod 4$, $a_1/d \equiv 1 \pmod 2$, and $d \equiv 1 \pmod 2$. Then G consists of d connected components isomorphic to graph $C_{n/d}(a_1/d, n/(2d))$. Since $a_1/d \equiv 1 \pmod 2$, it follows that $C_{n/d}(a_1/d, n/(2d))$ is isomorphic to Möbius ladder $M_{n/d}$. As $n/d \equiv 2 \pmod 4$, then, by Sedláček [230], $M_{n/d}$ is supermagic. Ivančo [136] proved that if H is a supermagic graph decomposable into odd number of edge-disjoint δ-regular factors, then kH (k disjoint copies of H) is supermagic for every odd positive integer k. The graph $C_{n/d}(a_1/d, n/(2d))$ is decomposable into three edge-disjoint 1-factors. Since $d \equiv 1 \pmod 2$, then the graph $C_n(a_1, n/2)$ is also supermagic. □

Hartsfield and Ringel [124] proved that if a bipartite graph is decomposable into even number of Hamilton cycles, then it is supermagic. In [16] it is proved that if the generators of a circulant graph $C_n(a_1, a_2, \ldots, a_{2k})$ satisfy the condition $\gcd(a_{2j-1}, a_{2j}, n) = 1$, for every $j \in \{1, 2, \ldots, k\}$, then $C_n(a_1, a_2, \ldots, a_{2k})$ has a Hamiltonian decomposition.

According to these assertions we have

Theorem 2.8.20 ([234]) *Let $G = C_n(a_1, a_2, \ldots, a_{2k})$ be a circulant graph. Let $n \equiv 0 \pmod 2$, $a_i \equiv 1 \pmod 2$ and $\gcd(a_{2j-1}, a_{2j}, n) = 1$, for every $i \in \{1, 2, \ldots, 2k\}$, $j \in \{1, 2, \ldots, k\}$. Then G is supermagic.*

Immediately from the previous theorem we obtain

Corollary 2.8.2 ([139]) *If n, k, a are positive integers, where a is odd and $2a + 4k - 2 \le n \equiv 0 \pmod 2$, then the $4k$-regular circulant graph $C_n(a, a+2, \ldots, a+2(k-1))$ is supermagic.*

In [143] Ivančo and Semaničová proved the following result.

Theorem 2.8.21 ([143]) *If G is a $4k$-regular circulant graph of odd order, then the Cartesian product of graphs G and K_2 is supermagic.*

Since for n even the graph $C_n(2, 4, \ldots, 4k, n/2)$ is isomorphic to the Cartesian product of graphs $C_{n/2}(1, 2, \ldots, 2k)$ and K_2, we have immediately

Corollary 2.8.3 ([139]) *If n, k are positive integers, $8k+2 \le n \equiv 2 \pmod 4$, then the circulant graph $C_n(2, 4, \ldots, 4k, n/2)$ is supermagic.*

Ivančo et al. [139] proved

Lemma 2.8.5 ([139]) *Let $C_n(a, a+b)$ be a 4-regular circulant graph and let v be its vertex. If $\gcd(n, b) = 1$, then there exists a labeling $f : E(C_n(a, a+b)) \to \{1, 2, \ldots, 2n\}$ such that, for every vertex $u \in V(C_n(a, a+b))$,*

$$f^{\star}(u) = \begin{cases} 3n+3 & \text{if } u = v \\ 4n+3 & \text{if } u \neq v. \end{cases}$$

Proof Since b and n are coprime, there exists a positive integer c (obviously, $c \equiv ab^{-1} \pmod n$) such that the graph $C_n(a, a+b)$ is isomorphic to the graph $C_n(c, c+$

1). Without loss of generality, we can assume to have the graph $C_n(c, c+1)$ with vertex set $\{v_1, v_2, \ldots, v_n\}$, where $v = v_1$. Define a mapping $f : E(C_n(c, c+1)) \to \{1, 2, \ldots, 2n\}$ by

$$f(v_i, v_{i+c+1}) = i \qquad \text{if } 1 \le i \le n$$

$$f(v_i, v_{i+c}) = \begin{cases} 2n+2-i & \text{if } 2 \le i \le n \\ n+1 & \text{if } i = 1. \end{cases}$$

It is easy to check that

$$f^\star(v_i) = \begin{cases} 3n+3 & \text{if } i = 1 \\ 4n+3 & \text{if } 2 \le i \le n. \end{cases}$$

Thus, the labeling f has the required properties. □

Theorem 2.8.22 ([139]) *Let $G = C_n(a, a+b, c, c+d)$ be a circulant graph of degree* 8. *If* $\gcd(b, n) = 1$ *and* $\gcd(d, n) = 1$, *then G is a supermagic graph.*

Proof The graph G is decomposable into two edge-disjoint 4-factors $G_1 = C_n(a, a+b)$ and $G_2 = C_n(c, c+d)$. According to Lemma 2.8.5, there exist labelings $f : E(G_1) \to \{1, 2, \ldots, 2n\}$ and $g : E(G_2) \to \{1, 2, \ldots, 2n\}$ such that

$$f^\star(v_1) = g^\star(v_1) = 3n+3,$$
$$f^\star(v_2) = f^\star(v_3) = \cdots = f^\star(v_n) = g^\star(v_2) = g^\star(v_3) = \cdots = g^\star(v_n) = 4n+3.$$

Consider a mapping $h : E(G) \to \{1, 2, \ldots, 4n\}$ defined by

$$h(e) = \begin{cases} f(e) & \text{if } e \in E(G_1) \\ 4n+1-g(e) & \text{if } e \in E(G_2). \end{cases}$$

Obviously, $h^\star(v) = 16n+4$, for every $v \in V(G)$, and thus h is a supermagic labeling of G. □

By choosing $b = d = 1$ and $c = a+2$, we get

Corollary 2.8.4 ([139]) *Let n, a be positive integers. The* 8*-regular circulant graph $C_n(a, a+1, a+2, a+3)$ is supermagic for all $n > 2a+6$.*

By induction it immediately follows that

Corollary 2.8.5 ([139]) *Let n, a, k be positive integers. The $8k$-regular circulant graph $C_n(a, a+1, \ldots, a+4k-1)$ is supermagic for all $n > 2a+8k-2$.*

In [139] it is proved that

Theorem 2.8.23 ([139]) *The circulant graph $C_n(1, 2, 3)$ has a supermagic labeling for all $n \geq 7$.*

Proof For any integer $n \geq 3$, let $H_n = H_n(1, 2, 3)$ be a graph with the vertex set $\{w_1, w_2, \ldots, w_{n+3}\}$ and edge set $\bigcup_{i=1}^{n}\{w_i w_{i+1}, w_{i+1} w_{i+3}, w_i w_{i+3}\}$. Notice that the vertices of the graph H_n are all of even degree, see Figs. 2.22, 2.23, 2.24, and 2.25. The vertices w_1, w_{n+2}, and w_{n+3} are of degree 2, for $n \geq 3$ the vertices w_2, w_3, and w_{n+1} are of degree 4, and for $n \geq 4$ all the remaining vertices are of degree 6. It is easy to observe that, by identifying the pairs w_1 and w_{n+1}, w_2 and w_{n+2}, w_3, and w_{n+3}, we obtain, for $n \geq 7$, from H_n the circulant graph $C_n(1, 2, 3)$.

The construction of the required labeling is done in two steps. First we find a labeling λ_n, called the *auxiliary labeling*, of H_n, for any $n \geq 3$. Then we show how to obtain from λ_n a supermagic labeling f of $C_n(1, 2, 3)$.

Since the graph H_n is a subgraph of a graph H_m, for $3 \leq n \leq m$, we can define the auxiliary labeling $\lambda_n : E(H_n) \to \{1, 2, \ldots, 3n\}$ recursively as follows. The labelings of the graphs H_3, H_4, H_5, and H_6 are given in Figs. 2.22, 2.23, 2.24, and

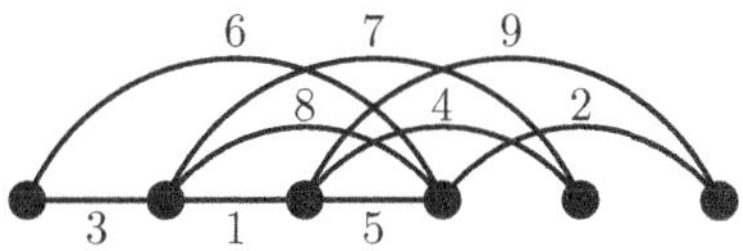

Fig. 2.22 Auxiliary labeling of $H_3(1, 2, 3)$

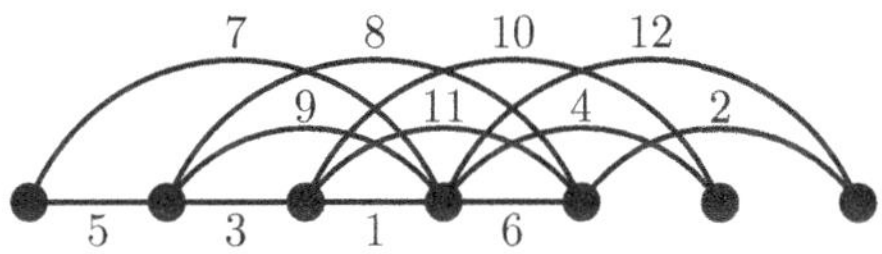

Fig. 2.23 Auxiliary labeling of $H_4(1, 2, 3)$

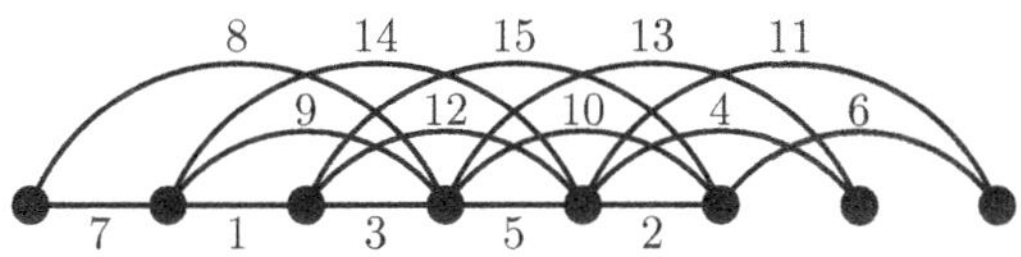

Fig. 2.24 Auxiliary labeling of $H_5(1, 2, 3)$

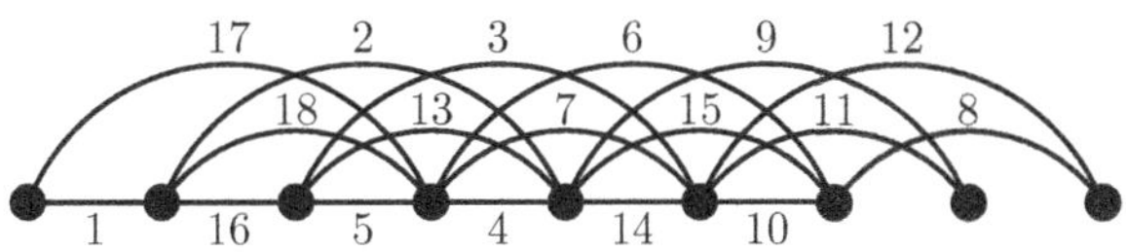

Fig. 2.25 Auxiliary labeling of $H_6(1, 2, 3)$

2.25. For larger n, the auxiliary labeling is defined by

$$\lambda_{n+4}(e) = \begin{cases} \lambda_n(e) + 6 & \text{for } e \in E(H_n) \\ 1 & \text{for } e = w_{n+3}w_{n+4} \\ 2 & \text{for } e = w_{n+5}w_{n+7} \\ 3 & \text{for } e = w_{n+2}w_{n+3} \\ 4 & \text{for } e = w_{n+4}w_{n+6} \\ 5 & \text{for } e = w_{n+1}w_{n+2} \\ 6 & \text{for } e = w_{n+4}w_{n+5} \\ 3n+7 & \text{for } e = w_{n+1}w_{n+4} \\ 3n+8 & \text{for } e = w_{n+2}w_{n+5} \\ 3n+9 & \text{for } e = w_{n+2}w_{n+4} \\ 3n+10 & \text{for } e = w_{n+3}w_{n+6} \\ 3n+11 & \text{for } e = w_{n+3}w_{n+5} \\ 3n+12 & \text{for } e = w_{n+4}w_{n+7}. \end{cases}$$

It is easy to verify that for every $n \in \{3, 4, 5, 6\}$ and all $i = 1, 2, \dots, n+3$,

$$\lambda_n^{\star}(w_i) = \begin{cases} \frac{1}{2}\deg(w_i)(3n+1) - 1 & \text{for } i = 1, 2, 3 \\ \frac{1}{2}\deg(w_i)(3n+1) + 1 & \text{for } i = n+1, n+2, n+3 \\ \frac{1}{2}\deg(w_i)(3n+1) & \text{otherwise.} \end{cases} \tag{2.28}$$

Similarly, for the first three vertices of H_{n+4},

$$\lambda_{n+4}^{\star}(w_1) = \lambda^{\star}(w_1) + 6\deg_{H_n}(w_1) = 3n + 12 = 3(n+4)$$
$$\lambda_{n+4}^{\star}(w_2) = \lambda^{\star}(w_2) + 6\deg_{H_n}(w_2) = (6n+1) + 24 = 6(n+4) + 1$$
$$\lambda_{n+4}^{\star}(w_3) = \lambda^{\star}(w_3) + 6\deg_{H_n}(w_3) = (6n+1) + 24 = 6(n+4) + 1.$$

For the vertices w_i, $i = 4, 5, \dots, n$, we have

$$\lambda_{n+4}^{\star}(w_i) = \lambda^{\star}(w_i) + 6\deg_{H_n}(w_i) = (9n+3) + 36 = 9(n+4) + 3.$$

Now for the remaining vertices in $V(H_n)$,

$$\lambda_{n+4}^{\star}(w_{n+1}) = \lambda^{\star}(w_{n+1}) + 6\deg_{H_n}(w_{n+1}) + 5 + (3n+7) = 9(n+4) + 3$$
$$\begin{aligned}\lambda_{n+4}^{\star}(w_{n+2}) &= \lambda^{\star}(w_{n+2}) + 6\deg_{H_n}(w_{n+2}) + 3 + 5 + (3n+8) + (3n+9) \\ &= 9(n+4) + 3\end{aligned}$$
$$\begin{aligned}\lambda_{n+4}^{\star}(w_{n+3}) &= \lambda^{\star}(w_{n+3}) + 6\deg_{H_n}(w_{n+3}) + 1 + 3 + (3n+10) + (3n+11) \\ &= 9(n+4) + 3.\end{aligned}$$

Finally, for the four last vertices in $V(H_{n+4})$,

$$\lambda^\star_{n+4}(w_{n+4}) = 1+4+6+(3n+7)+(3n+9)+(3n+12) = 9(n+4)+3$$
$$\lambda^\star_{n+4}(w_{n+5}) = 2+6+(3n+8)+(3n+11) = 6(n+4)+3$$
$$\lambda^\star_{n+4}(w_{n+6}) = 4+(3n+10) = 3(n+4)+2$$
$$\lambda^\star_{n+4}(w_{n+7}) = 2+(3n+12) = 3(n+4)+2.$$

Thus, the equations in (2.28) hold for every vertex of H_n and for all $n \geq 3$.

For $n \geq 7$ we obtain the 6-regular circulant graph $C_n(1,2,3)$ from the graph H_n by identifying vertices w_1 and w_{n+1}, w_2 and w_{n+2}, and w_3 and w_{n+3}. More precisely, the mapping $\xi : V(H_n) \to V(C_n(1,2,3))$, given by $\xi(w_i) = v_i$, for all $i = 1,2,\ldots,n$, and $\xi(w_{n+j}) = v_j$, for $j = 1,2,3$, is a homomorphism of graphs H_n and $C_n(1,2,3)$. Moreover, the homomorphism ξ induces a bijective mapping ξ_E from $E(H_n)$ to $E(C_n(1,2,3))$. Consider the labeling $f : E(C_n(1,2,3)) \to \{1,2,\ldots,3n\}$ defined by $f(e) = \lambda_n(\xi_E^{-1}(e))$. Evidently,

$$f^\star(v_i) = \begin{cases} \lambda^\star_n(w_i) + \lambda^\star_n(w_{n+i}) & \text{for } 1 \leq i \leq 3 \\ \lambda^\star_n(w_i) & \text{for } 4 \leq i \leq n. \end{cases}$$

Therefore, $f^\star(v) = 9n+3$, for every vertex $v \in V(C_n(1,2,3))$. Thus, f is a supermagic labeling of the circulant graph $C_n(1,2,3)$. □

According to Corollary 2.8.2, the circulant graph $C_n(1,3)$ is supermagic for every even integer $n \geq 8$. However, we can extend this claim.

Theorem 2.8.24 ([139]) *The circulant graph $C_n(1,3)$ has a supermagic labeling for all $n \geq 7$.*

Proof For any integer $n \geq 4$, let $H_n = H_n(1,3)$ be a graph with the vertex set $\{w_1, w_2, \ldots, w_{n+3}\}$ and the edge set $\bigcup_{i=1}^{n}\{w_i w_{i+3}\} \cup \bigcup_{i=3}^{n}\{w_i w_{i+1}\} \cup \{w_1 w_2, w_{n+2} w_{n+3}\}$. Notice that the vertices of the graph H_n are all of even degree, see Figs. 2.26, 2.27, 2.28, and 2.29. The vertices w_1, w_2, w_3, w_{n+1}, w_{n+2}, and w_{n+3} are of degree 2 and all the remaining vertices are of degree 4. It is easy to observe that by identifying the pairs w_1 and w_{n+1}, w_2 and w_{n+2}, and w_3 and w_{n+3}, we obtain, for $n \geq 7$, from H_n the circulant graph $C_n(1,3)$.

Since the graph H_n is a subgraph of a graph H_m, for $4 \leq n \leq m-2$, we can define the auxiliary labeling $\lambda_n : E(H_n) \to \{1,2,\ldots,2n\}$ recursively for $n \neq 5$ as follows. The labelings of the graphs H_4, H_6, H_7, and H_9 are given in Figs. 2.26,

Fig. 2.26 Auxiliary labeling of $H_4(1,3)$

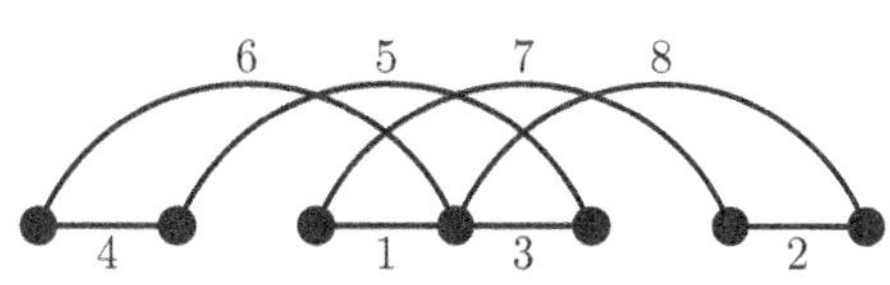

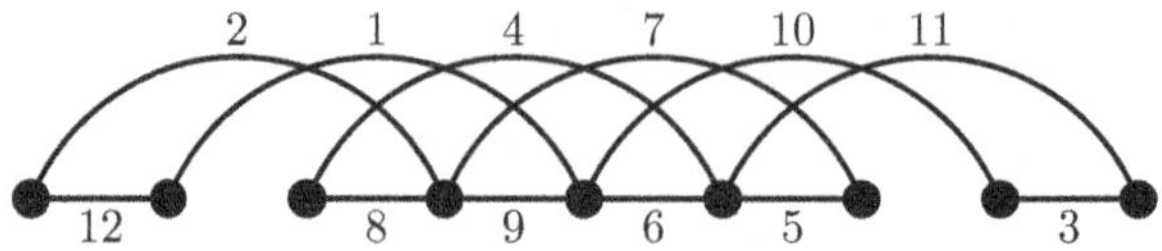

Fig. 2.27 Auxiliary labeling of $H_6(1, 3)$

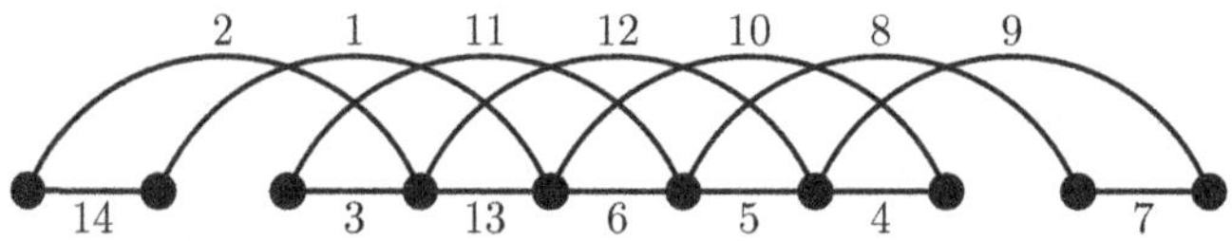

Fig. 2.28 Auxiliary labeling of $H_7(1, 3)$

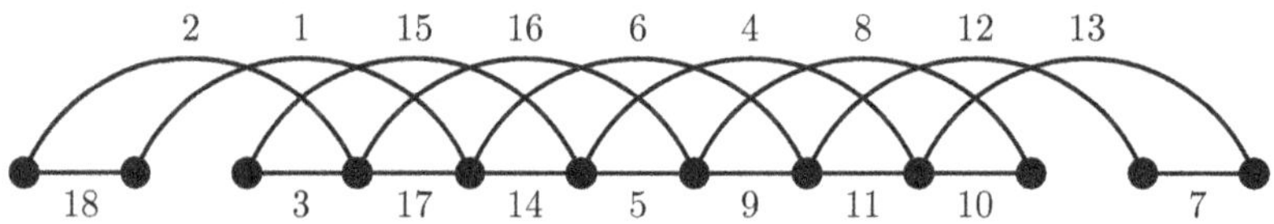

Fig. 2.29 Auxiliary labeling of $H_9(1, 3)$

2.27, 2.28, and 2.29. For $n = 8$ and for $n \geq 10$, the auxiliary labeling is defined by

$$\lambda_{n+4}(e) = \begin{cases} \lambda_n(e) + 4 & \text{for } e \in E(H_n) \\ 1 & \text{for } e = w_{n+3}w_{n+4} \\ 2 & \text{for } e = w_{n+6}w_{n+7} \\ 3 & \text{for } e = w_{n+4}w_{n+5} \\ 4 & \text{for } e = w_{n+1}w_{n+2} \\ 2n+5 & \text{for } e = w_{n+2}w_{n+5} \\ 2n+6 & \text{for } e = w_{n+1}w_{n+4} \\ 2n+7 & \text{for } e = w_{n+3}w_{n+6} \\ 2n+8 & \text{for } e = w_{n+4}w_{n+7}. \end{cases}$$

As in the proof of Theorem 2.8.23, it is easy to verify that for every vertex of H_n, $n \geq 6$,

$$\lambda_n^\star(w_i) = \begin{cases} \frac{1}{2}\deg(w_i)(2n+1) - 1 & \text{for } i = 1, n+3 \\ \frac{1}{2}\deg(w_i)(2n+1) + 1 & \text{for } i = 3, n+1 \\ \frac{1}{2}\deg(w_i)(2n+1) & \text{otherwise.} \end{cases}$$

For $n \geq 7$ we obtain the 4-regular circulant graph $C_n(1, 3)$ from the graph H_n by identifying vertices w_1 and w_{n+1}, w_2 and w_{n+2}, and w_3 and w_{n+3}. More precisely, the mapping $\xi : V(H_n) \to V(C_n(1, 3))$, given by $\xi(w_i) = v_i$, for all $i = 1, 2, \ldots, n$, and $\xi(w_{n+j}) = v_j$, for $j = 1, 2, 3$, is a homomorphism of the graphs H_n and $C_n(1, 3)$. Moreover, the homomorphism ξ induces a bijective mapping ξ_E from $E(H_n)$ to $E(C_n(1, 3))$. Now the edge labeling f, given by $f(e) = \lambda_n(\xi_E^{-1}(e))$, similarly as in Theorem 2.8.23, is a supermagic labeling of the circulant graph $C_n(1, 3)$ with index $4n + 2$. □

Theorem 2.8.25 ([139]) *The circulant graph $C_n(1, 2, 3, 4, 6)$ has a supermagic labeling for all $n \geq 13$.*

Proof For any integer $n \geq 4$, let $H_n = H_n(1, 2, 3, 4, 6)$ be a graph given by

$$V(H_n) = \{w_1, w_2, \ldots, w_{n+6}\}$$

$$E(H_n) = \bigcup_{i=1}^{n} \{w_{i+2}w_{i+3}, w_{i+2}w_{i+4}, w_{i+3}w_{i+6}, w_i w_{i+4}, w_i w_{i+6}\}.$$

All vertices of the graph H_n are of even degree, see Figs. 2.30, 2.31, 2.32, and 2.33. It is easy to observe that, for $n \geq 13$, we obtain from H_n the circulant graph $C_n(1, 2, 3, 4, 6)$ by identifying the pairs w_i and w_{n+i}, for $1 \leq i \leq 6$.

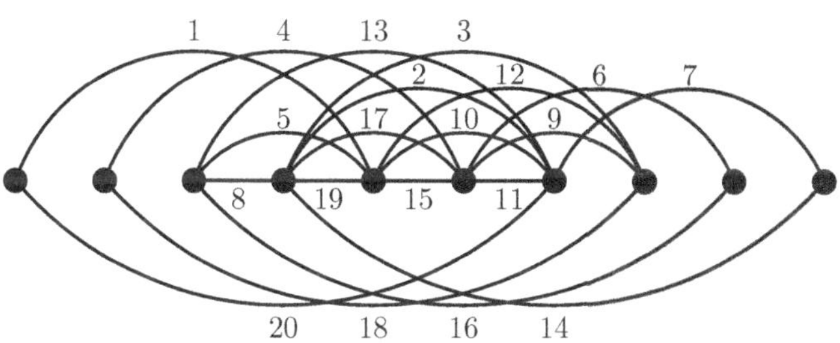

Fig. 2.30 Auxiliary labeling of $H_4(1, 2, 3, 4, 6)$

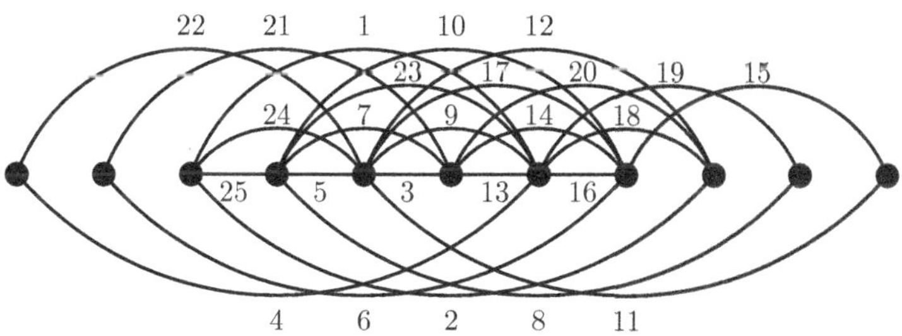

Fig. 2.31 Auxiliary labeling of $H_5(1, 2, 3, 4, 6)$

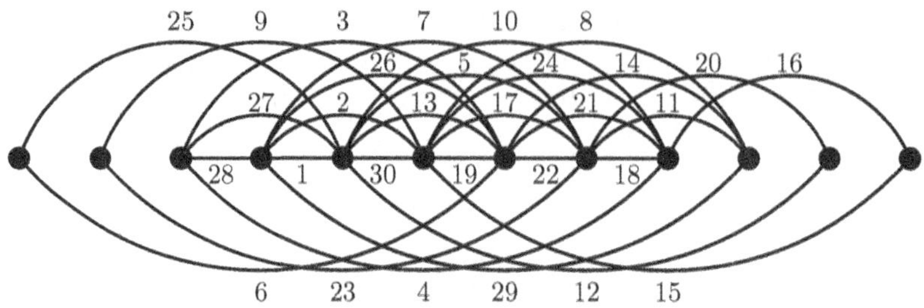

Fig. 2.32 Auxiliary labeling of $H_6(1, 2, 3, 4, 6)$

Since the graph H_n is a subgraph of a graph H_m, for $4 \leq n \leq m$, we can define auxiliary labeling $\lambda_n : E(H_n) \to \{1, 2, \ldots, 5n\}$ recursively as follows. The labelings of the graphs H_4, H_5, H_6, and H_7 are given in Figs. 2.30, 2.31, 2.32, and 2.33. For $n \geq 8$ the auxiliary labeling is defined by

$$\lambda_{n+4}(e) = \begin{cases} \lambda_n(e) + 10 & \text{for } e \in E(H_n) \\ 1 & \text{for } e = w_{n+1}w_{n+5} \\ 2 & \text{for } e = w_{n+4}w_{n+7} \\ 3 & \text{for } e = w_{n+4}w_{n+8} \\ 4 & \text{for } e = w_{n+2}w_{n+6} \\ 5 & \text{for } e = w_{n+3}w_{n+5} \\ 6 & \text{for } e = w_{n+6}w_{n+9} \\ 7 & \text{for } e = w_{n+7}w_{n+10} \\ 8 & \text{for } e = w_{n+3}w_{n+4} \\ 9 & \text{for } e = w_{n+6}w_{n+8} \\ 10 & \text{for } e = w_{n+5}w_{n+7} \\ 5n+11 & \text{for } e = w_{n+6}w_{n+7} \\ 5n+12 & \text{for } e = w_{n+5}w_{n+8} \\ 5n+13 & \text{for } e = w_{n+3}w_{n+7} \\ 5n+14 & \text{for } e = w_{n+4}w_{n+10} \\ 5n+15 & \text{for } e = w_{n+5}w_{n+6} \\ 5n+16 & \text{for } e = w_{n+3}w_{n+9} \\ 5n+17 & \text{for } e = w_{n+4}w_{n+6} \\ 5n+18 & \text{for } e = w_{n+2}w_{n+8} \\ 5n+19 & \text{for } e = w_{n+4}w_{n+5} \\ 5n+20 & \text{for } e = w_{n+1}w_{n+7}. \end{cases}$$

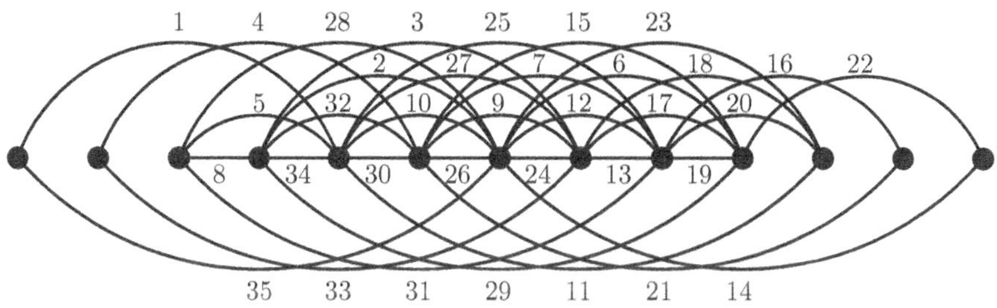

Fig. 2.33 Auxiliary labeling of $H_7(1, 2, 3, 4, 6)$

As in the proof of Theorem 2.8.23, it is easy to verify that for every vertex of H_n, $n \geq 4$,

$$\lambda_n^\star(w_i) = \begin{cases} \frac{1}{2}\deg(w_i)(5n+1) - 1 & \text{for } i = 2, n+5 \\ \frac{1}{2}\deg(w_i)(5n+1) + 1 & \text{for } i = 5, n+2 \\ \frac{1}{2}\deg(w_i)(5n+1) & \text{otherwise.} \end{cases}$$

For $n \geq 13$, we obtain the 10-regular circulant graph $C_n(1, 2, 3, 4, 6)$ from the graph H_n by identifying vertices w_i and w_{n+i}, for $i = 1, 2, 3, 4, 5, 6$. More precisely, the mapping $\xi : V(H_n) \to V(C_n(1, 2, 3, 4, 6))$, given by $\xi(w_i) = v_i$, for all $i = 1, 2, \ldots, n$, and $\xi(w_{n+j}) = v_j$, for $j = 1, 2, 3, 4, 5, 6$, is a homomorphism of the graphs H_n and $C_n(1, 2, 3, 4, 6)$. Moreover, the homomorphism ξ induces a bijective mapping ξ_E from $E(H_n)$ to $E(C_n(1, 2, 3, 4, 6))$. Now the edge labeling f, given by $f(e) = \lambda_n(\xi_E^{-1}(e))$, as in Theorem 2.8.23, is a supermagic labeling of the circulant graph $C_n(1, 2, 3, 4, 6)$ with the index $25n + 5$. □

Ivančo et al. [139] characterized all pairs n, r for which an r-regular supermagic graph of order n exists.

Theorem 2.8.26 ([139]) *Let r, n be positive integers, $n \geq r + 1$. There exists an r-regular supermagic graph of order n if and only if one of the following statements holds.*

(i) $r = 1$ *and* $n = 2$.
(ii) $3 \leq r \equiv 1 \pmod 2$ *and* $n \equiv 2 \pmod 4$.
(iii) $4 \leq r \equiv 0 \pmod 2$ *and* $n > 5$.

Proof The necessity of the Conditions (i), (ii), or (iii) follows from Theorem 2.5.2.

Sufficiency we examine case by case. Let k be a nonnegative integer such that $0 \leq r - 8k \leq 7$. Consider the following cases.

Case A Obviously the only graph satisfying (i) is K_2 and it is the only 1-regular supermagic graph.

Case B Suppose Condition (ii) holds.

Case B1 For $r = 8k + 1$, $k \geq 1$, and $n \equiv 2 \pmod 4$, let us consider the r-regular graph $G = C_n(1, 2, \ldots, 4k, n/2)$ of order n. According to Corollaries 2.8.3 and 2.8.2, the graphs $G_1 = C_n(2, 4, \ldots, 4k, n/2)$ and $G_2 = C_n(1, 3, \ldots, 4k - 1)$ are supermagic for all $n > r$. Since G can be decomposed into two edge-disjoint factors G_1 and G_2, the graph G is supermagic by Theorem 2.5.6.

Case B2 For $r = 3$, let us consider the Möbius ladder $C_n(1, n/2)$. By Theorem 2.8.17, it is supermagic.

For $r = 8k + 3$, $k \geq 1$, let us consider the r-regular circulant graph $G = C_n(1, 2, \ldots, 4k + 1, n/2)$ on $n > r$ vertices. By Theorem 2.8.17, there exists a 3-regular supermagic graph $G_1 = C_n(1, n/2)$, for all feasible values of n. According to Corollary 2.8.5, the graph $G_2 = C_n(2, 3, \ldots, 4k + 1)$ is supermagic. Since G can be decomposed into two edge-disjoint factors G_1 and G_2, the existence of a supermagic labeling of G is guaranteed by Theorem 2.5.6.

Case B3 For $r = 8k + 5$ and $n = 8k + 6$, there exists a unique graph, namely, the complete graph K_n; it is supermagic by Theorem 2.6.6.

For $r = 5$ and $n > 6$, let us consider the circulant graph $C_n(2, 4, n/2)$; it is supermagic by Corollary 2.8.3.

For $r = 8k + 5$ and $k \geq 1$, let us consider the r-regular circulant graph $G = C_n(1, 2, \ldots, 4k+1, 4k+3, n/2)$ of order n, $n > 8k+6$. The graph G is decomposable into two edge-disjoint factors $C_n(2, 4, \ldots, 4k, n/2)$ and $C_n(1, 3, \ldots, 4k + 1, 4k + 3)$. Thus it is supermagic according to Corollaries 2.8.3 and 2.8.2, and Theorem 2.5.6.

Case B4 For $r = 7$ and $n = 10$, let us consider the graph $C_{10}(1, 2, 4, 5)$. The graph $C_{10}(1, 5)$ is supermagic by Theorem 2.8.17. In [136] it is proved that the graph $2K_5$ (isomorphic to $C_{10}(2, 4)$) is also supermagic. Then the graph $C_{10}(1, 2, 4, 5)$ is supermagic according to Theorem 2.5.6.

For $r = 7$ and $14 \leq n \equiv 2 \pmod 4$, let us consider the graph $C_n(1, 2, 6, n/2)$. The graph $C_n(1, n/2)$ is supermagic by Theorem 2.8.17. The graph $C_n(2, 6)$ is isomorphic to $2C_{n/2}(1, 3)$ and so is supermagic by Theorems 2.8.24 and 2.5.7. According to Theorem 2.5.6, the graph $C_n(1, 2, 6, n/2)$ is supermagic.

For $r = 15$ and $18 \leq n \equiv 2 \pmod 4$, let us consider the circulant graph $G_1 = C_n(1, 2, 3, 4, 5, 6, 8, n/2)$. The graphs $C_n(1, n/2)$ and $C_n(3, 5)$ are supermagic by Theorem 2.8.17 and Corollary 2.8.2. The graph $C_n(2, 4, 6, 8)$, isomorphic to $2C_{n/2}(1, 2, 3, 4)$, is supermagic by Corollary 2.8.4 and Theorem 2.5.7. Therefore, according to Theorem 2.5.6, the graph G_1 is supermagic.

For $r = 8k + 7$, $k > 1$, $n > 8(k + 1)$, let us consider the circulant graph $G = C_n(1, 2, 3, 4, 5, 6, 8, 9, \ldots, 8(k + 1), n/2)$. The graph G is decomposable into G_1 (from the previous paragraph) and $G_2 = C_n(9, 10, \ldots, 8(k + 1))$. Since G_2 is a supermagic graph by Corollary 2.8.5, the graph G is also supermagic by Theorem 2.5.6.

Case C Suppose Condition (iii) holds.

Case C1 For $r = 8k$, let us consider the circulant graph $C_n(1, 2, \ldots, 4k)$ which is supermagic by Corollary 2.8.5.

Case C2 For $r = 8k + 2$ and $n = 8k + 3$ we have the complete graph K_n which is supermagic due to Theorem 2.6.6.

The unique graph for $r = 8k+2$ and $n = 8k+4$ is the complete $(4k+2)$-partite graph $K_{4k+2[2]}$ which is supermagic by Theorem 2.8.1.

For $r = 10$ and $n \geq 13$, let us consider the circulant graph $C_n(1, 2, 3, 4, 6)$ which is supermagic by Theorem 2.8.25.

Finally, for $r = 8k + 2$, $k > 1$, $n \geq 8k + 5$, let us consider the circulant graph $G = C_n(1, 2, 3, 4, 6, 7, \ldots, 4k + 2)$ which is decomposable into two edge-disjoint factors $C_n(1, 2, 3, 4, 6)$ and $C_n(7, 8, \ldots, 4k + 2)$. These factors are supermagic by Theorem 2.8.25 and Corollary 2.8.5. Therefore, G is supermagic due to Theorem 2.5.6.

Case C3 For $r = 8k + 4$ and $n = 8k + 5 > 5$, we have the complete graph K_n which is supermagic due to Theorem 2.6.6.

The only graph for $r = 8k + 4$ and $n = 8k + 6$ is the complete $(4k + 3)$-partite graph $K_{4k+3[2]}$ which is supermagic by Theorem 2.8.1.

For $r = 4$ and $n \geq 7$, we have the circulant graph $C_n(1, 3)$ which is supermagic by Theorem 2.8.24.

For $r = 8k + 4$, $k \geq 1$, $n \geq 8k + 7$, let us consider the circulant graph $G = C_n(1, 3, 4, \ldots, 4k + 3)$ which is decomposable into two edge-disjoint factors $C_n(1, 3)$ and $C_n(4, 5, \ldots, 4k+3)$. These factors are supermagic by Theorem 2.8.24 and Corollary 2.8.5. Therefore, G is supermagic due to Theorem 2.5.6.

Case C4 For $r = 6$ and $n \geq 7$, we have the circulant graph $C_n(1, 2, 3)$ which is supermagic by Theorem 2.8.23.

Finally, for $r = 8k + 6 > 6$ and $n > r$ we can construct an r-regular circulant graph $C_n(1, 2, \ldots, 4k + 3)$ on n vertices by Corollary 2.8.5 and Theorem 2.5.6, from the circulant graphs $C_n(1, 2, 3)$ and $C_n(4, 5, \ldots, 4k + 3)$.

Therefore, there is a supermagic r-regular circulant graph of order n for an arbitrary pair (n, r) satisfying (i), (ii), or (iii). This completes the proof. □

2.8.6 *Constructions Using Vertex-Antimagic Graphs*

In [143] Ivančo and Semaničová introduced constructions of supermagic graphs using some vertex-antimagic graphs. Although vertex-antimagic labelings are discussed in Chap. 5. For self-contained reading of these constructions let us introduce in a general way a vertex-antimagic edge labeling.

By an (a, d)*-vertex-antimagic edge* (VAE) *labeling* of a graph G we mean a one-to-one mapping from $E(G)$ onto $\{1, 2, \ldots, |E(G)|\}$ such that the set of all vertex-weights in G is $\{a, a+d, \ldots, a+(|V(G)|-1)d\}$, where $a > 0$ and $d \geq 0$ are two fixed integers. A graph that allows an (a, d)-vertex-antimagic edge labeling will be often called an (a, d)-VAE graph. For more details on (a, d)-VAE graphs see Sect. 5.1.

For any graph G we define a graph $G^{\bowtie}$ by $V(G^{\bowtie}) = \bigcup_{v\in V(G)}\{v^0, v^1\}$ and $E(G^{\bowtie}) = E_1(G^{\bowtie}) \cup E_2(G^{\bowtie})$, where the set $E_1(G^{\bowtie}) = \bigcup_{vu\in E(G)}\{v^0u^1, v^1u^0\}$ and $E_2(G^{\bowtie}) = \bigcup_{v\in V(G)}\{v^0v^1\}$.

Theorem 2.8.27 ([143]) *Let G be an $(a, 1)$-VAE $2r$-regular graph. Then $G^{\bowtie}$ is a supermagic graph.*

Proof Put $n = |V(G)|$. Since G is a $2r$-regular graph, every component is Eulerian. Therefore, there exists a digraph $\vec{G}$ obtained from G by an orientation of its edges in such a way that the outdegree (and also the indegree) of every vertex of $\vec{G}$ is equal to r. By $[u, v]$ we denote an arc of $\vec{G}$ and by $N^+(v)$ (resp. $N^-(v)$) we denote the outneighborhood (resp. inneighborhood) of a vertex v in $\vec{G}$.

Let $f : E(G) \to \{1, 2, \ldots, rn\}$ be an $(a, 1)$-VAE labeling of G. Consider the bijection $g : E_1(G^{\bowtie}) \to \{1, 2, \ldots, 2rn\}$ given by

$$g(u^iv^j) = \begin{cases} f(uv) & \text{if } i = 0, j = 1 \\ f(uv) + rn & \text{if } i = 1, j = 0, \end{cases}$$

for every arc $[u, v]$ of $\vec{G}$.

For its index-mapping we have

$$\begin{aligned} g^\star(v^0) &= \sum_{w\in N^+(v)} g(v^0w^1) + \sum_{u\in N^-(v)} g(u^1v^0) \\ &= \sum_{w\in N^+(v)} f(vw) + \sum_{u\in N^-(v)} (f(uv) + rn) = f^\star(v) + r^2n \end{aligned}$$

for every vertex $v^0 \in V(G^{\bowtie})$. Similarly, we have $g^\star(v^1) = f^\star(v) + r^2n$ for every vertex $v^1 \in V(G^{\bowtie})$. Thus $g^\star(v^0) = g^\star(v^1) = f^\star(v) + r^2n$ for every vertex $v \in V(G)$. Since f is an $(a, 1)$-VAE labeling, the set $\{f^\star(v) : v \in V(G)\}$ consists of consecutive integers. This means that the bijection $h : E(G^{\bowtie}) \to \{1, 2, \ldots, (2r+1)n\}$, given by

$$\begin{aligned} h(u^iv^j) &= g(u^iv^j), & &\text{for } u^iv^j \in E_1(G^{\bowtie}) \\ h(v^0v^1) &= \frac{2rn(r+1) + (2r+1)(n+1)}{2} - f^\star(v), & &\text{for } v \in V(G) \end{aligned}$$

is a supermagic labeling of $G^{\bowtie}$. □

Note that $C_n^{\bowtie}$ is isomorphic to either the Möbius ladder M_{2n}, for n odd, or to the graph of n-sided prism $\mathcal{S}_n$, for n even. Moreover, for the disjoint union of graphs G_1 and G_2, it holds that $(G_1 \cup G_2)^{\bowtie} = G_1^{\bowtie} \cup G_2^{\bowtie}$. The following corollary is proved in [143].

Corollary 2.8.6 ([143]) *Let k, n and m be positive integers. For k odd the following graphs are supermagic.*

(i) kM_{2n} *when* $3 \le n \equiv 1 \pmod 2$.
(ii) $k(M_6 \cup \mathcal{S}_n)$ *when* $6 \le n \equiv 0 \pmod 2$.
(iii) $k(\mathcal{S}_4 \cup M_{2n})$ *when* $5 \le n \equiv 1 \pmod 2$.
(iv) $k(M_{10} \cup \mathcal{S}_n)$ *when* $4 \le n \equiv 0 \pmod 2$.
(v) $k(\mathcal{S}_m \cup M_{2n})$ *when* $6 \le m \equiv 0 \pmod 2$, $n \equiv 1 \pmod 2$, $n \ge m/2 + 2$.

Corollary 2.8.7 ([143]) *Let G be a $2r$-regular graph of odd order n. If G is circulant, Hamiltonian, or $n < 4r$, then $G^{\bowtie}$ is a supermagic graph.*

It is easy to see that $G^{\bowtie}$ is isomorphic to the Cartesian product $G \square K_2$ whenever G is a bipartite graph. However, a regular bipartite graph of even degree is never $(a, 1)$-VAE. In the next theorem we describe another construction of supermagic Cartesian products.

Theorem 2.8.28 ([143]) *Let G be an $(a, 1)$-VAE graph decomposable into two edge-disjoint r-factors. Then $G \square K_2$ is a supermagic graph.*

Proof Suppose that F^1, F^2 are edge-disjoint r-factors which form a decomposition of G and $f : E(G) \to \{1, 2, \ldots, rn\}$, where $n = |V(G)|$ is an $(a, 1)$-VAE labeling of G.

We denote the vertices of $G \square K_2$ by v_i, $i \in \{1, 2\}$, $v \in V(G)$, in such a way that the vertices $\{v_i : v \in V(G)\}$ induce a subgraph G_i isomorphic to G. Then $G \square K_2$ consists of subgraphs G_1, G_2, and n edges $v_1 v_2$, for all $v \in V(G)$. By F_i^j, $i \in \{1, 2\}$, $j \in \{1, 2\}$, we denote the factor of G_i corresponding to F^j.

Consider the bijection $g : E(G_1 \cup G_2) \to \{1, 2, \ldots, 2rn\}$ given by

$$g(e) = \begin{cases} f(e) & \text{if } e \in F_1^1 \text{ or } e \in F_2^2 \\ f(e) + rn & \text{if } e \in F_2^1 \text{ or } e \in F_1^2. \end{cases}$$

For its index-mapping we have

$$\begin{aligned} g^\star(v_1) &= \sum_{v_1 u_1 \in E(G_1)} g(v_1 u_1) = \sum_{v_1 u_1 \in E(F_1^1)} g(v_1 u_1) + \sum_{v_1 w_1 \in E(F_1^2)} g(v_1 w_1) \\ &= \sum_{vu \in E(F^1)} f(vu) + \sum_{vw \in E(F^2)} (f(vw) + rn) \\ &= \sum_{vu \in E(G)} f(uv) + r^2 n = f^\star(v) + r^2 n, \end{aligned}$$

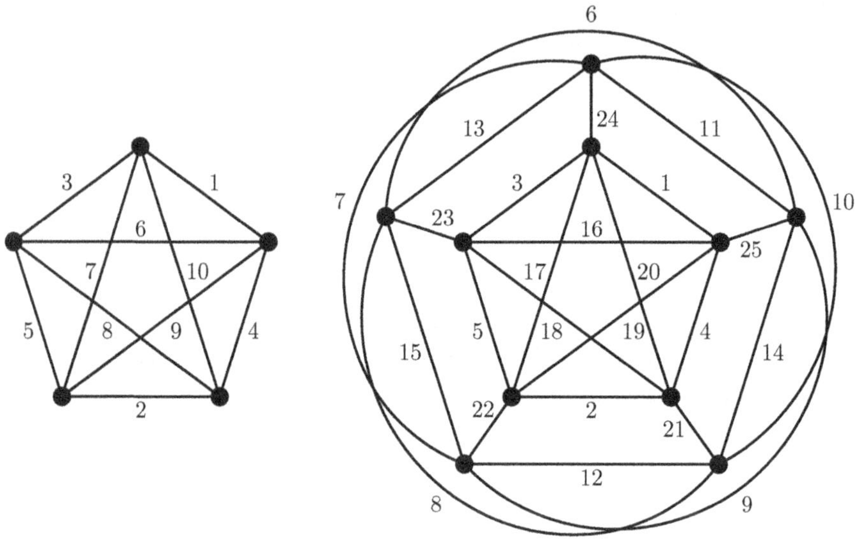

Fig. 2.34 An $(a, 1)$-VAE labeling of K_5 and the corresponding supermagic labeling of $K_5\Box K_2$

for every vertex $v_1 \in V(G_1)$. Similarly, $g^\star(v_2) = f^\star(v)+r^2n$, for every vertex $v_2 \in V(G_2)$. Thus $g^\star(v_1) = g^\star(v_2) = f^\star(v) + r^2n$, for every vertex $v \in V(G)$. Since f is an $(a, 1)$-VAE labeling, the set $\{f^\star(v) : v \in V(G)\}$ consists of consecutive integers.

This means that the bijection $h : E(G\Box K_2) \to \{1, 2, \ldots, (2r + 1)n\}$ given by

$$h(e) = g(e), \qquad \text{for every } e \in E(G_1 \cup G_2)$$

$$h(v_1v_2) = \frac{2rn(r + 1) + (2r + 1)(n + 1)}{2} - f^\star(v), \quad \text{for every } v \in V(G)$$

is a supermagic labeling of $G\Box K_2$. □

Figure 2.34 illustrates the use of the construction mentioned in the proof of Theorem 2.8.28. The supermagic labeling of $K_5\Box K_2$ is constructed from the $(a, 1)$-VAE labeling of K_5.

Corollary 2.8.8 ([143]) *Let G be a $4r$-regular graph of odd order n. If G is circulant, Hamiltonian, or $n < 8r$, then $G\Box K_2$ is a supermagic graph.*

2.8.7 Constructions Using Double-Consecutive Labeling

In [138] Ivančo defines a new type of edge labeling, the so-called double-consecutive labeling. Let U_1, U_2 be the subsets of the vertex set of a graph G such that $|U_1| = |U_2|$, $U_1 \cup U_2 = V(G)$ and $U_1 \cap U_2 = \emptyset$. An injective mapping f from the edge set $E(G)$ into positive integers is called a *double-consecutive labeling with respect to* (U_1, U_2), *DC-labeling* for short, if the sets of the vertex-weights in U_1 and U_2 are the same and they form a consecutive integer sequence, i.e.,

$$\{w(v) : v \in U_1\} = \{w(v) : v \in U_2\} = \{a, a+1, \ldots, a+|U_1|-1\},$$

for some positive integer a.

In [138] Ivančo gave some constructions of supermagic graphs that are decomposed into some special subgraphs having appropriate DC-labelings. Using this he constructed some supermagic complements of bipartite graphs. The *complement* $\overline{G}$ *of a graph* G is the graph on the same vertex set such that two vertices of $\overline{G}$ are adjacent if and only if they are not adjacent in G.

Theorem 2.8.29 ([138]) *Let G be an r-regular bipartite graph of order $8k$. Then $\overline{G}$ is a supermagic graph if and only if r is odd.*

Theorem 2.8.30 ([138]) *Let G be an r-regular bipartite graph of order $2n$, where n is odd and r is even. Then $\overline{G}$ is a supermagic graph if and only if $(n, r) \neq (3, 2)$.*

Theorem 2.8.31 ([138]) *Let G be an r-regular bipartite graph of order $2n$ with parts U_1 and U_2. If $n \geq 5$ and r are odd and $\overline{G}$ is a Hamiltonian graph, then $\overline{G}$ is a supermagic graph.*

Corollary 2.8.9 ([138]) *Let G be an r-regular bipartite graph of order $2n$. If $2r < n$ and $5 \leq n \equiv r \equiv 1 \pmod 2$, then $\overline{G}$ is a supermagic graph.*

2.8.8 Constructions Using Degree-Magic Labeling

A bijection f from $E(G)$ into $\{1, 2, \ldots, |E(G)|\}$ is called a *degree-magic labeling* (or just *d-magic labeling*) of a graph G if its index-mapping $f^\star$ satisfies

$$f^\star(v) = \frac{1 + |E(G)|}{2} \deg(v), \quad \text{for all } v \in V(G),$$

where $\deg(v)$ is the degree of a vertex v. We say that a graph G is *degree-magic* (or simply *d-magic*) when there exists a degree-magic labeling of G.

The concept of degree-magic graphs was introduced in [62] as an extension of supermagic regular graphs. In [62] it was proved that the family of degree-magic graphs is closed under edge-bijective homomorphism and the family of balanced degree-magic graphs is closed under edge-disjoint union. Some other properties of degree-magic graphs and characterizations of some classes of degree-magic and balanced degree-magic graphs were described in [63] and [64].

For regular graphs Bezegová and Ivančo [62] described the relationship between supermagic and degree-magic labelings.

Theorem 2.8.32 ([62]) *Let G be a regular graph. Then G is supermagic if and only if it is degree-magic.*

Using this extension of supermagic labeling, Bezegová and Ivančo [62] proved that there is no forbidden subgraph characterization of supermagic graphs.

Theorem 2.8.33 ([62]) *For any graph G there is a supermagic regular graph which contains an induced subgraph isomorphic to G.*

In [63] Bezegová and Ivančo introduced some constructions of supermagic (and also balanced degree-magic) labelings for a large family of graphs. Degree-magic labelings allow us to construct supermagic labelings for the disjoint union of some regular non-isomorphic graphs.

Theorem 2.8.34 ([63]) *Let $\delta > 4$ be an even integer. Let G be a δ-regular graph whose each component is a complete multipartite graph of even size. Then G is a supermagic graph. Moreover, for any δ-regular supermagic graph H, the union of disjoint graphs H and G is also a supermagic graph.*

Theorem 2.8.35 ([63]) *Let $\delta \equiv 0 \pmod 8$ be a positive integer. Let G be a δ-regular graph whose each component is a circulant graph. Then G is a supermagic graph. Moreover, for any δ-regular supermagic graph H, the union of disjoint graphs H and G is also a supermagic graph.*

Theorem 2.8.36 ([63]) *Let $k, n_1, n_2, \ldots, n_k$ be positive integers such that $k \equiv 1 \pmod 4$ and $11 \leq n_i \equiv 3 \pmod 8$, for all $i \in \{1, 2, \ldots, k\}$. Then the complement of the union of disjoint cycles $C_{n_1} \cup C_{n_2} \cup \cdots \cup C_{n_k}$ is supermagic.*

Theorem 2.8.37 ([61]) *Let G be an r-regular bipartite graph of order $8k$ and let H be an $(8k-r-1)$-regular supermagic graph. If r is odd, then $\overline{G} \cup H$ is a supermagic graph.*

2.9 Non-regular Magic and Supermagic Graphs

2.9.1 Insertion of a New Edge

Let G be a graph not isomorphic to K_n. Let $G + \{e\}$ be a graph formed from G by inserting a new edge e. Trenkler [277] showed that

Theorem 2.9.1 ([277]) *If $G + \{e\}$ is formed from a magic graph G by inserting a new edge e which belongs to a (1-2)-factor of $G + \{e\}$, then $G + \{e\}$ is magic.*

In [152] Jeurissen proved

Theorem 2.9.2 ([152]) *A magic graph stays magic if an edge is inserted into one of its components, unless this turns a bipartite component into non-bipartite one.*

Theorem 2.9.3 ([21]) *Let $G = V_1V_2$ be a magic balanced connected bipartite graph. If at least two edges are inserted into V_1 (V_2) and at least one edge into V_2 (V_1), then the resulting graph will be magic.*

Trenkler [279] and also Jeurissen [152] proved

Theorem 2.9.4 ([279, 152]) *Let G be a connected non-bipartite magic graph. Then by inserting a new edge into G we obtain a magic graph.*

An unbalanced complete bipartite graph $G = V_1V_2$ is not magic, see Theorem 2.3.5. By inserting extra edges into this graph we can obtain a magic graph. Let $|V_1| = n > |V_2| = m \geq 2$. Bača [21] proved

Theorem 2.9.5 ([21]) *Let $G = V_1V_2$ be unbalanced complete bipartite graph. If $\lfloor (n-m)/2 \rfloor + 1$ independent edges are inserted into V_1 in G, then the resulting graph will be magic.*

By $\mathcal{G}$ we denote a non-bipartite graph, which is obtained from an unbalanced complete bipartite graph $G = V_1V_2$ by inserting $\lfloor (n-m)/2 \rfloor + 1$ independent edges into V_1.

Theorem 2.9.6 ([21]) *Let $V(H) = V(\mathcal{G})$. If the graph $\mathcal{G}$ is a subgraph of graph H, then H is magic.*

As a corollary of this result we obtain that a complete graph K_n is magic if $n \geq 5$ [263]. The converse of Theorem 2.9.6 is not necessarily true. The graph in Fig. 2.35 is magic but does not contain $\mathcal{G}$ as a subgraph with the same vertex set.

The *ith power* G^i, $i \geq 2$, of a graph G is a graph with the same vertex set as G and such that two vertices of G^i are adjacent if and only if the distance between these vertices in G is at most i. An I-graph is a graph with a 1-factor whose every edge is incident with a vertex of degree 1. An example of an I-graph is given in Fig. 2.36.

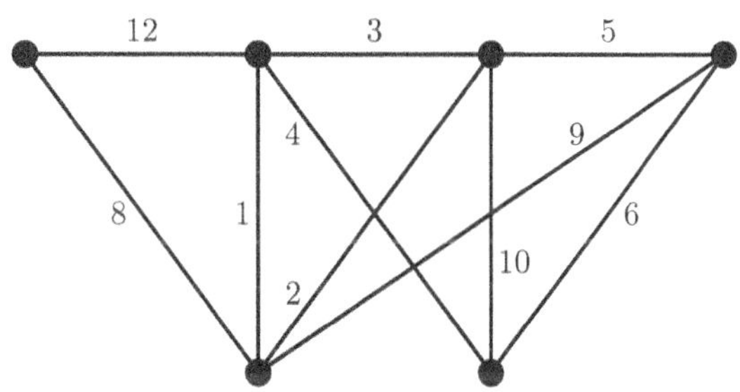

Fig. 2.35 A magic graph that does not contain $\mathcal{G}$ as a subgraph with the same vertex set

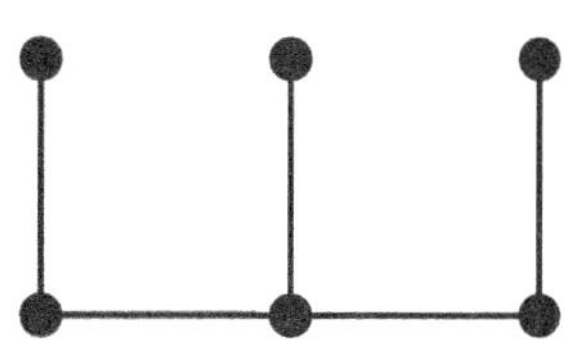

Fig. 2.36 Example of an I-graph

Trenkler and Vetchý [282] characterized magic powers of graphs.

Theorem 2.9.7 ([282]) *Let a graph G have order $n \geq 5$. The graph G^2 is magic if and only if G is not an I-graph and it is not the path P_6. The graph G^i is magic for all $i \geq 3$.*

2.9.2 Deletion of an Edge

Let $G - \{e\}$ be a graph obtained from G by deleting an edge $e \in E(G)$.

Bača [21] proved

Theorem 2.9.8 ([21]) *Let G be a regular connected magic graph of degree $r \geq 3$. Then the graph stays magic if an arbitrary edge is deleted.*

Hartsfield and Ringel [124] presented a simple construction of non-regular supermagic graphs. They proved

Theorem 2.9.9 ([124]) *Let G be a regular supermagic graph. Then there exists an edge $e \in E(G)$ such that $G - \{e\}$ is a supermagic graph.*

Theorem 2.9.10 ([124]) *If G is regular and $G - \{e\}$ is supermagic, then G is supermagic.*

Drajnová et al. [91] proved that the complete graph on n vertices with deleted one edge, $K_n - \{e\}$, is a supermagic graph for every positive integer $n \geq 6$; see also Theorem 2.6.7.

In [233] Semaničová dealt with the problem of whether the complete graph without two edges, $K_n - \{e, f\}$, is supermagic or not.

Theorem 2.9.11 ([233]) *Let $n \geq 6$ be a positive integer. If $K_n - \{e, f\}$ is a supermagic graph, then $n \not\equiv 0 \pmod 8$.*

Theorem 2.9.12 ([233]) *Let $10 \leq n \equiv 2 \pmod 4$ and e, f be two nonadjacent edges in K_n. Then $K_n - \{e, f\}$ is a supermagic graph.*

Proof Let $10 \leq n \equiv 2 \pmod 4$. Let e, f be two nonadjacent edges in the complete graph K_n. Denote the vertices of K_n by $v_1, v_2, \ldots, v_n$ in such a way that $e = v_n v_{n-2}$, $f = v_{n-1} v_{n-3}$.

Let G be a subgraph of K_n induced by the set $\{v_1, v_2, \ldots, v_{n-4}\}$. The graph G is isomorphic to K_{n-4} and, by Theorem 2.6.6, there exists a supermagic labeling g from $E(G)$ into $\{1, 2, \ldots, \binom{n-4}{2}\}$. Clearly, $g^\star(v_i) = \left(\binom{n-4}{2} + 1\right)(n-5)/2$ for all i, $1 \leq i \leq n-4$.

Let H be a subgraph of K_n which is isomorphic to a complete bipartite graph with the partition $V_1 = \{v_1, v_2, \ldots, v_{n-4}\}$, $V_2 = \{v_{n-3}, v_{n-2}, v_{n-1}, v_n\}$. Note that the graph H is isomorphic to the graph $K_{4,n-4}$. Consider a mapping h from $E(H)$

into $\{1, 2, \ldots, 4(n-4)\}$ defined by

$$h(v_i v_j) = \begin{cases} 4n - 15 - 2i & \text{for } j = n-3, i \equiv 1 \pmod 2 \\ 2i - 2 & \text{for } j = n-3, i \equiv 0 \pmod 2 \\ 2i - 1 & \text{for } j = n-2, i \equiv 1 \pmod 2 \\ 4n - 12 - 2i & \text{for } j = n-2, i \equiv 0 \pmod 2 \\ 4n - 16 - 2i & \text{for } j = n-1, i \equiv 1 \pmod 2 \\ 2i - 1 & \text{for } j = n-1, i \equiv 0 \pmod 2 \\ 2i + 2 & \text{for } j = n, i \equiv 1 \pmod 2 \\ 4n - 15 - 2i & \text{for } j = n, i \equiv 0 \pmod 2. \end{cases}$$

It is easy to check that for its index-mapping

$$h^\star(v_i) = \begin{cases} 2\,(4(n-4)+1) & \text{for } 1 \le i \le n-4 \\ \frac{n-4}{2}\,(4(n-4)+1) & \text{for } n-3 \le i \le n. \end{cases}$$

Put a positive integer $a = (n^3 + 8n^2 + 29n + 78)/4$ and define a mapping f : $E(K_n - \{e, f\}) \to \{a, a+1, \ldots, a + n(n-1)/2 - 3\}$ by

$$f(v_i v_j) = \begin{cases} a - 1 + g(v_i v_j) & \text{for } 1 \le i \le n-4 \text{ and } 1 \le j \le n-4 \\ a + \binom{n-4}{2} - 1 + h(v_i v_j) & \text{for } (i,j) \ne (1, n-3), (i,j) \ne (1, n-1) \\ & \text{and } 1 \le i \le n-4 \text{ and } n-3 \le j \le n \\ a + \binom{n-4}{2} + 2 & \text{for } i = 1,\ j = n-3 \\ a + \binom{n-4}{2} + 1 & \text{for } i = 1,\ j = n-1 \\ a + \binom{n-4}{2} + 4(n-4) & \text{for } i = n-3,\ j = n-2 \\ a + \binom{n-4}{2} + 4(n-4) + 3 & \text{for } i = n-2,\ j = n-1 \\ a + \binom{n-4}{2} + 4(n-4) + 1 & \text{for } i = n-1,\ j = n \\ a + \binom{n-4}{2} + 4(n-4) + 2 & \text{for } i = n,\ j = n-3. \end{cases}$$

It is easy to see that the mapping f is a bijection and for its index-mapping we get

$$f^\star(v_i) = \tfrac{1}{4}(n^4 + 9n^3 + 24n^2 + 92n - 48), \quad \text{for } 1 \le i \le n.$$

Thus, f is a supermagic labeling and $K_n - \{e, f\}$ is a supermagic graph. □

Let $G-$(1-factor) be a graph obtained from G by deleting the edges of a 1-factor. Doob [90] proved.

Theorem 2.9.13 ([90]) *The complete graph K_{2n} with a 1-factor deleted is magic if and only if $n \geq 3$.*

Hartsfield and Ringel [124] showed

Theorem 2.9.14 ([124]) *If $n \equiv 0 \pmod 2$, $n \neq 2$, then $K_{n,n}$ − (1-factor) is not supermagic.*

Theorem 2.9.15 ([124]) *If $n \equiv 1 \pmod 4$, then $K_{n,n}$ − (1-factor) is supermagic.*

The problem for $n \equiv 3 \pmod 4$ is not yet solved. Hartsfield and Ringel [124] proposed the following open problem

Open Problem 2.9.1 ([124]) Is the graph $K_{n,n}$ − (1-factor) supermagic when $n \equiv 3 \pmod 4$?

$K_n - C_n$ is a graph obtained from a complete graph on n vertices with a Hamiltonian cycle deleted. Doob [90] showed

Theorem 2.9.16 ([90]) *The graph $K_n - C_n$ is magic if and only if $n \geq 7$.*

2.9.3 Contraction of an Edge

Ivančo and Semaničová [144] dealt with supermagic graphs obtained from a regular graph by contraction of an edge.

Theorem 2.9.17 ([144]) *Let G be a 3-regular triangle-free supermagic graph. Then there exists an edge $e \in E(G)$ such that the graph obtained from G by contraction of the edge e is supermagic.*

Proof Let G be a 3-regular supermagic graph of order n. In [136] it is proved that $n \equiv 2 \pmod 4$ and there exists a supermagic labeling $f : E(G) \to \{1, 2, \dots, 3n/2\}$ of G for an index $(9n + 6)/4$. Let $u_1u_2 \in E(G)$ be the edge of G such that $f(u_1u_2) = 3n/2$. By H we denote the graph obtained from G by the contraction of the edge u_1u_2. Let w denote the vertex in $V(H)$ which arose by the identification of u_1 and u_2. Consider the bijection $g : E(H) \to \{1 + 3(n-2)/4, 2 + 3(n-2)/4, \dots, 3n/2 + 3(n-2)/4\}$ given by

$$g(e) = f(e) + \tfrac{3(n-2)}{4}, \quad \text{for every } e \in E(H).$$

For its index-mapping we get

$$\begin{aligned} g^\star(w) &= f^\star(u_1) + f^\star(u_2) - 2f(u_1u_2) + 3(n-2) \\ &= 3\left(1 + \tfrac{3n}{2}\right) - 3n + 3(n-2) = \tfrac{3(3n-2)}{2} \end{aligned}$$

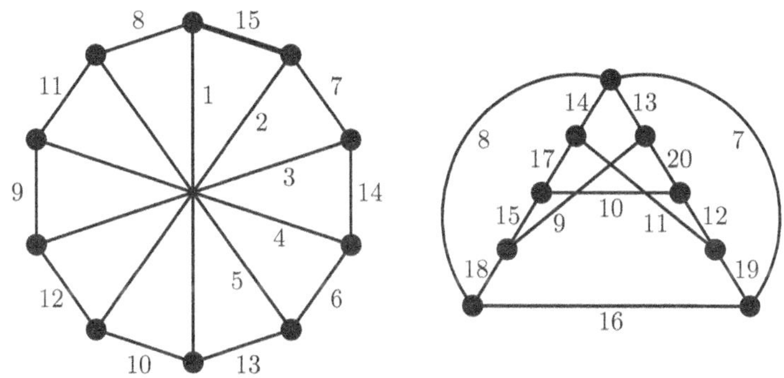

Fig. 2.37 A supermagic labeling of a graph obtained from a supermagic graph by the contraction of the edge with the largest value

and

$$g^{\star}(v) = f^{\star}(v) + 3\frac{3(n-2)}{4} = \frac{3}{2}\left(1 + \frac{3n}{2}\right) + \frac{9(n-2)}{4} = \frac{3(3n-2)}{2},$$

for every vertex $v \in V(H) - w$.

Thus g is a supermagic labeling of H. □

Figure 2.37 illustrates the construction of a non-regular supermagic graph obtained from a supermagic graph by the contraction of the edge with the largest value.

Let $\mathcal{A}_n$ be a graph isomorphic to a Cartesian product of C_n and K_2, in which one edge joining two edge-disjoint cycles of length n is contracted. Thus, the graph $\mathcal{A}_n$ has the vertex set

$$V(\mathcal{A}_n) = \{u, v_2^1, v_3^1, \dots, v_n^1, v_2^2, v_3^2, \dots, v_n^2\}.$$

For the sake of clarity, let $u = v_1^1 = v_1^2$, and the edge set

$$E(\mathcal{A}_n) = \bigcup_{i=1}^{n} \{v_i^1 v_{i+1}^1, v_i^2 v_{i+1}^2\} \cup \bigcup_{i=2}^{n} \{v_i^1 v_i^2\},$$

where subscripts are taken modulo n.

Theorem 2.9.18 ([144]) *The graph $\mathcal{A}_n$ is supermagic for every positive integer $n \geq 3$.*

Proof If $n \geq 3$ is an odd positive integer, then $\mathcal{A}_n$ is isomorphic to a graph obtained from the Möbius ladder M_{2n} by contracting one chord. In [230] there is a construction of a supermagic labeling of M_{2n}, $3 \leq n \equiv 1 \pmod 2$, with the

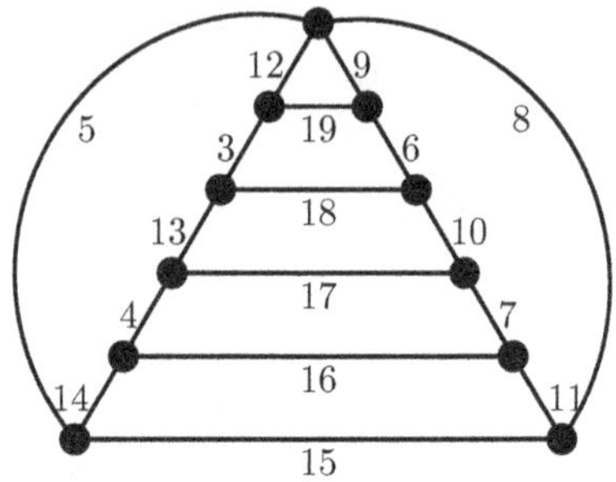

Fig. 2.38 Supermagic labeling of $\mathcal{A}_6$

smallest value on the chord. Then we consider the dual labeling to this supermagic labeling and, according to Theorem 2.9.17, we get that $\mathcal{A}_n$ is a supermagic graph.

If $n \geq 4$ is an even positive integer, we consider a mapping $f : E(\mathcal{A}_n) \to \{n/2, n/2+1, \ldots, 7n/2-2\}$ defined by

$$f(v_i^1 v_{i+1}^1) = \begin{cases} 2n + \frac{i-1}{2} & \text{if } i \equiv 1 \pmod 2 \\ \frac{n}{2} + \frac{i-2}{2} & \text{if } i \equiv 0 \pmod 2 \end{cases}$$

$$f(v_i^2 v_{i+1}^2) = \begin{cases} \frac{3n}{2} + \frac{i-1}{2} & \text{if } i \equiv 1 \pmod 2 \\ n + \frac{i-2}{2} & \text{if } i \equiv 0 \pmod 2 \end{cases}$$

$$f(v_i^1 v_i^2) = \frac{7n}{2} - i \qquad \text{if } i \geq 2.$$

It is easy to check that f is a bijection and

$$f^\star(v) = 6n - 2, \quad \text{for every } v \in V(\mathcal{A}_n).$$

Thus f is a supermagic labeling of $\mathcal{A}_n$. □

Figure 2.38 depicts a supermagic labeling of $\mathcal{A}_6$.

2.9.4 Splitting a Vertex and Adding an Edge

Another construction of supermagic non-regular graphs is provided by the following theorem.

Theorem 2.9.19 ([144]) *Let f be a supermagic labeling of a 4-regular graph G such that there exists a vertex $v \in V(G)$ such that*

$$f(vu_1) + f(vu_2) = f(vu_3) + f(vu_4),$$

where u_i, $i = 1, 2, 3, 4$, are the vertices adjacent to v. Let H be a graph with the vertex set $V(H) = (V(G) - v) \cup \{v^1, v^2\}$ and the edge set $E(H) = \left(E(G) - \bigcup_{i=1}^{4}\{vu_i\}\right) \cup \{v^1u_1, v^1u_2, v^2u_3, v^2u_4, v^1v^2\}$. Then H is a supermagic graph.

Proof Let f be a supermagic labeling of 4-regular graph G such that for the vertex $v \in V(G)$,

$$f(vu_1) + f(vu_2) = f(vu_3) + f(vu_4).$$

This expression is equal to $\lambda/2$, where λ is the index of f.

Consider a bijection $g : E(H) \to \{1, 2, \ldots, |E(G)|, |E(G)| + 1\}$ defined by

$$g(e) = \begin{cases} f(e) & \text{if } e \in E(G) \\ f(u_iv) & \text{if } e = u_iv^j, (i, j) \in \{(1, 1), (2, 1), (3, 2), (4, 2)\} \\ |E(G)| + 1 & \text{if } e = v^1v^2. \end{cases}$$

For the index-mapping of g we have

$$\begin{aligned} g^\star(v^1) &= g(v^1u_1) + g(v^1u_2) + g(v^1v^2) = f(vu_1) + f(vu_2) + 1 + |E(G)| \\ &= 1 + |E(G)| + 1 + |E(G)| = 2(1 + |E(G)|) \\ g^\star(v^2) &= g(v^2u_3) + g(v^2u_4) + g(v^1v^2) = f(vu_3) + f(vu_4) + 1 + |E(G)| \\ &= 1 + |E(G)| + 1 + |E(G)| = 2(1 + |E(G)|) \\ g^\star(u) &= f^\star(u) = 2(1 + |E(G)|) \quad \text{for every } u \in V(H) - \{v^1, v^2\}. \end{aligned}$$

Thus g is a supermagic labeling of H. □

Figure 2.39 gives an illustration of a supermagic labeling of a bipartite graph decomposable into two Hamilton cycles and the supermagic labeling of the corresponding graph obtained from the original by splitting a vertex and adding an edge. The vertex which is split in the original graph, and the added edge in the corresponding derived graph, are specially marked.

2.9.5 Disjoint Union of Regular Graphs

In [144] Ivančo and Semaničová dealt with the disjoint union of two regular graphs.

Theorem 2.9.20 ([144]) *For $i = 1, 2$ let G_i be an r_i-regular supermagic graph of order n_i. If $r_1 > r_2$ and*

$$p = \frac{n_2r_2^2 - n_1r_1^2 + 2r_1r_2n_1}{4(r_1 - r_2)} - \frac{1}{2}$$

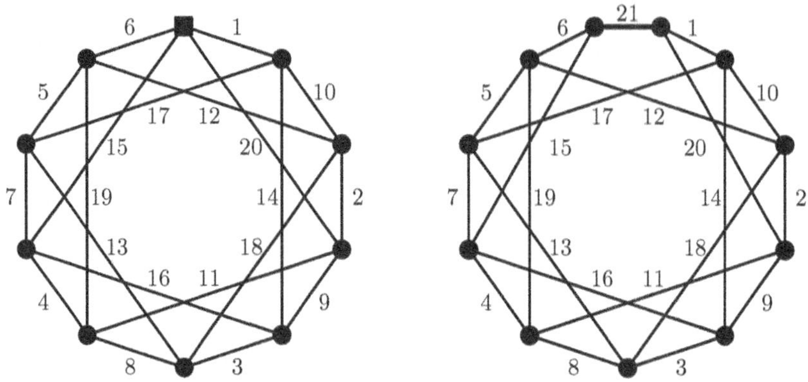

Fig. 2.39 Supermagic graph obtained from an original supermagic graph by splitting a vertex and adding an edge. Illustration of Theorem 2.9.19

is a nonnegative integer, then the disjoint union of the graphs G_1 and G_2 is a supermagic graph.

Proof Since G_1 is an r_1-regular supermagic graph, there exists a supermagic labeling $f_1 : E(G_1) \to \{1, 2, \ldots, r_1n_1/2\}$ for the index $\lambda_1 = r_1 (1 + r_1n_1/2)/2$. Analogously, there exists a supermagic labeling $f_2 : E(G_2) \to \{1, 2, \ldots, r_2n_2/2\}$ of G_2 for the index $\lambda_2 = r_2 (1 + r_2n_2/2)/2$.

If $p = (n_2r_2^2 - n_1r_1^2 + 2r_1r_2n_1)/(4r_1 - 4r_2) - 1/2$ is a nonnegative integer, then we consider a bijection $g : E(G_1 \cup G_2) \to \{1+p, 2+p, \ldots, (r_1n_1 + r_2n_2)/2+p\}$, defined by

$$g(e) = \begin{cases} f_1(e) + p & \text{if } e \in E(G_1) \\ f_2(e) + \frac{r_1n_1}{2} + p & \text{if } e \in E(G_2). \end{cases}$$

For its index-mapping we get

$$g^\star(v) = \begin{cases} \lambda_1 + r_1p & \text{if } v \in V(G_1) \\ \lambda_2 + r_2(\frac{r_1n_1}{2} + p) & \text{if } v \in V(G_2). \end{cases}$$

Thus $g^\star(v) = r_1r_2(r_1n_1 + r_2n_2)/(4r_1 - 4r_2)$, for every vertex $v \in V(G_1 \cup G_2)$ and so g is a supermagic labeling of $G_1 \cup G_2$. □

Figure 2.40 depicts a supermagic labeling of $K_{3,3} \cup K_{4,4}$ obtained by using the construction described in Theorem 2.9.20.

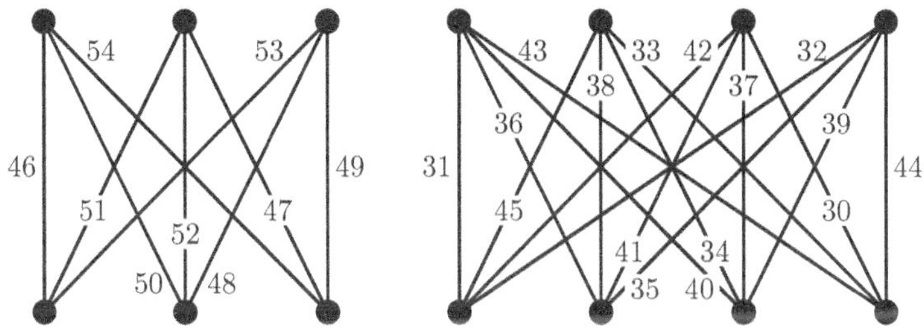

Fig. 2.40 Supermagic labeling of $K_{3,3} \cup K_{4,4}$

2.9.6 Constructions Using $(a, 1)$-Vertex-Antimagic Edge Graphs

In [144] Ivančo and Semaničová described two constructions of non-regular supermagic graphs using $(a, 1)$-VAE graphs.

Theorem 2.9.21 ([144]) *Let G_1, G_2, G_3 be 2-regular $(a, 1)$-VAE graphs, each of order n. Then there exists a supermagic graph G which is decomposable into two edge-disjoint spanning subgraphs F_1 and F_2, where F_1 is isomorphic to the disjoint union of G_1, G_2, G_3 (i.e., $F_1 \cong G_1 \cup G_2 \cup G_3$) and F_2 is isomorphic to n copies of the path on 3 vertices (i.e., $F_2 \cong nP_3$).*

Proof Let G_1, G_2, G_3 be 2-regular $(a, 1)$-VAE graphs of the same order n. Then $n \equiv 1 \pmod 2$. Moreover, for $j = 1, 2, 3$ there exists an $(a, 1)$-VAE labeling $f_j : E(G_j) \to \{1, 2, \dots, n\}$ of G_j, such that its index-mapping $f_j^\star$ satisfies

$$\{f_j^\star(v) : v \in V(G_j)\} = \left\{\tfrac{n+3}{2}, \tfrac{n+5}{2}, \dots, \tfrac{3n+1}{2}\right\}.$$

We denote the vertices of the graph G_j, $j = 1, 2, 3$, by $v_1^j, v_2^j \dots, v_n^j$ in such a way that

$$f_1^\star(v_i^1) = \frac{n+1}{2} + i \qquad \text{for } i = 1, 2, \dots, n$$

$$f_2^\star(v_i^2) = \begin{cases} \frac{n-1}{2} + 2i & \text{for } i = 1, 2, \dots, \frac{n+1}{2} \\ 2i - \frac{n+1}{2} & \text{for } i = \frac{n+3}{2}, \frac{n+5}{2}, \dots, n \end{cases}$$

$$f_3^\star(v_i^3) = \begin{cases} n+i & \text{for } i = 1, 2, \dots, \frac{n+1}{2} \\ i & \text{for } i = \frac{n+3}{2}, \frac{n+5}{2}, \dots, n. \end{cases}$$

Evidently it holds that

$$f_1^\star(v_1^1) < f_1^\star(v_2^1) < \cdots < f_1^\star(v_n^1)$$

$$f_2^\star(v_1^2) < f_2^\star(v_{\frac{n+3}{2}}^2) < f_2^\star(v_2^2) < f_2^\star(v_{\frac{n+5}{2}}^2) < \cdots < f_2^\star(v_{\frac{n+1}{2}}^2)$$

$$f_3^\star(v_{\frac{n+3}{2}}^3) < f_3^\star(v_{\frac{n+5}{2}}^3) < \cdots < f_3^\star(v_n^3) < f_3^\star(v_1^3) < f_3^\star(v_2^3) < \cdots < f_3^\star(v_{\frac{n+1}{2}}^3).$$

Let G be a graph obtained from the disjoint union of G_1, G_2, G_3 with the added edges $v_i^1 v_i^2,\ v_i^2 v_i^3,\ i = 1, 2, \dots, n$. Now we consider a mapping $g : E(G) \to \{(7n+1)/2, (7n+3)/2, \dots, (17n-1)/2\}$ defined by

$$g(e) = \begin{cases} f_1(e) + \frac{15n-1}{2} & \text{for } e \in E(G_1) \\ f_2(e) + \frac{9n-1}{2} & \text{for } e \in E(G_2) \\ f_3(e) + \frac{13n-1}{2} & \text{for } e \in E(G_3) \\ \frac{9n+1}{2} - i & \text{for } e = v_i^1 v_i^2, i = 1, 2, \dots, n \\ 6n + 1 - i & \text{for } e = v_i^2 v_i^3, i = 1, 2, \dots, \frac{n+1}{2} \\ 7n + 1 - i & \text{for } e = v_i^2 v_i^3, i = \frac{n+3}{2}, \frac{n+5}{2}, \dots, n. \end{cases}$$

It is easy to check that g is a bijection and for its index-mapping we get

$$g^\star(v_i^1) = f_1^\star(v_i^1) + 2(\tfrac{15n-1}{2}) + g(v_i^1 v_i^2) = \tfrac{n+1}{2} + i + 15n - 1 + \tfrac{9n+1}{2} - i = 20n,$$

for every $i = 1, 2, \dots, n$.

Analogously, for every $j = 2, 3$ and every $i = 1, 2, \dots, n$, we have

$$g^\star(v_i^j) = 20n.$$

Thus g is a supermagic labeling of G. □

Let $n \geq 3$ be an odd positive integer. Let $\mathcal{H}_n$ be a graph with the vertex set

$$V(\mathcal{H}_n) = \{v_1^1, v_2^1, \dots, v_n^1, v_1^2, v_2^2, \dots, v_n^2, v_1^3, v_2^3, \dots, v_n^3\}$$

and the edge set

$$E(\mathcal{H}_n) = \bigcup_{i=1}^{n} \{v_i^1 v_{i+1}^1, v_i^2 v_{i+\frac{n+1}{2}}^2, v_i^3 v_{i+1}^3, v_i^1 v_i^2,\ v_i^2 v_i^3\},$$

where subscripts are taken modulo n.

It is easy to see that $\mathcal{H}_n$ is decomposable into two edge-disjoint spanning subgraphs F_1 and F_2, where F_1 is isomorphic to three copies of odd cycle C_n (i.e., $F_1 \cong 3C_n$) and F_2 is isomorphic to n copies of the path on 3 vertices (i.e.,

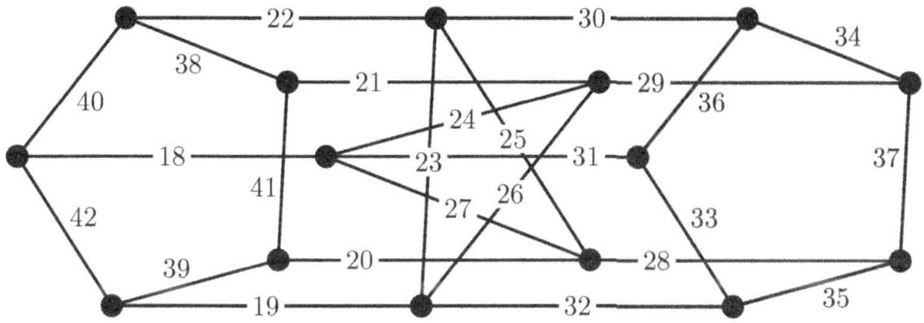

Fig. 2.41 Supermagic labeling of $\mathcal{H}_5$

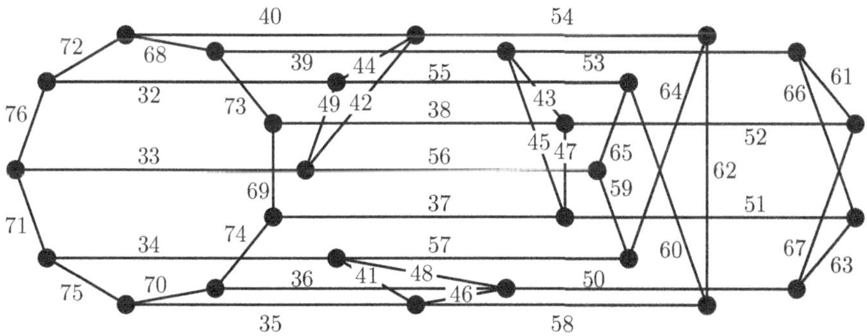

Fig. 2.42 Supermagic labeling of a non-regular graph

$F_2 \cong nP_3$). It is known, see [72], that there exists an $(a, 1)$-VAE labeling of odd cycle $C_n = v_1v_2 \dots v_nv_1$ such that $f^\star(v_i) = (n+1)/2 + i$ for $i = 1, 2, \dots, n$.

Corollary 2.9.1 ([144]) *For every odd positive integer $n \geq 3$, the graph $\mathcal{H}_n$ is supermagic.*

Figure 2.41 depicts a supermagic labeling of $\mathcal{H}_5$.

Using Theorem 2.9.21 we can find many others supermagic non-regular graphs. In Fig. 2.42 a supermagic labeling of one such graph is depicted.

Theorem 2.9.22 ([144]) *Let k be a positive integer. Let G_1 (G_2) be an r-regular ($(r + 2k)$ regular) $(a, 1)$-VAE graph of order n. Suppose that*

$$p_1 = \tfrac{nr^2}{4k} - \tfrac{nk+1}{2} \quad or \quad p_2 = \tfrac{nr(r+2)}{4k} - \tfrac{nk+1}{2}$$

is a nonnegative integer. Then there exists a supermagic graph G which is decomposable into two edge-disjoint spanning subgraphs F_1 and F_2, where F_1 is isomorphic to the disjoint union of G_1 and G_2 (i.e., $F_1 \cong G_1 \cup G_2$) and F_2 is isomorphic to n copies of K_2 (i.e., $F_2 \cong nK_2$).

Proof Since G_1 is an r-regular $(a_1, 1)$-VAE graph of order n, there exists an $(a_1, 1)$-VAE labeling $f_1 : E(G_1) \to \{1, 2, \ldots, nr/2\}$ of G_1 such that for its index-mapping $f_1^\star$ we have

$$\{f_1^\star(v) : v \in V(G_1)\} = \{a_1, a_1 + 1, \ldots, a_1 + n - 1\}.$$

Clearly, $a_1 = (nr^2 + 2r - 2n + 2)/4$ in this case.

Analogously, there exists an $(a_2, 1)$-VAE labeling $f_2 : E(G_2) \to \{1, 2, \ldots, n(r + 2k)/2\}$ of G_2. For its index-mapping we have

$$\{f_2^\star(v) : v \in V(G_2)\} = \{a_2, a_2 + 1, \ldots, a_2 + n - 1\},$$

where $a_2 = (n(r + 2k)^2 + 2(r + 2k) - 2n + 2)/4$.

We denote the vertices of G_j, $j = 1, 2$, by $v_1^j, v_2^j \ldots, v_n^j$ in such a way that

$$f_1^\star(v_i^1) = a_1 - 1 + i, \quad \text{for } i = 1, 2, \ldots, n$$
$$f_2^\star(v_i^1) = a_2 - 1 + i, \quad \text{for } i = 1, 2, \ldots, n.$$

Let G be a graph obtained from the disjoint union of G_1 and G_2 with added edges $v_i^1 v_i^2$, $i = 1, 2, \ldots, n$.

If $p_1 = nr^2/(4k) - (nk + 1)/2$ is a nonnegative integer, then we consider a bijection $g_1 : E(G) \to \{1 + p_1, 2 + p_1, \ldots, n(r + k + 1) + p_1\}$, defined by

$$g_1(e) = \begin{cases} f_1(e) + \frac{n(r+2k)}{2} + p_1 & \text{for } e \in E(G_1) \\ f_2(e) + p_1 & \text{for } e \in E(G_2) \\ n(r + k + 1) + 1 - i + p_1 & \text{for } e = v_i^1 v_i^2, i = 1, 2, \ldots, n. \end{cases}$$

For its index-mapping we get

$$\begin{aligned} g_1^\star(v_i^1) &= f_1^\star(v_i^1) + r(\tfrac{n(r+2k)}{2} + p_1) + g(v_i^1 v_i^2) \\ &= a_1 - 1 + i + r(\tfrac{n(r+2k)}{2} + p_1) + n(r + k + 1) + 1 - i + p_1 \\ &= \tfrac{n(3r^2+4r+2)}{4} + \tfrac{kn(r+1)}{2} + \tfrac{nr^2(r+1)}{4k} \\ g_1^\star(v_i^2) &= f_2^\star(v_i^2) + (r + 2k)p_1 + g(v_i^1 v_i^2) \\ &= a_2 - 1 + i + (r + 2k)p_1 + n(r + k + 1) + 1 - i + p_1 \\ &= \tfrac{n(3r^2+4r+2)}{4} + \tfrac{kn(r+1)}{2} + \tfrac{nr^2(r+1)}{4k}, \end{aligned}$$

for $i = 1, 2, \ldots, n$.

Thus g_1 is a supermagic labeling of G.

If $p_2 = nr(r+2)/(4k) - (nk+1)/2$ is a nonnegative integer, then we consider a bijection $g_2 : E(G) \to \{1 + p_2, 2 + p_2, \ldots, n(r+k+1) + p_2\}$, defined by

$$g_2(e) = \begin{cases} f_1(e) + \frac{n(r+2k+2)}{2} + p_2 & \text{for } e \in E(G_1) \\ f_2(e) + p_2 & \text{for } e \in E(G_2) \\ \frac{n(r+2k+2)}{2} + 1 - i + p_2 & \text{for } e = v_i^1 v_i^2, i = 1, 2, \ldots, n. \end{cases}$$

For its index-mapping we get

$$g_2^\star(v) = \frac{n(3r^2+6r+2)}{4} + \frac{kn(r+1)}{2} + \frac{nr(r+1)(r+2)}{4k}, \qquad \text{for } v \in V(G).$$

Thus g_2 is a supermagic labeling of G in this case. □

For r even and $k = 1$ in the previous theorem, we immediately obtain

Corollary 2.9.2 ([144]) *Let G_1 be a $2r$-regular $(a, 1)$-VAE graph of order n and let G_2 be a $2(r+1)$-regular $(a, 1)$-VAE graph of order n. Then there exists a supermagic graph G which is decomposable into two edge-disjoint spanning subgraphs F_1 and F_2, where F_1 is isomorphic to the disjoint union of G_1 and G_2 (i.e., $F_1 \cong G_1 \cup G_2$) and F_2 is isomorphic to n copies of K_2 (i.e., $F_2 \cong nK_2$).*

Proof Since the graphs G_1 and G_2 are $(a, 1)$-VAE graphs of even degree, then $n \equiv 1 \pmod 2$. Thus $p_1 = n(2r)^2/4 - (n+1)/2 = nr^2 - (n+1)/2$ is a positive integer and, according to Theorem 2.9.22, there exists a desired supermagic graph G. □

The construction described in Theorem 2.9.22 is illustrated in Fig. 2.43.

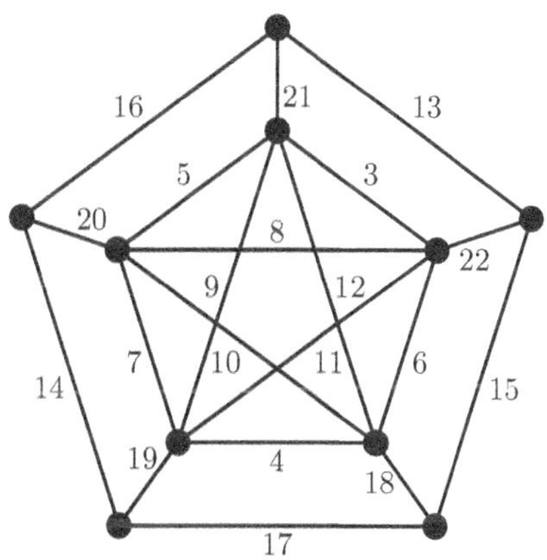

Fig. 2.43 Illustration of the construction described in Theorem 2.9.22

2.9.7 Join of Graphs

The *join* $G \oplus H$ of the disjoint graphs G and H is the graph $G \cup H$ together with all the edges joining vertices of $V(G)$ and vertices of $V(H)$. A vertex v of graph G is called a *saturated vertex* if it is adjacent to all the others vertices. The graph H with saturated vertex v is isomorphic to the graph $(H - v) \oplus K_1$. A graph with saturated vertex is also denoted by $G \oplus K_1$.

In [235] Semaničová characterized magic graphs with a saturated vertex.

Theorem 2.9.23 ([235]) *Let G be a graph of order at least 4. The graph $G \oplus K_1$ is magic if and only if the minimum degree of G is at least 1, G does not contain isolated edge, and $|\Gamma_G(S)| \geq |S|$, for every independent subset S of $V(G)$.*

Proof Let G be a graph of order at least 4. Let G be a graph with minimum degree at least 1 such that G does not contain an isolated edge and $|\Gamma_G(S)| \geq |S|$, for every independent subset S of $V(G)$.

The graph $G \oplus K_1$ is non-bipartite. Suppose that v is a saturated vertex of $G \oplus K_1$.

Suppose $G \oplus K_1$ is of type $\mathcal{A}$. As the saturated vertex v is incident with all other vertices, it is easy to see that $G \oplus K_1$ has 4 vertices, thus $|V(G)| = 3$, a contradiction.

Suppose that $G \oplus K_1$ is of type $\mathcal{B}$. Let H be the component of $(G \oplus K_1) - \{e_1, e_2\}$ which is a balanced bipartite graph. Consider two cases.

Let $v \in V(H)$. It is easy to see that G is isomorphic to K_2. This means G contains an isolated edge, a contradiction.

Let $v \notin V(H)$. As saturated vertex is incident with all vertices of $V(G)$, $|V(G)| = 2$, and H is a component of G. Thus G contains an isolated edge, a contradiction.

Let S be an independent set of vertices in $V(G \oplus K_1)$. Then either $S = \{v\}$ or S is an independent set of vertices in G. If $S = \{v\}$, then $\Gamma_{G \oplus K_1}(S) = V(G)$ and $|S| = 1 < |V(G)| = |\Gamma_{G \oplus K_1}(S)|$. If S is an independent set of vertices in G, then $|\Gamma_{G \oplus K_1}(S)| = |\Gamma_G(S) \cup \{v\}| = |\Gamma_G(S)| + 1 \geq |S| + 1 > |S|$.

According to Theorem 2.3.6, we get that $G \oplus K_1$ is a magic graph.

The converse implication is obvious. □

Let I denote the graph in Fig. 2.36.

From Theorem 2.9.23 we immediately obtain the bound for the smallest possible number of edges in graph G.

Corollary 2.9.3 ([235]) *Let $G \oplus K_1$ be a magic graph. Then for every nonnegative integer k the following statements hold.*

(i) If $|V(G)| = 5 + 4k$, then $|E(G)| \geq 5 + 3k$.
(ii) If $|V(G)| = 6 + 4k$, then $|E(G)| \geq 5 + 3k$.
(iii) If $|V(G)| = 7 + 4k$, then $|E(G)| \geq 6 + 3k$.
(iv) If $|V(G)| = 8 + 4k$, then $|E(G)| \geq 6 + 3k$.

Moreover, the equality holds if and only if G is a graph isomorphic to (i) $kP_4 \cup C_5$, (ii) $kP_4 \cup I$ or $kP_4 \cup P_6$, (iii) $(k + 1)P_4 \cup C_3$, or (iv) $(k + 2)P_4$.

Proof According to Theorem 2.6.8, proved in [279], in a magic graph the number of vertices of degree 2 is less than the number of vertices of degree at least 3. This means that if the graph $G \oplus K_1$ is magic, then the number of vertices of degree 1 in G, denoted by n_1, is less than or equal to the number of vertices of degree at least 2 in G. Let $V_{\geq 2}$ be the set of vertices of degree at least 2 in G, and $n_{\geq 2}$ be the cardinality of $V_{\geq 2}$. Then

$$n_1 \leq n_{\geq 2}. \tag{2.29}$$

According to Theorem 2.9.23, G contains no isolated vertex, so that

$$|V(G)| = n_1 + n_{\geq 2}. \tag{2.30}$$

For the number of edges we have

$$2|E(G)| = \sum_{v \in V(G)} \deg(v) = n_1 + \sum_{v \in V_{\geq 2}} \deg(v) \geq n_1 + 2n_{\geq 2} = 2|V(G)| - n_1. \tag{2.31}$$

Combining (2.29), (2.30), and (2.31), we get

$$|E(G)| \geq \frac{3}{4}|V(G)|, \tag{2.32}$$

which implies the bounds for the number of edges in G.

Consider Case (iv), i.e., $|V(G)| = 8 + 4k$. If the equality holds, and because G has no isolated vertex and no isolated edge, then every component of G has exactly 4 vertices and 3 edges. Moreover, G contains a (1-2)-factor, thus its every component is isomorphic to P_4. The other cases are analogous. □

Using Stewart's characterization of supermagic complete graphs, see Theorem 2.6.6, we have that the graph $K_n \oplus K_1$ is supermagic if and only if $n = 1$ or $4 < n \not\equiv 3 \pmod 4$.

In [235] Semaničová established an upper bound for the number of vertices n in supermagic graph $G \oplus K_1$, where G is an r-regular graph.

Lemma 2.9.1 ([235]) *Let G be an r-regular graph of order n, $2 \leq r < n - 1$. Let $G \oplus K_1$ be a supermagic graph with supermagic labeling $f : E(G \oplus K_1) \to \{a, a+1, \ldots, a + n(r+2)/2 - 1\}$. Then*

$$a \leq \frac{\frac{n(r+2)^2}{2} - n^2 - r - 1}{2(n - r - 1)}.$$

Proof Let G be an r-regular graph of order n, with $2 \le r < n-1$. Let $f : E(G \oplus K_1) \to \{a, a+1, \ldots, a+n(r+2)/2-1\}$ be a supermagic labeling of $G \oplus K_1$ for an index λ. Then

$$(n+1)\lambda = 2 \sum_{e \in E(G \oplus K_1)} f(e) = 2\left(a + (a+1) + \cdots + \left(a + \frac{n(r+2)}{2} - 1\right)\right)$$

$$= \left(2a + \frac{n(r+2)}{2} - 1\right)\frac{n(r+2)}{2}.$$

Thus

$$\lambda = \frac{(2a + \frac{n(r+2)}{2} - 1)\frac{n(r+2)}{2}}{n+1}. \tag{2.33}$$

Since the degree of the saturated vertex v is n, then

$$\lambda = f^{\star}(v) \ge a + (a+1) + \cdots + (a+n-1). \tag{2.34}$$

According to (2.33),

$$\frac{(2a+n-1)n}{2} \le \frac{(2a + \frac{n(r+2)}{2} - 1)\frac{n(r+2)}{2}}{n+1}. \tag{2.35}$$

After some manipulations, we obtain

$$2a(n-r-1) \le \frac{n(r+2)^2}{2} - n^2 - r - 1. \tag{2.36}$$

As $r < n-1$, we get the required bound for a. □

Using the previous lemma we get the following theorem.

Theorem 2.9.24 ([235]) *Let G be an r-regular graph of order $n \ge 4$, $2 \le r < n-1$. If $G \oplus K_1$ is a supermagic graph, then*

$$n \le \frac{\frac{r^2}{2} + 2r + \sqrt{(\frac{r^2}{2} + 2r)^2 + 4(r+1)}}{2}. \tag{2.37}$$

Proof Let G be an r-regular graph of order $n \ge 4$, $2 \le r < n-1$. Let $G \oplus K_1$ be a supermagic graph and suppose

$$n > \frac{\frac{r^2}{2} + 2r + \sqrt{(\frac{r^2}{2} + 2r)^2 + 4(r+1)}}{2}. \tag{2.38}$$

Then

$$n^2 - n(\frac{r^2}{2} + 2r) - r - 1 > 0. \tag{2.39}$$

After some manipulations we get

$$\frac{n(r+2)^2}{2} - n^2 - r - 1 < 2(n - r - 1). \tag{2.40}$$

This means that

$$\frac{\frac{n(r+2)^2}{2} - n^2 - r - 1}{2(n - r - 1)} < 1. \tag{2.41}$$

According to Lemma 2.9.1, we have $a < 1$, a contradiction. □

Corollary 2.9.4 ([235]) *Let G be a 2-regular graph. The graph $G \oplus K_1$ is supermagic if and only if G is isomorphic to C_4 or C_5.*

Proof Let G be a 2-regular graph of order n. Let $G \oplus K_1$ be a supermagic graph. According to Theorems 2.9.23 and 2.9.24, we have $4 \leq n \leq 6$.

If $n = 6$ then, by Lemma 2.9.1, we have $a \leq 3/2$, and thus $a = 1$. In this case, by (2.33), we get $\lambda = (2 + 11)12/7 = 13 \cdot 12/7$. This is impossible since λ must be an integer.

If $4 \leq n \leq 5$, then G is connected which means that G is isomorphic to a cycle C_n.

The graph $C_n \oplus K_1$ is isomorphic to the wheel with n spokes. Stewart [262] proved that wheel with n spokes is supermagic if and only if $n = 4$ or $n = 5$. The corresponding supermagic labelings of W_4 and W_5 are depicted in Fig. 2.44. □

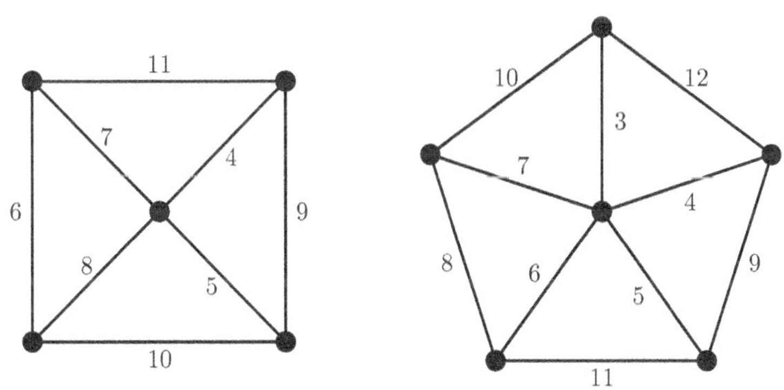

Fig. 2.44 Supermagic labelings of W_4 and W_5

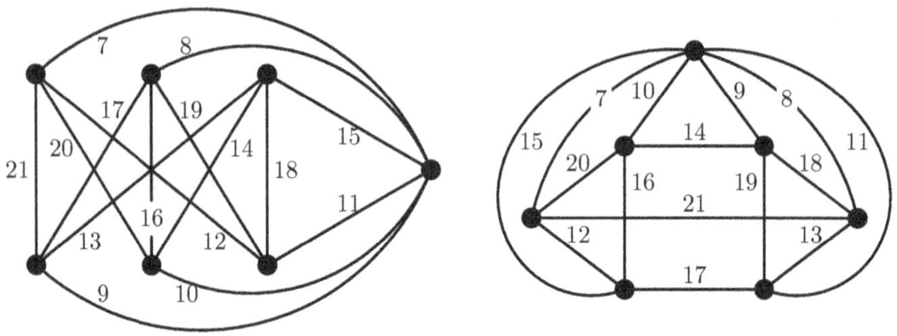

Fig. 2.45 Supermagic labelings of $G \oplus K_1$ for 3-regular graphs G of order 6

Corollary 2.9.5 ([235]) *Let G be a 3-regular graph of order n. The graph $G \oplus K_1$ is supermagic if and only if $n = 6$ or $n = 8$.*

Proof Let G be a 3-regular graph of order n, $4 \leq n \equiv 0 \pmod 2$. If $G \oplus K_1$ is a supermagic graph, then, according to Theorem 2.9.24, we have $n \leq 10$.

If $n = 10$, then, by Lemma 2.9.1, we have $a \leq 7/4$, and thus $a = 1$. By (2.33), we get $\lambda = (2 + 24)25/11 = 26 \cdot 25/11$ which is not an integer, a contradiction.

If $n = 4$, then G is a complete graph and $G \oplus K_1$ is isomorphic to K_5. According to Theorem 2.6.6, K_5 is not supermagic. Thus we have $n = 6$ or $n = 8$.

There exist exactly two non-isomorphic 3-regular graphs of order 6 and exactly six non-isomorphic 3-regular graphs of order 8; Figs. 2.45 and 2.46 display supermagic labelings of $G \oplus K_1$ for all these graphs. □

Corollary 2.9.6 ([235]) *Let G be 4-regular graph of order n. If $G \oplus K_1$ is a supermagic graph, then $5 \leq n \leq 15$.*

In [143] Ivančo and Semaničová described another construction of supermagic joins $G \oplus K_1$.

Theorem 2.9.25 ([143]) *Let G be an $(a, 1)$-VAE r-regular graph of order n. If $(n - r - 1)$ is a divisor of the nonnegative integer $a + n(1 + r - (n + 1)/2)$, then the join $G \oplus K_1$ is a supermagic graph.*

Proof Put $\lambda_1 = a + n(1 + r)$ and $\lambda_2 = n(n + 1)/2$. According to the assumption there is a nonnegative integer p such that $\lambda_1 - \lambda_2 = p(n - r - 1)$ (thus $(r+1)p + \lambda_1 = np + \lambda_2$). Let f be an $(a, 1)$-VAE labeling of G. The join $G \oplus K_1$ is obtained from G by adding the vertex w and the edges wv for all $v \in V(G)$.

Consider the mapping h from $E(G \oplus K_1)$ into positive integers given by

$$h(e) = \begin{cases} p + n + f(e) & \text{if } e \in E(G) \\ p + n + a - f^{\star}(v) & \text{if } e = wv \text{ for } v \in V(G). \end{cases}$$

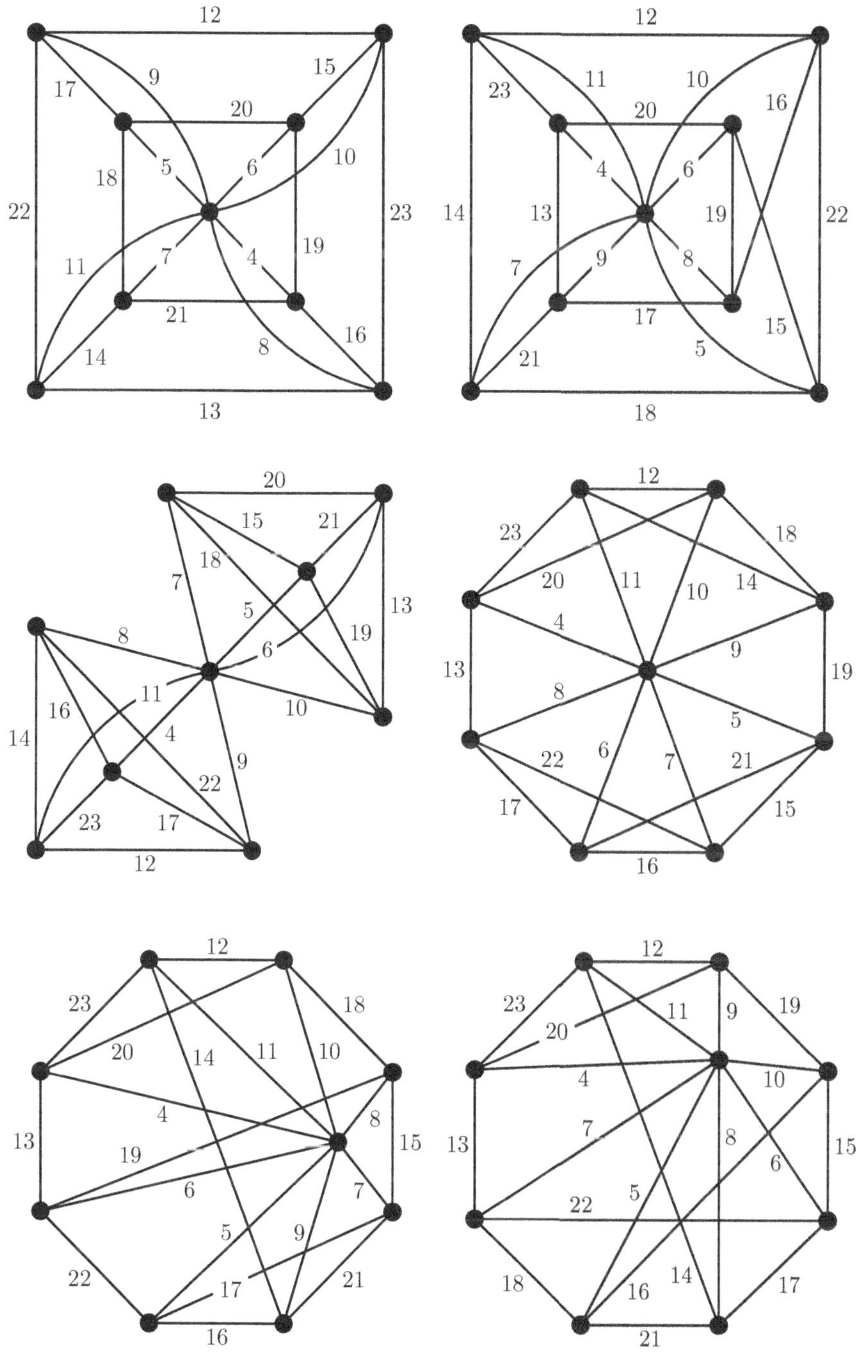

Fig. 2.46 Supermagic labelings of $G \oplus K_1$ for 3-regular graphs G of order 8

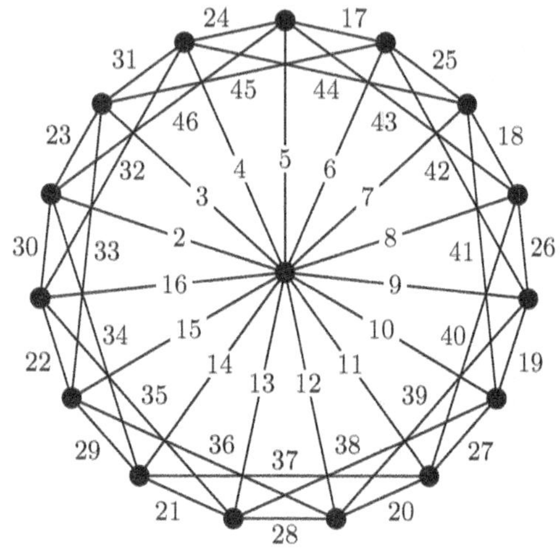

Fig. 2.47 The supermagic labeling of $G \oplus K_1$, where G is isomorphic to the circulant graph $C_{15}(1, 3)$

Evidently, $\{h(wv) : v \in V(G)\} = \{p + 1, p + 2, \ldots, p + n\}$ and $\{h(e) : e \in E(G)\} = \{p + n + 1, p + n + 2, \ldots, p + n + |E(G)|\}$. Thus, the set $\{h(e) : e \in E(G \oplus K_1)\}$ consists of consecutive positive integers. Moreover, $h^\star(w) = np + \lambda_2$ and $h^\star(v) = (r + 1)p + \lambda_1$ for all $v \in V(G)$. Therefore, h is a supermagic labeling of $G \oplus K_1$. □

Figure 2.47 displays the supermagic labeling of $G \oplus K_1$, where G is a circulant graph $C_{15}(1, 3)$.

Using divisibility, it is not difficult to check the assumptions of Theorem 2.9.25 for given values of n and r. Thus

Corollary 2.9.7 ([143]) *Let n and r be positive integers such that one of the following conditions is satisfied.*

(i) $5 \leq n \equiv 1 \pmod 2$ *and* $r = n - 3$.
(ii) $11 \leq n \equiv 1 \pmod 2$ *and* $r = n - 7$.
(iii) $8 \leq n \equiv 0 \pmod 4$ *and* $r = \frac{n}{2} - 1$.
(iv) $11 \leq n \equiv 3 \pmod 8$ *and* $r = n - 5$.
(v) $12 \leq n \equiv 4 \pmod 8$ *and* $r = n - 3$.
(vi) $12 \leq n \equiv 4 \pmod 8$ *and* $r = n - 7$.
(vii) $13 \leq n \equiv 5 \pmod 8$ *and* $r = n - 5$.

If G is an $(a, 1)$-VAE r-regular graph of order n, then the join $G \oplus K_1$ is supermagic.

Corollary 2.9.8 ([143]) *Let G be any $(n - 3)$-regular ($(n - 7)$-regular) graph of odd order $n \geq 7$ ($n \geq 15$). Then $G \oplus K_1$ is a supermagic graph.*

Ivančo and Polláková [142] characterized supermagic graphs $G \oplus K_1$ by using a vertex-magic total labeling which is defined in the following way.

A *vertex-magic total labeling* (or VMT labeling for short) of a graph G is a bijection from the set $V(G) \cup E(G)$ to the set of integers, $\{1, 2, \ldots, |V(G)| +$

$|E(G)|\}$, with the property that the sum of the labels of a vertex v and of all the edges incident to v is equal to a certain magic constant, for every vertex v of G. A more formal definition can be found in Chap. 3.

Theorem 2.9.26 ([142]) *Let G be a d-regular graph of order n. The graph $G \oplus K_1$ is supermagic if and only if G admits a VMT labeling with magic constant h such that $(n-d-1)$ is a divisor of the nonnegative integer $(n+1)h - n(d+2)/2(n(d+2)/2+1)$.*

Applying Theorem 2.9.26 to regular graphs of even order they immediately obtained the following corollaries.

Corollary 2.9.9 ([142]) *Let G be a $2(k-1)$-regular graph of order $2k$. The graph $G \oplus K_1$ is supermagic if and only if G admits a VMT labeling with magic constant h such that $(2k+1)h \geq 2k^2(2k^2+1)$.*

Corollary 2.9.10 ([142]) *Let G be a k-regular graph of order $2k$. The graph $G \oplus K_1$ is supermagic if and only if G admits a VMT labeling with magic constant h such that the nonnegative integer $(2k+1)h - k(k+2)(k+1)^2$ is an integral multiple of $(k-1)$.*

Ivančo and Polláková [142] also characterized several supermagic complete multipartite graphs.

Theorem 2.9.27 ([142]) *Let C be a Hamiltonian cycle of the complete bipartite graph $K_{n,n}$, where $n \geq 5$. Then $K_{n,n} - E(C)$ is a supermagic graph.*

Corollary 2.9.11 ([142]) *The complete tripartite graph $K_{1,n,n}$ is supermagic if and only if $n \geq 2$.*

Theorem 2.9.28 ([142]) *The complement of the circulant graph $C_{2n}(1, n)$ is supermagic for any integer $n \geq 4$.*

Corollary 2.9.12 ([142]) *The complete multipartite graph $K_{1,n[2]}$ is supermagic if and only if $n \geq 2$.*

In [138] Ivančo presented some constructions of supermagic joins of graphs using DC-labelings.

Theorem 2.9.29 ([138]) *Let G_1, G_2 be disjoint r-regular Hamiltonian graphs of order n. If $r \geq 4$ is even and n is odd, then the join $G_1 \oplus G_2$ is a supermagic graph.*

Corollary 2.9.13 ([138]) *Let G_1, G_2 be disjoint r-regular Hamiltonian graphs of order n. If $2r \geq 4$, $5 \leq n \equiv 1 \pmod 2$, and $4 \leq r \equiv 0 \pmod 2$, then the join $G_1 \oplus G_2$ is supermagic.*

Theorem 2.9.30 ([138]) *Let G_i, $i \in \{1, 2\}$, be an r_i-regular Hamiltonian graph of order n. If $4 \leq r_1 \equiv 0 \pmod 4$, $r_1 = r_2 + 2$ and n is odd, then the join $G_1 \oplus G_2$ is a supermagic graph.*

For dense graphs

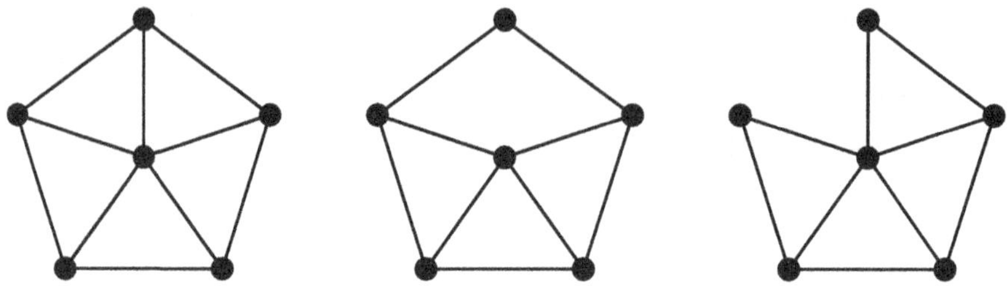

Fig. 2.48 The wheel W_5, basket B_5, and fan F_5

Corollary 2.9.14 ([138]) *Let G_i, $i \in \{1, 2\}$, be an r_i-regular Hamiltonian graph of order n. If $n \leq 2r_2$, $r_1 = r_2 + 2$, and $4 \leq r_1 \equiv 0 \pmod 4$, then the join $G_1 \oplus G_2$ is a supermagic graph.*

2.9.8 Other Magic and Supermagic Non-regular Graphs

A *wheel* W_n with n spokes is isomorphic to the graph $C_n \oplus K_1$. The vertex of degree n is called the *central vertex*, or the *hub vertex*, and the remaining vertices are called the *rim vertices*. The edges incident to the central vertex are called *spokes* and the edges not incident to the central vertex are called *rim edges*. If one spoke is removed from W_n, the resulting graph is called a *basket*, denoted by B_n. If one rim edge is removed from W_n, the resulting graph is called a *fan*, denoted by F_n. Alternatively, a fan F_n is isomorphic to the graph $P_n \oplus K_1$. Figure 2.48 depicts a wheel W_5, a basket B_5, and a fan F_5.

In [262] Stewart proved

Theorem 2.9.31 ([262]) *A wheel W_n is magic for $n \geq 4$.*

Stewart [262] proved that wheel W_n is supermagic if and only if $n = 4$ or $n = 5$.

Theorem 2.9.32 ([262]) *A basket B_n is magic for $n = 4$ and all $n \geq 6$.*

Theorem 2.9.33 ([262]) *A fan F_n is magic if and only if n is odd and $n \geq 3$.*

Stewart [262] also proved that out of the graphs corresponding to the five Platonic solids only the octahedron is supermagic.

2.10 Related Topics

Magic graphs in which each edge label is a prime number are called *prime-magic*. The notion of a prime-magic graph was introduced by Stewart in [262]. Except for the trivial K_2, the simplest prime-magic graph is depicted in Fig. 2.49.

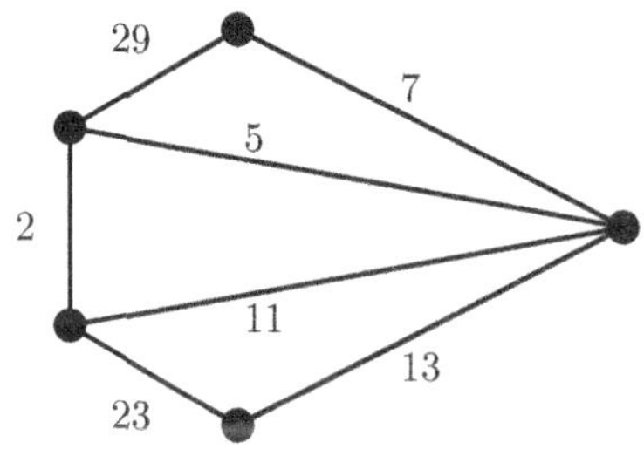

Fig. 2.49 Prime-magic labeling of a graph

There are infinitely many graphs that are magic, but not prime-magic. This answers, in the negative, a conjecture of Sedláček that every magic graph is also prime-magic. Stewart additionally proposed the more plausible conjecture.

Conjecture 2.10.1 ([262]) Every regular magic graph is prime-magic.

The difficulty surrounding any general statement about prime-magic graphs is obviously due to the irregular distribution of primes.

For relatively prime positive integers a and b, define a set of integers: $\Pi(a, b) = \{k : k \geq 0, ak + b \text{ is a prime}\}$. According to Dirichlet, the set $\Pi(a, b)$ is infinite. Stewart proved

Theorem 2.10.1 ([262]) *If G is regular and if G is magic under a labeling λ whose values are in $\Pi(a, b)$, then G is prime-magic.*

For example, complete graph K_5 is magic under a labeling λ, whose values 1, 2, 3, 4, 5, 9, 11, 12, 13, 15 are all in $\Pi(30, 7)$, see Fig. 2.50. We can apply Theorem 2.10.1 to obtain prime-magic labeling of K_5, see Fig. 2.51.

In the theory of numbers, magic squares consisting of prime numbers are known. Sierpiński [252] gives examples of magic squares with three (four) rows formed only from prime numbers, where the sum of the numbers along every row, every column, and every main diagonal is 1077 (798). Stewart [262] shows that complete bipartite graph $K_{3,3}$ is prime-magic with index $\sigma = 139$, see Fig. 2.52.

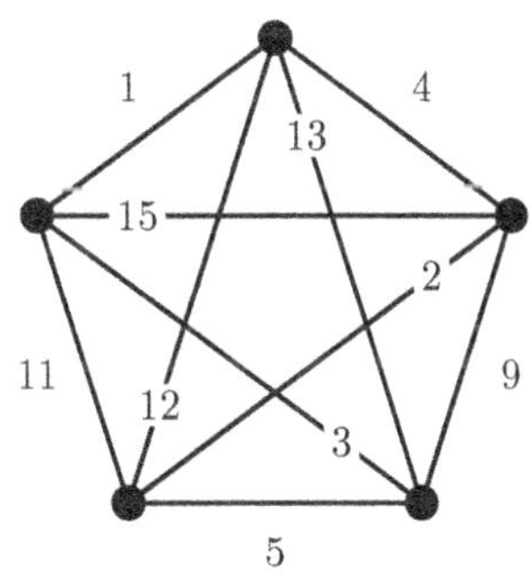

Fig. 2.50 Magic labeling of K_5

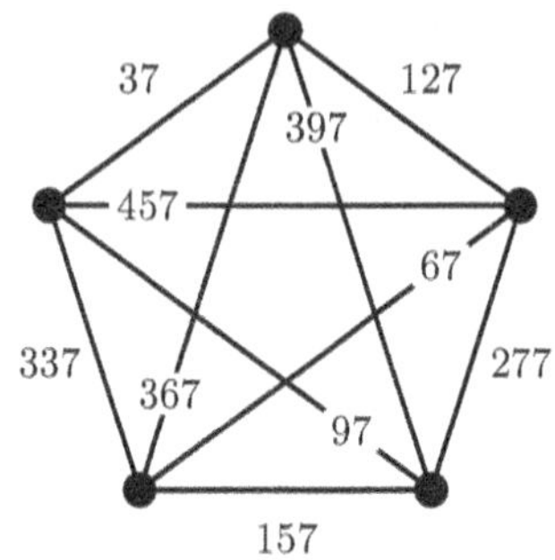

Fig. 2.51 Prime-magic labeling of K_5

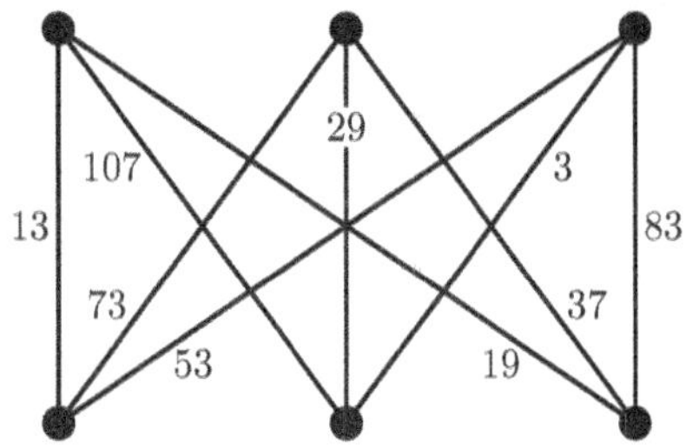

Fig. 2.52 Prime-magic labeling of $K_{3,3}$ with index $\sigma = 139$

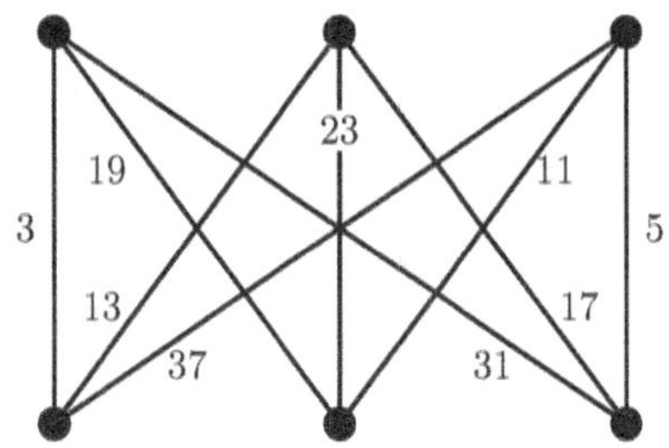

Fig. 2.53 Prime-magic labeling of $K_{3,3}$ with index $\sigma = 53$

Note that a prime-magic labeling of $K_{n,n}$ is not necessarily a magic square of order n because the sums along the main diagonals are not needed for prime-magic labeling of $K_{n,n}$.

What minimum value of index σ can be assigned to the prime-magic graph $K_{n,n}$, $n \geq 3$?

Sedláček showed that

Theorem 2.10.2 ([231]) *The smallest value of index σ which can be assigned to the prime-magic graph $K_{3,3}$ is $\sigma = 53$.*

Figure 2.53 illustrates the prime-magic labeling of $K_{3,3}$ with $\sigma = 53$.

Bača and Holländer have shown that

Theorem 2.10.3 ([31]) *The minimum value of index σ which can be assigned to the prime-magic graph $K_{4,4}$ is $\sigma = 114$.*

Proof Let $A = (f(e_{ij}))$ be a square matrix, formed only from prime numbers, describing a labeling f of graph $K_{4,4}$ and denote $S = \sum_{i=1}^{4}\sum_{j=1}^{4} f(e_{ij})$, where e_{ij} is the edge joining the ith vertex and jth vertex of $K_{4,4}$. Clearly, $S \geq 438$. It is required that $\sigma \equiv 0 \pmod 2$ and $S = 4\sigma$. The first S satisfying these conditions is $S = 440$. Then $\sigma = 110 \equiv -1 \pmod 3$ and so each 4-tuple of prime numbers $(f(e_{ij}) \neq 2)$ besides $f(e_{11}) = 3$ must have two elements $\equiv -1$ and one $\equiv 1 \pmod 3$ and each 4-tuple of prime numbers $(f(e_{ij}) \neq 2, 3)$ must have either all four elements $\equiv -1 \pmod 3$ or three elements $\equiv 1$ and one $\equiv -1 \pmod 3$. There exists exactly one modified $(0, 1, -1)$-matrix, see the first matrix in Fig. 2.54, such that the sum of every row and every column is $\equiv -1 \pmod 3$. However, the sum of three and ten prime numbers congruent to $+1 \pmod 3$ and five prime numbers congruent to $-1 \pmod 3$ (which formed the matrix) is greater than 440, and thus we have a contradiction.

Let us discuss the case where $\sigma = 112 \equiv 1 \pmod 3$. In this case each 4-tuple of prime numbers $(f(e_{ij}) \neq 2)$, besides $f(e_{11}) = 3$, must have two elements $\equiv +1$ and one $\equiv -1 \pmod 3$ and each 4-tuple of prime numbers $(f(e_{ij}) \neq 2, 3)$ must have either all four elements $\equiv 1 \pmod 3$ or three elements $\equiv -1$ and one $\equiv 1 \pmod 3$. We get the unique modified $(0, 1, -1)$-matrix for $\sigma \equiv 1 \pmod 3$, see the second matrix in Fig. 2.54.

The sum of three and five prime numbers congruent to $+1 \pmod 3$ and ten prime numbers congruent to $-1 \pmod 3$ is greater than 448. This means that for $\sigma = 112$ there does not exist a prime-magic graph $K_{4,4}$.

Similar to above, it can be shown that for $\sigma = 114$ there exist exactly two modified $(0, 1, -1)$-matrices. The one on the left in Fig. 2.55 yields a contradiction and the one on the right corresponds to a matrix of prime numbers and it shows that $\sigma = 114$ is the minimum value of the index. □

$$\begin{bmatrix} 0 & 1 & -1 & -1 \\ 1 & -1 & 1 & 1 \\ -1 & 1 & 1 & 1 \\ -1 & 1 & 1 & 1 \end{bmatrix} \qquad \begin{bmatrix} 0 & -1 & 1 & 1 \\ -1 & 1 & -1 & -1 \\ 1 & -1 & -1 & -1 \\ 1 & -1 & -1 & -1 \end{bmatrix}$$

Fig. 2.54 Modified matrix for $\sigma = 110$ and for $\sigma = 112$

$$\begin{bmatrix} 0 & 1 & 1 & 1 \\ 1 & -1 & 1 & -1 \\ 1 & 1 & -1 & -1 \\ 1 & -1 & -1 & 1 \end{bmatrix} \qquad \begin{bmatrix} 0 & -1 & -1 & -1 \\ -1 & 1 & -1 & 1 \\ -1 & 1 & 1 & -1 \\ -1 & -1 & 1 & 1 \end{bmatrix}$$

Fig. 2.55 Two modified matrices for $\sigma = 114$

Fig. 2.56 Square matrix describing prime-magic labeling of $K_{4,4}$

$$\begin{bmatrix} 3 & 11 & 47 & 53 \\ 23 & 37 & 41 & 13 \\ 29 & 61 & 7 & 17 \\ 59 & 5 & 19 & 31 \end{bmatrix}$$

Fig. 2.57 Square matrix describing prime-magic labeling of $K_{5,5}$

$$\begin{bmatrix} 3 & 107 & 7 & 37 & 79 \\ 17 & 19 & 67 & 71 & 59 \\ 43 & 61 & 89 & 11 & 29 \\ 73 & 41 & 23 & 83 & 13 \\ 97 & 5 & 47 & 31 & 53 \end{bmatrix}$$

The corresponding square matrix for prime-magic labeling of $K_{4,4}$ with $\sigma = 114$ is shown in Fig. 2.56.

Let $\sigma(S_n)$ denote the smallest natural number ($\sigma(S_n) \leq \sigma$) for which there exists S_n, where S_n is the sum of n^2 pairwise distinct prime numbers. Clearly, if n is odd, then $\sigma(S_n) \not\equiv 0 \pmod 2$, and if n is even, then $\sigma(S_n) \equiv 0 \pmod 2$, and furthermore, $n\sigma(S_n) = S_n$.

Bača and Holländer formulated the following conjecture.

Conjecture 2.10.2 ([31]) If $n \geq 5$, then the minimum value of index σ which can be assigned to the prime-magic graph $K_{n,n}$ is $\sigma(S_n)$.

They have proved that, for $5 \leq n \leq 17$, the conjecture holds. For instance, for $n = 5$, the sum of the first 25 prime numbers ($\neq 2$) is $S_5 = 1159$ but $S_5/5$ is not an integer. The smallest natural number greater than $S_5/5$ is $\sigma(S_5) = 233$ and the corresponding square matrix, see Fig. 2.57, shows the required prime-magic labeling of $K_{5,5}$.

Trenkler [281, 280] extended the definition of supermagic graphs to include hypergraphs and proved that the complete k-uniform n-partite hypergraph is supermagic if $n \neq 2$ or 6 and $k \geq 2$.

Chapter 3
Vertex-Magic Total Labelings

A *vertex-magic total labeling* (or VMT labeling for short) of a (p, q) graph G with p vertices and q edges is a bijection λ from the set $V(G) \cup E(G)$ to the set of integers, $\{1, 2, \ldots, p+q\}$, with the property that the sum of the labels of a vertex v and of all the edges incident to v is equal to a certain magic constant k, for every vertex v of G. This sum is called the *vertex-weight* of the vertex v and is denoted by $wt(v)$. More formally,

$$wt(v) = \lambda(v) + \sum_{u \in N(v)} \lambda(vu) = k. \tag{3.1}$$

Different VMT labelings of wheel W_4 with the same magic constant $k = 26$ are displayed in Fig. 3.1.

A VMT labeling λ is called *super* if $\lambda(V(G)) = \{1, 2, \ldots, p\}$. That is, in a super VMT labeling the smallest labels are assigned to the vertices. A graph that allows a (super) VMT labeling will be often called a (super) VMT graph.

VMT labeling was first suggested by MacDougall, Miller, Slamin, and Wallis, and introduced in [179] and [181].

Let S_V and S_E denote the sum of all vertex labels and all edge labels, respectively, in a VMT labeling λ of a (p, q) graph G. The sum of all labels is

$$S_V + S_E = 1 + 2 + \cdots + p + q = \frac{(p+q+1)(p+q)}{2}. \tag{3.2}$$

If we sum all the vertex-weights, we count each vertex label once and each edge label twice; therefore, we have the equality

$$S_V + 2S_E = pk. \tag{3.3}$$

M. Bača et al., *Magic and Antimagic Graphs*, Developments in Mathematics 60,
https://doi.org/10.1007/978-3-030-24582-5_3

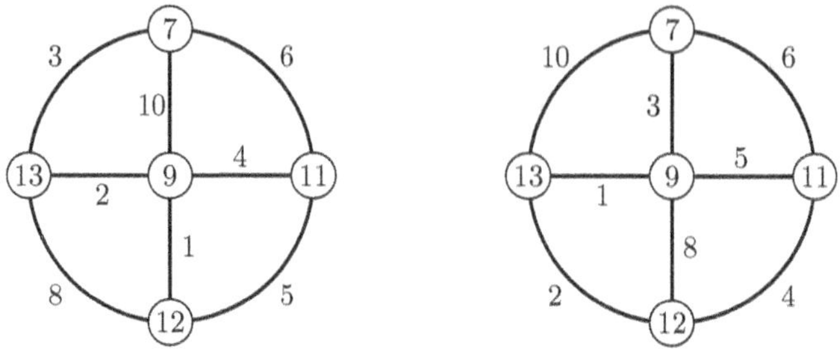

Fig. 3.1 Two non-isomorphic VMT labelings of W_4

Now, if we combine Eqs. (3.2) and (3.3) we get

$$S_E + \frac{(p+q+1)(p+q)}{2} = pk. \tag{3.4}$$

The edges could conceivably receive either the q smallest labels or, at the other extreme, the q largest labels, or anything in between. Consequently, we have

$$\frac{q(q+1)}{2} \leq S_E \leq \frac{(p+q+1)(p+q)}{2} - \frac{p(p+1)}{2}. \tag{3.5}$$

Combining (3.4) and (3.5), we get the inequality

$$\frac{(p+q+1)(p+q)+q(q+1)}{2p} \leq k \leq \frac{2(p+q+1)(p+q)-p(p+1)}{2p}. \tag{3.6}$$

This inequality gives the range for the feasible values of the magic constant k.

It is clear from Eq. (3.1) that when the value of k and the labels of edges are given, then the vertex labels are determined. So in this case the VMT labeling is completely described by the edge labels. Note however that the vertex labels do not completely determine the VMT labeling. Figure 3.1 shows two VMT labelings of the wheel W_4 that use the same vertex labels and have the same magic constant $k = 26$, but the edge labels are different.

3.1 Vertex-Magic Total Labelings of Regular Graphs

If a regular (p, q) graph G possesses a VMT labeling with magic constant k, we can create a new labeling λ' from λ by setting

$$\begin{aligned} \lambda'(u) &= p+q+1-\lambda(u), && \text{for any vertex } u \in V(G) \\ \lambda'(uv) &= p+q+1-\lambda(uv), && \text{for any edge } uv \in E(G). \end{aligned} \tag{3.7}$$

Clearly, like λ, the labeling λ' is also a bijection from the set $V(G) \cup E(G)$ to $\{1, 2, \ldots, p+q\}$ and we will call λ' the *dual* of λ. If r is the degree of each vertex of G, then

$$k' = (r+1)(p+q+1) - k$$

is the new magic constant.

The general problem of whether or not one can use a labeling of a graph G to produce a labeling of some subgraph of G appears to be very difficult. The next theorem answers a very special case of this question for regular graphs.

Theorem 3.1.1 ([179, 287]) *Let G be a regular graph having a VMT labeling in which the label 1 is assigned to some edge e. Then the graph $G - e$ has a VMT labeling.*

3.1.1 Cycles

The easiest regular graphs to deal with are the cycles. Let C_n be a cycle on n vertices and let P_n be a path on n vertices. MacDougall et al. [179] proved

Theorem 3.1.2 ([179]) *The cycle C_n has a VMT labeling for any $n \geq 3$.*

The VMT labelings of C_n, $n \geq 3$, from [179] assign the label 1 to some edge e. According to Theorem 3.1.1, the graph $C_n - e$ has a VMT labeling. Therefore we have

Corollary 3.1.1 ([179]) *The path P_n has a VMT labeling for any $n \geq 3$.*

3.1.2 Complete Graphs

MacDougall et al. [179] obtained a result concerning VMT labelings for complete graphs K_n when n is odd.

Theorem 3.1.3 ([179]) *There is a VMT labeling of K_n for all odd n.*

VMT labelings for complete graphs K_n when n is even have been given in [190] and [175].

Theorem 3.1.4 ([190]) *There is a VMT labeling of K_n, for all $n \equiv 2 \pmod 4$, $n \geq 6$.*

Theorem 3.1.5 ([175]) *There is a VMT labeling of K_n, for all $n \equiv 0 \pmod 4$, $n \geq 4$.*

Gray et al. [123] used the existence of magic rectangles to present a simpler proof that all complete graphs are VMT. Krishnappa et al. [168] gave another proof that all complete graphs are vertex-magic total.

The following conjecture was proposed in [179].

Conjecture 3.1.1 ([179]) For each $n \geq 5$ there is a VMT labeling of K_n, for every feasible value of k.

For complete graph K_n with $p = n$ and $q = n(n-1)/2$, Inequality (3.6) gives

$$\frac{n(n^2+3)}{4} \leq k \leq \frac{n(n+1)^2}{4}. \tag{3.8}$$

Any integer between these bounds is called a *feasible value* for k. McQuillan and Smith [186] developed a new technique that constructs a VMT labeling of complete graphs K_n, for n odd, for all values of k in the feasible range. The following theorem is a partial solution to Conjecture 3.1.1.

Theorem 3.1.6 ([186]) *If n is an odd integer, then K_n has a VMT labeling with magic constant k if and only if*

$$\frac{n(n^2+3)}{4} \leq k \leq \frac{n(n+1)^2}{4}.$$

Armstrong and McQuillan [18] proved that if $n \equiv 2 \pmod 4$, $n \geq 6$, then K_n has a VMT labeling with magic constant k for any $(n^3+6n)/4 \leq k \leq (n^3+2n^2-2n)/4$. Moreover if $n \equiv 2 \pmod 8$ then K_n has a VMT labeling with magic constant k for any $(n^3+4n)/4 \leq k \leq (n^3+2n^2)/4$.

Gómez [115] dealt with super VMT labelings for K_n. The value of the magic constant k of a super VMT graph is the largest possible. Thus, from (3.8) it follows that the magic constant for a super VMT labeling of K_n is

$$k = \frac{n(n+1)^2}{4}. \tag{3.9}$$

The construction in the proof of Theorem 3.1.3 uses a pair of orthogonal latin squares and there is a scope within the application of latin squares to provide labelings with a variety of different magic constants, including the largest feasible values. Thus, this construction proves that K_n has a super VMT labeling for any odd n. According to (3.9) and due to the fact that $n(n+1)^2/4$ is not an integer, when $n \equiv 2 \pmod 4$, the complete graph K_n, $n \equiv 2 \pmod 4$, does not have a super VMT labeling. A systematic search for VMT labelings on K_4, see [179], shows that K_4 does not have a VMT labeling with $k = 25$. Since according to (3.9), super VMT labeling of K_4 would have $k = 25$, it follows that K_4 does not have a super VMT labeling.

The case that completes the study of super VMT labeling of K_n corresponds to $n \equiv 0 \pmod 4$, $n \geq 5$. Gómez proposed a new method to obtain the super VMT labelings in [115]. He applied this method to the complete graphs and proved

Theorem 3.1.7 ([115]) *If $n \equiv 0 \pmod 4$, $n \geq 5$, then K_n has a super VMT labeling.*

3.1.3 Generalized Petersen Graphs

The *generalized Petersen graph* $P(n, m)$, $n \geq 3$ and $1 \leq m \leq \lfloor (n-1)/2 \rfloor$, consists of an outer n-cycle $u_0, u_1, \dots, u_{n-1}$, a set of n spokes $u_i v_i$, $0 \leq i \leq n-1$, and n edges $v_i v_{i+m}$, $0 \leq i \leq n-1$, with indices taken modulo n. The generalized Petersen graph was introduced by Watkins in [293].

The next theorem, proved in [23], shows that $P(n, m)$ admits an $(a, 1)$-VAE labeling which was defined in Sect. 2.8.6.

Theorem 3.1.8 ([23]) *A generalized Petersen graph $P(n, m)$ is $(a, 1)$-VAE if and only if n is even, $n \geq 4$, $1 \leq m \leq n/2 - 1$, and $a = (7n + 4)/2$.*

$P(n, m)$ is regular of degree $r = 3$ with $p = 2n$ vertices and $q = 3n$ edges. Trivially, there exists a vertex labeling of $P(n, m)$ with values in the set $\{q + 1, q + 2, \dots, q + p\}$ which together with an $(a, 1)$-VAE labeling combine to a VMT labeling of $P(n, m)$ with the magic constant $k = a + p + q = 17n/2 + 2$. By duality, there exists a VMT labeling of $P(n, m)$ with the new magic constant $k' = (r + 1)(p + q + 1) - k = 23n/2 + 2$. From (3.6) it follows that the feasible values of the magic constant k for the generalized Petersen graph $P(n, m)$ are

$$\frac{17n}{2} + 2 \leq k \leq \frac{23n}{2} + 2. \tag{3.10}$$

Since the magic constant of a dual VMT labeling is the largest possible, then from Theorem 3.1.8 it follows that

Corollary 3.1.2 *For $n \geq 4$, n even and $1 \leq m \leq n/2 - 1$, the generalized Petersen graph $P(n, m)$ has a super VMT labeling.*

According to the upper bound in (3.10) and due to the fact that $23n/2 + 2$ is not an integer for n odd, the generalized Petersen graph $P(n, m)$ does not have a super VMT labeling for n odd.

Slamin and Miller [258] described the VMT labeling of $P(n, m)$ when n and m are coprime. They showed that

Theorem 3.1.9 ([258]) *For $n \geq 3$, $1 \leq m \leq \lfloor (n-1)/2 \rfloor$ and $\gcd(m, n) = 1$, every generalized Petersen graph $P(n, m)$ has a VMT labeling with the magic constants $k = 9n + 2$, $k = 10n + 2$ and $k = 11n + 2$.*

Figure 3.2 shows the VMT labelings of the generalized Petersen graphs $P(7, 2)$ and $P(7, 3)$ with magic constant $k = 72$.

Bača, Miller, and Slamin proved the existence of VMT labeling of $P(n, m)$ for $n \geq 3$ and all feasible values of m.

Theorem 3.1.10 ([44]) *For $n \geq 3$, $1 \leq m \leq \lfloor (n-1)/2 \rfloor$, every generalized Petersen graph $P(n, m)$ has a VMT labeling with the magic constants $k = 9n + 2$, $k = 10n + 2$, and $k = 11n + 2$.*

Figure 3.3 illustrates VMT labeling for the Petersen graph $P(5, 2)$ with magic constant $k = 52$.

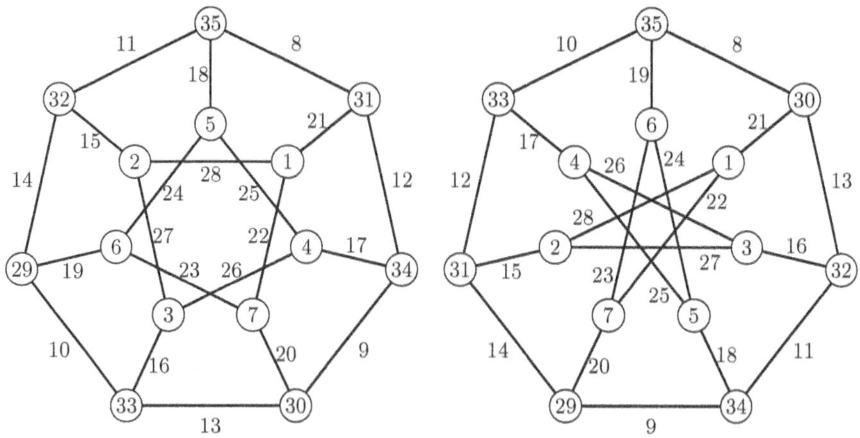

Fig. 3.2 VMT labeling of $P(7, 2)$ and $P(7, 3)$

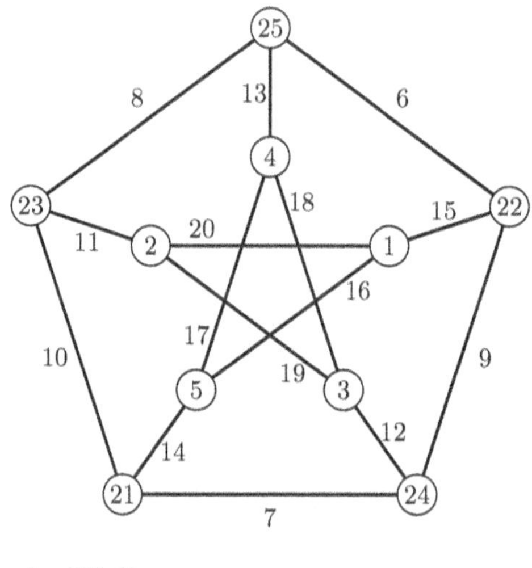

Fig. 3.3 VMT labeling for $P(5, 2)$

3.1.4 Two Families of Convex Polytopes

We shall investigate graphs of two families of convex polytopes. First we consider the graph R_n consisting of $2n$ 5-sided faces, n 6-sided faces and a pair of n-sided faces, embedded in the plane and labeled as in Fig. 3.4.

Using Inequality (3.6), where $p = 6n$ and $q = 9n$, we can determine the feasible values of the magic constant k for the graph R_n as

$$\frac{51n}{2} + 2 \leq k \leq \frac{69n}{2} + 2. \tag{3.11}$$

The following notation will be useful to define the VMT labeling.

$$\delta(x) = \begin{cases} 0 & \text{if } x \text{ is even} \\ 1 & \text{if } x \text{ is odd} \end{cases} \tag{3.12}$$

$$\rho(x, y) = \begin{cases} 0 & \text{if } x > y \\ 1 & \text{if } x \leq y. \end{cases} \tag{3.13}$$

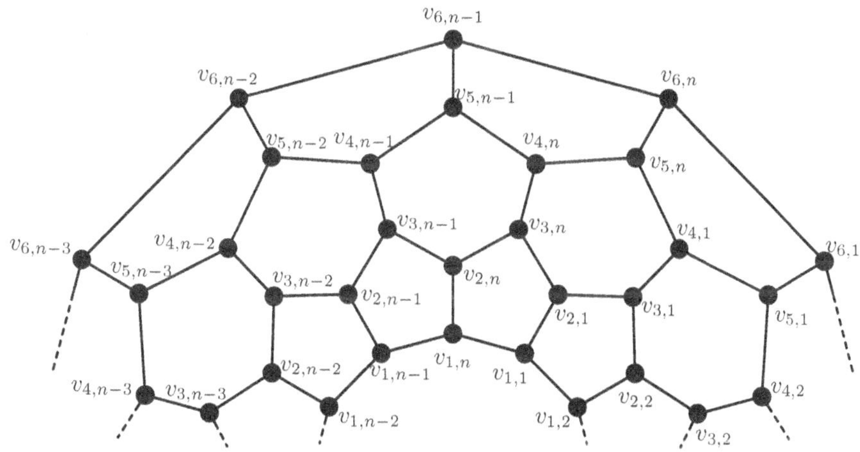

Fig. 3.4 The convex polytope R_n

To obtain a VMT labeling of R_n, $i = 1, 2, \ldots, n$, we construct an edge labeling $\lambda_1 : E(R_n) \to \{1, 2, \ldots, 9n\}$ as follows.

$$\lambda_1(v_{1,i}v_{1,i+1}) = ((8n+i)\delta(i) + (n-i)\delta(i+1))\rho(i, n-1) + n\rho(n, i)$$

$$\lambda_1(v_{1,i}v_{2,i}) = \left(\frac{5n}{2} + i\right)\delta(i) + \left(\frac{5n}{2} - i + 1\right)\delta(i+1)$$

$$\lambda_1(v_{2,i}v_{3,i}) = \left(\frac{15n}{2} - i + 1\right)\delta(i) + \left(\frac{9n}{2} - i + 2\right)\delta(i+1)$$

$$\lambda_1(v_{3,i}v_{2,i+1}) = \left(\frac{13n}{2} + i\right)\delta(i) + \left(\frac{9n}{2} - i + 1\right)\delta(i+1)$$

$$\lambda_1(v_{3,i}v_{4,i}) = (n+1)\rho(i, 1) + \left(\frac{3n-i+3}{2}\delta(i) + \left(\frac{9n}{2} + i\right)\delta(i+1)\right)\rho(2, i)$$

$$\lambda_1(v_{4,i}v_{5,i}) = \frac{11n+i+1}{2}\delta(i) + \left(\frac{11n}{2} - i + 1\right)\delta(i+1)$$

$$\lambda_1(v_{5,i}v_{4,i+1}) = \frac{13n-i+1}{2}\delta(i) + \frac{15n+i}{2}\delta(i+1)$$

$$\lambda_1(v_{5,i}v_{6,i}) = \left(\frac{5n}{2} - i + 1\right)\delta(i) + \left(\frac{5n}{2} + i\right)\delta(i+1)$$

$$\lambda_1(v_{6,i}v_{6,i+1}) = (8n+i+1)\delta(i) + (n-i+1)\delta(i+1).$$

It is a matter of routine checking to see that the labeling λ_1 is an $(a, 1)$-VAE with vertex-weights $21n/2+2, 21n/2+3, \ldots, 33n/2+1$. Trivially, there exists a vertex labeling of R_n with values in the set $\{|E(R_n)| + 1, |E(R_n)| + 2, \ldots, |E(R_n)| + |V(R_n)|\} = \{9n+1, 9n+2, \ldots, 15n\}$ which together with $(a, 1)$-VAE labeling λ_1 combine to a VMT labeling with the magic constant $k = 51n/2 + 2$. Thus we have

Theorem 3.1.11 ([189]) *For $n \geq 4$, n even, the plane graph R_n has a VMT labeling with $k = 51n/2 + 2$.*

Since R_n is regular, by duality we have

Corollary 3.1.3 *For $n \geq 4$, n even, the plane graph R_n has a super VMT labeling.*

The *antiprisms* A_n, $n \geq 3$, is a family of planar graphs that are regular of degree 4. These are Archimedean convex polytopes and, in particular, A_3 is the octahedron.

We will denote the vertex set of A_n by $V(A_n) = \{u_i, v_i : 1 \leq i \leq n\}$ and the edge set $E(A_n) = \{u_iu_{i+1} : 1 \leq i \leq n\} \cup \{v_iv_{i+1} : 1 \leq i \leq n\} \cup \{u_iv_i : 1 \leq i \leq n\} \cup \{v_iu_{i+1} : 1 \leq i \leq n\}$, with indices taken modulo n, see Fig. 3.5.

From (3.6) we get the range of feasible values for k

$$\frac{26n+5}{2} \leq k \leq \frac{34n+5}{2}. \tag{3.14}$$

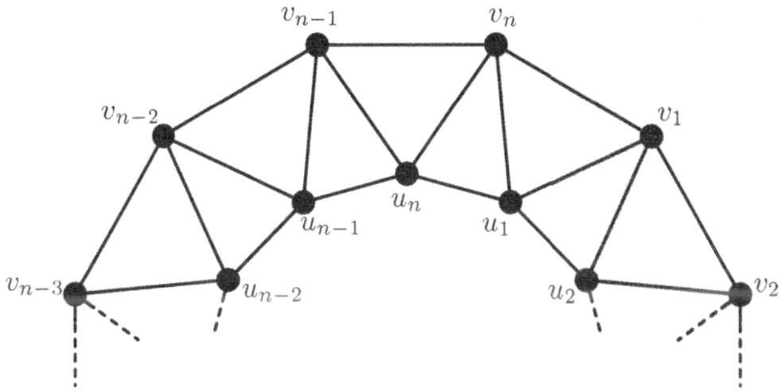

Fig. 3.5 The antiprism A_n

Theorem 3.1.12 ([189]) *For $n \geq 4$, n even, the antiprism A_n has a VMT labeling with $k = 15n + 2$.*

Proof We construct an edge labeling λ_2 of A_n, $n = 2m$, $m \geq 2$, in the following way:

$$\begin{aligned}
\lambda_2(u_i u_{i+1}) =& 6n\rho(i, 1) + ((5n + i - 1)\delta(i) + i\delta(i + 1))\rho(2, i) \\
\lambda_2(v_i v_{i+1}) =& ((5n + i)\delta(i) + (2n + i + 1)\delta(i + 1))\rho(i, n - 1) + (2n + 1)\rho(n, i) \\
\lambda_2(u_i v_i) =& (3n + 1)\alpha(1, i, 1) + (5n - 2i + 3)\alpha(2, i, m + 1) \\
&+ (3n + 3)\alpha(m + 2, i, m + 2) + (5n - 2i + 3)\alpha(m + 3, i, n - 1) \\
&+ (4n - 1)\alpha(n, i, n) \\
\lambda_2(v_i u_{i+1}) =& 2n - 2i + 1,
\end{aligned}$$

for $i = 1, 2, \ldots, n$, where

$$\alpha(x, y, z) = \begin{cases} 1 & \text{if } x \leq y \leq z \\ 0 & \text{otherwise.} \end{cases} \tag{3.15}$$

The edge labeling λ_2 is a bijection from $E(A_n)$ onto the set $\{i : 1 \leq i \leq n\} \cup \{n + 2j - 1 : 1 \leq j \leq 2n\} \cup \{5n + i : 1 \leq i \leq n\}$. The weights of the vertices under the edge labeling λ_2 constitute the set $\{10n + 2j : 1 \leq j \leq 2n\}$. If λ_3 is a vertex labeling with values in the set $\{n + 2j : j = 1, 2, \ldots, 2n\}$ then the labelings λ_2 and λ_3 combine to give a VMT labeling of A_n with the magic constant $k = 15n + 2$. □

Again, by duality we have

Corollary 3.1.4 *For $n \geq 4$, n even, the antiprism A_n has a VMT labeling with $k = 15n + 3$.*

Figure 3.6 depicts a VMT labeling of the antiprism A_4.

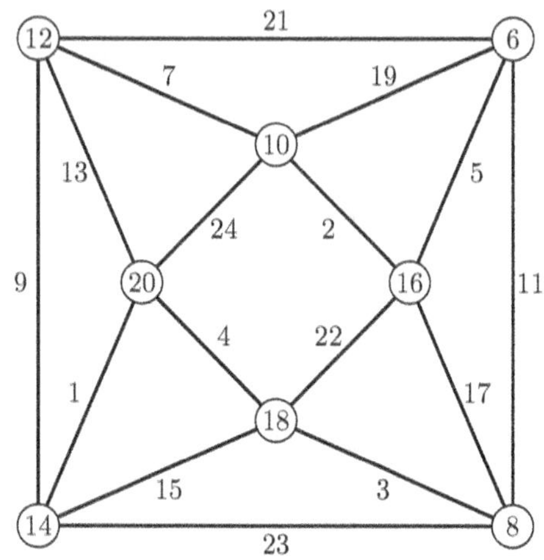

Fig. 3.6 VMT labeling of the antiprism A_4

Miller et al. [189] were unable to find a construction that will produce a VMT labeling for the plane graph R_n for n odd. However, they suggest the following:

Conjecture 3.1.2 ([189]) There is a VMT labeling for the plane graph R_n, for every $n \geq 3$.

For antiprism A_n they propose

Open Problem 3.1.1 ([189]) Find a VMT labeling for the antiprism A_n, for all odd $n \geq 3$.

3.1.5 Cartesian Product of Graphs

The construction of VMT labelings of Cartesian products of certain r-regular VMT graphs and certain s-regular supermagic graphs shown in the paper [166] is based on a VMT labeling of copies of an r-regular VMT graph and on a supermagic labeling of copies of a $2s$-regular graph, which can be factored into two s-regular factors. Wallis [288] proved the following theorem

Theorem 3.1.13 ([288]) *Suppose G is a regular graph of degree $\Delta \geq 1$, which has a VMT labeling. The following statements hold.*

(i) If Δ is even, then nG has a VMT labeling whenever n is an odd positive integer.
(ii) If Δ is odd, then nG has a VMT labeling for every positive integer n.

Theorem 3.1.13 gives a general method for constructing VMT labelings for n copies of certain regular VMT graphs. The following theorem gives a similar result for copies of certain $2s$-regular supermagic graphs.

Theorem 3.1.14 ([166]) *Let s be a positive integer. Let G be a $2s$-regular supermagic graph, which can be factorized into two s-regular factors. Then nG is also a supermagic graph.*

Taking a graph G_0 on v vertices, which satisfies the conditions of Theorem 3.1.13, and taking a graph H_0 on u vertices, which satisfies the conditions of Theorem 3.1.14, one can get a graph uG_0 on uv vertices, which is VMT and a graph vH_0 on uv vertices, which is supermagic. Next theorem shows that in both cases we can combine the graphs to get a VMT graph.

Theorem 3.1.15 ([166]) *Let G be an r-regular VMT graph on uv vertices, which consists of u copies of G_0. Let λ_G be a VMT labeling of G with the magic constant h. Let H be an s-regular supermagic graph on uv vertices, which consists of v copies of H_0. Let λ_H be a supermagic labeling of H with the magic constant k. Then there exists a VMT labeling of $G_0 \square H_0$ with the magic constant $h + k + s(2 + r)/2$.*

The construction given in the proof of Theorem 3.1.15 allows to build Cartesian products for several families of graphs. For G in the theorem we can take any graph with a VMT labeling, which satisfies Theorem 3.1.13, e.g., K_n, $K_{n,n}$, C_n, Petersen graph $P(n,k)$, prisms $C_n \square P_2$ if the necessary condition of being even regular or odd regular with even number of copies is satisfied. For H we have a variety of graph classes to choose from. Among graphs, which are proven to have a supermagic labeling and are also factorable into two s-regular factors are, e.g., K_n for $n \not\equiv 0 \pmod 4$, $K_{n,n}$, Q_n for n even.

Frončel, Kovář, and Kovářová presented in [104] a construction of VMT labelings of Cartesian products of cycles $C_m \square C_n$ for $m, n \geq 3$ and n odd. The construction is based on a generalized (a,d)-vertex-antimagic total labeling of cycles in which nonconsecutive integers are used.

An injection $\lambda : V(G) \cup E(G) \to \mathbb{N}$ is called a *generalized* (a,d)*-vertex-antimagic total labeling* of G if, for every $v \in V(G)$, the set of sums

$$\lambda(v) + \sum_{u \in N(v)} \lambda(vu),$$

forms an arithmetic progression $\{a, a+d, \ldots, a + (|V(G)| - 1)d\}$.

The generalized (a,d)-vertex-antimagic total labeling of G allows the set of labels to be distinct integers, not necessarily $1, 2, \ldots, |V(G)| + |E(G)|$. The generalized (a,d)-vertex antimagic total labeling of G is called (a,d)-vertex-antimagic total ((a,d)-VAT for short) if the set of labels is exactly $\{1, 2, \ldots, |V(G)| + |E(G)|\}$. Hence every (a,d)-VAT labeling of a graph is the generalized (a,d)-vertex-antimagic total labeling. For more details on (a,d)-VAT labeling see Sect. 5.2.

The existence of two types of generalized $(a,2)$-vertex-antimagic total labelings of cycles C_n is shown in [104].

Theorem 3.1.16 ([104]) *Let a, b, and n be positive integers, $n \geq 3$. Then there exists a generalized $(a + 2b + 2(n-1), 2)$-vertex-antimagic total labeling of C_n,* *where* $a, a+2, \ldots, a+2(n-1)$ *are the vertex labels and* $b, b+2, \ldots, b+2(n-1)$ *are the edge labels.*

Taking $a = 1$ and $b = 2$, we get a $(2n+3, 2)$-VAT labeling for every C_n with labels $1, 2, \ldots, 2n$. This special case is already known, see [28]. Taking $a = 2$ and $b = 1$, we get another $(2n+2, 2)$-VAT labeling of C_n.

Theorem 3.1.17 ([104]) *Let* a *and* b *be positive integers and let* $n \geq 3$ *be odd. Then there exists a generalized* $(a+2b+(n-1)/2, 2)$*-vertex-antimagic total labeling of* C_n*, where* $a, a+1, \ldots, a+n-1$ *are the vertex labels and* $b, b+1, \ldots, b+n-1$ *are the edge labels.*

Taking $a = 1$ and $b = n+1$, we get a $((5n+5)/2, 2)$-VAT labeling for every C_n with labels $1, 2, \ldots, 2n$. Taking $a = n+1$ and $b = 1$, we get a $((3n+5)/2, 2)$-VAT labeling of C_n. These labelings are already known, see [28].

These types of generalized $(s, 2)$-vertex-antimagic total labelings of cycles C_n have been used to obtain the two methods for constructing VMT labelings of Cartesian products of cycles. The first method (which proves Theorem 3.1.18) is more general, for Cartesian products of cycles of any length with odd cycles. The second method (which proves Theorem 3.1.19) can be used only for Cartesian products of cycles of odd lengths.

Theorem 3.1.18 ([104]) *For each* $m, n \geq 3$ *and* n *odd, there exists a VMT labeling of* $C_m \square C_n$ *with the magic constant* $k = m(15n+1)/2 + 2$.

Theorem 3.1.19 ([104]) *For each* $m, n \geq 3$ *and* m, n *odd, there exists a VMT labeling of* $C_m \square C_n$ *with the magic constant* $k = (17mn+5)/2$.

Using both methods one can construct several different VMT labelings that give distinct magic constants for the same graph.

3.1.6 Knödel Graphs

The *Knödel graph*, $W_{\Delta,n}$, was introduced in 1975 by Knödel [159] and formally defined in [82]. The graph $W_{\Delta,n}$ is regular of even order $n \geq 2$ and degree Δ, $1 \leq \Delta \leq \lfloor \log_2 n \rfloor$. The vertices of $W_{\Delta,n}$ are the pairs (i, j) with $i = 1, 2$ and $0 \leq j \leq n/2 - 1$. For every j, $0 \leq j \leq n/2 - 1$, there is an edge between vertex $(1, j)$ and every vertex $(2, l)$, where $l \equiv (j + 2^k - 1) \pmod{n/2}$, for $k = 0, 1, \ldots, \Delta - 1$.

The Knödel graphs $W_{\Delta,n}$ have been studied in [97, 103].

From the definition of the Knödel graph, for $\Delta = 3$ and even $n \geq 8$, the vertex set is

$$V(W_{3,n}) = \{v_0, v_1, \ldots, v_{\frac{n}{2}-1}, u_0, u_1, \ldots, u_{\frac{n}{2}-1}\},$$

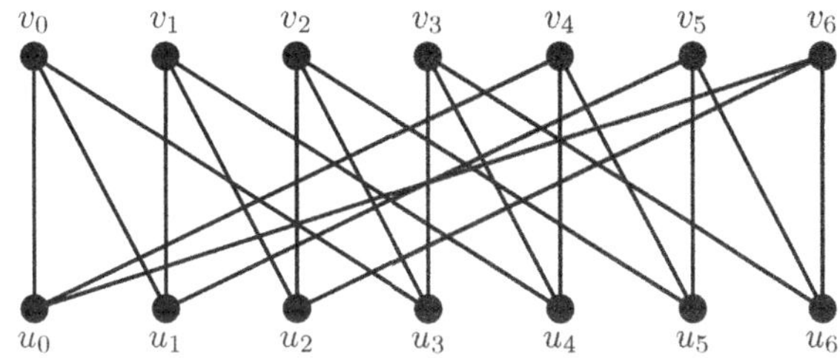

Fig. 3.7 Knödel graph $W_{3,14}$

and the edge set

$$E(W_{3,n}) = \bigcup_{i=0}^{\frac{n}{2}-1} \{v_i u_i, v_i u_{i+1}, v_i u_{i+3}\}.$$

Figure 3.7 illustrates Knödel graph $W_{3,14}$.

For VMT 3-regular graph $W_{3,n}$ with $p = n$ vertices and $q = 3n/2$ edges, Inequality (3.6) gives

$$\frac{17n+8}{4} \leq k \leq \frac{23n+8}{4}. \tag{3.16}$$

For super VMT graph the value of the magic constant k is the largest possible. Thus, from (3.16) it follows that the magic constant for $W_{3,n}$ is $k = (23n+8)/4$ and it is an integer only for $n \equiv 0 \pmod 4$.

Yue et al. [300] defined the edge labeling λ_4 of $W_{3,n}$, for $n \equiv 0 \pmod 4$ as follows:

$$\lambda_4(v_i u_i) = \begin{cases} n+1+\frac{i}{2} & \text{if } i \text{ is even } 0 \leq i \leq \frac{n}{2}-2 \\ \frac{3n-1+i}{2} & \text{if } i \text{ is odd } 1 \leq i \leq \frac{n}{2}-1 \end{cases}$$

$$\lambda_4(v_i u_{i+1}) = \begin{cases} \frac{5n}{4}+1+\frac{i}{2} & \text{if } i \text{ is even } 0 \leq i \leq \frac{n}{2}-4 \\ 2n-\frac{1+i}{2} & \text{if } i \text{ is odd } 1 \leq i \leq \frac{n}{2}-3 \\ \frac{7n}{4} & \text{if } i = \frac{n}{2}-2 \\ 2n+1 & \text{if } i = \frac{n}{2}-1 \end{cases}$$

$$\lambda_4(v_i u_{i+3}) = \begin{cases} \frac{5n-i}{2} & \text{if } i \text{ is even } 0 \leq i \leq \frac{n}{2}-2 \\ 2n+\frac{3+i}{2} & \text{if } i \text{ is odd } 1 \leq i \leq \frac{n}{2}-3 \\ 2n & \text{if } i = \frac{n}{2}-1. \end{cases}$$

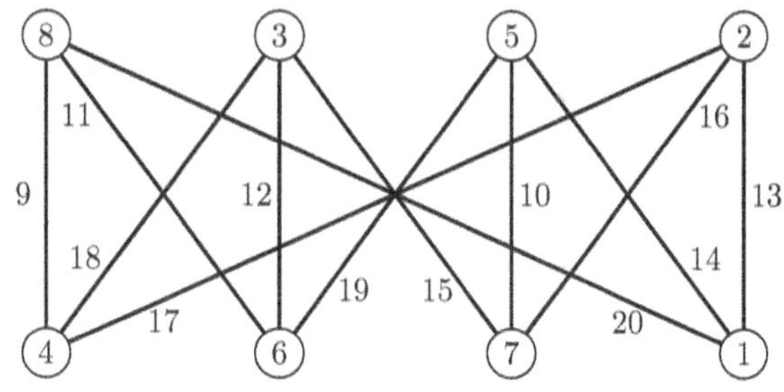

Fig. 3.8 A super VMT labeling of $W_{3,8}$

They verified that λ_4 is a bijection from the set $E(W_{3,n})$ onto the set $\{n+1, n+2, \ldots, 5n/2\}$ and showed that

$$g_{\lambda_4}(v) = k - \sum_{u \in N(v)} \lambda_4(vu),$$

for all vertices $v \in V(W_{3,n})$, gives the arithmetic progression $\{1, 2, \ldots, n\}$. Details can be found in [300]. Thus

Theorem 3.1.20 ([300]) *The Knödel graph $W_{3,n}$ admits a super VMT labeling for $n \equiv 0 \pmod 4$.*

Figure 3.8 shows a super VMT labeling of Knödel graph $W_{3,8}$.

3.1.7 General Results for Regular Graphs

For regular graphs, MacDougall [178] conjectured

Conjecture 3.1.3 ([178]) All regular graphs other than K_2 and $2K_3$ possess VMT labelings.

To date, while constructions have been derived for some families of regular graphs including those previously mentioned, no counterexamples to MacDougall's conjecture have been found. Gray [120] added significant further support to the conjecture by demonstrating that "almost all" regular graphs of odd order possess VMT labelings, as well as many graphs of even order.

Gray calls VMT labeling *strong* if the largest labels are assigned to the vertices. It is easy to see that if λ is a strong VMT labeling of a regular graph G, then the dual labeling λ', see (3.7), is a super VMT labeling.

In [120], it is shown that

Theorem 3.1.21 ([120]) *If G is a graph of order n with a spanning subgraph H which possesses a strong VMT labeling and $G - E(H)$ is even regular, then G also possesses a strong VMT labeling.*

Corollary 3.1.5 ([120]) *Every Hamiltonian regular graph of odd order admits a strong VMT labeling.*

It is shown in [74] that every 2-connected r-regular graph of order $p \leq 3r + 1$ is Hamiltonian so, clearly, Corollary 3.1.5 applies to all such graphs. In [222], it is shown that almost all regular graphs are Hamiltonian. However, this is an asymptotic result.

Gray [120] opines that one way of establishing that every regular graph of odd order has a VMT labeling would be to show that every regular graph of odd order greater than 7 possesses a 2-factor with a strong VMT labeling. But it is not known whether this is true. However, the following partial result is already known.

Corollary 3.1.6 ([120]) *Every regular graph of odd order with a spanning subgraph consisting of isomorphic cycles has a strong VMT labeling.*

Quasi-prism is defined as a cubic graph of order $2n$ which can be partitioned into two 2-factors, each of order n, with a 1-factor between them. Particular examples of quasi-prisms are prisms and generalized Petersen graphs. In [183], McQuillan provided a construction for VMT labelings of quasi-prisms.

Lemma 3.1.1 ([120]) *Let G be a $(2r + 1)$-regular graph of order $2m$ with a VMT labeling such that one of the following conditions is satisfied.*

(i) *Its vertices are assigned distinct labels from $\{(2r - 2)m + i, 2rm + i : i = 1, 2, \ldots, m\}$ and it has a 1-factor whose labels are distinct members of $\{(2r - 1)m + i : i = 1, 2, \ldots, m\}$.*
(ii) *Its vertices are assigned distinct labels from $\{(2r - 3)m + i, (2r - 1)m + i : i = 1, 2, \ldots, m\}$ and it has a 1-factor whose labels are distinct members of $\{2rm + i : i = 1, 2, \ldots, m\}$.*

If a 2-factor is added to the graph, the resulting $(2r + 3)$-regular graph has a VMT labeling.

Using McQuillan's construction and Lemma 3.1.1, Gray proved the following theorem.

Theorem 3.1.22 ([120]) *Every $(2r + 3)$-regular graph which has a quasi-prism as a spanning subgraph has a VMT labeling.*

Since every complete graph of even order has a quasi-prism as a spanning subgraph, Theorem 3.1.22 permits an alternative construction for a VMT labeling of K_{2n} to that found in [123].

The general question of whether all even regular graphs of even order possess VMT labelings seems to be much harder. As mentioned earlier, constructions are known for C_{2m}, K_{2m}, and $K_{2m,2m}$. In [120] it is shown how to construct families of $2r$-regular graphs of even order $2n$ with a VMT labeling, for all $3 \leq r \leq m - 1$.

From now on, we focus on r-regular graphs on n vertices with VMT labelings in which vertex labels are consecutive integers. First we observe that for such labelings there are restrictions on n and r.

Lemma 3.1.2 ([167]) *Let G be an r-regular graph on n vertices. If G has a VMT labeling such that the vertex labels constitute an arithmetic progression with odd difference, then either r is even and n is odd or r is odd and $n \equiv 0 \pmod 4$.*

The following results are based on the Petersen Theorem.

Theorem 3.1.23 (Petersen Theorem) *Let G be a $2r$-regular graph. Then there exists a 2-factor in G.*

Notice that, after removing edges of the 2-factor guaranteed by the Theorem 3.1.23, we have again an even regular graph. Thus, by induction, an even regular graph has a 2-factorization.

Lemma 3.1.2 states the limits for the constructions shown below. The constructions require vertex labels to be consecutive integers.

Theorem 3.1.24 ([167]) *Let G be a $(2+s)$-regular graph of order n such that it contains an s-regular factor G' which allows a VMT labeling with magic constant k and vertex labels being consecutive integers starting at h. Then G has VMT labelings with magic constants*

$$k = \frac{1}{4}(s+4)(n(s+4)+2) - \frac{1}{2}(n-1) - t,$$

where $t \in \{h, n(s+2)/2+1\}$.

Figure 3.9 shows a VMT labeling for a 3-regular graph on 8 vertices with consecutive vertex labels and with magic constant 36. Using Theorem 3.1.24 we can find VMT labeling of 5-regular graph of order 8, see Fig. 3.10, where a 2-factor is drawn in thick and the magic constant is 77.

Theorem 3.1.24 gives a recursive method for constructing VMT labelings of regular graphs. Hence the following theorem.

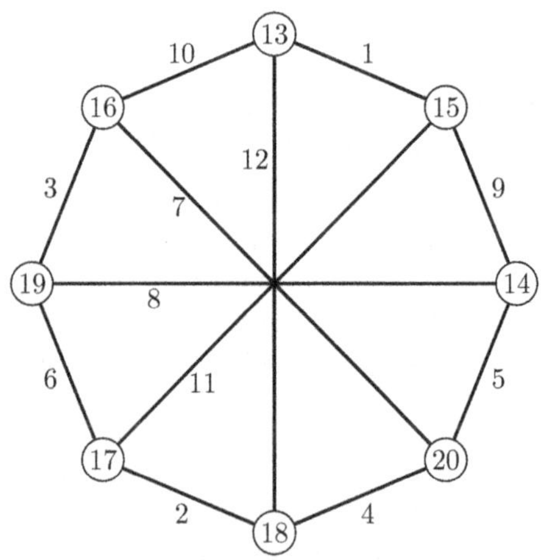

Fig. 3.9 A VMT labeling of 3-regular graph

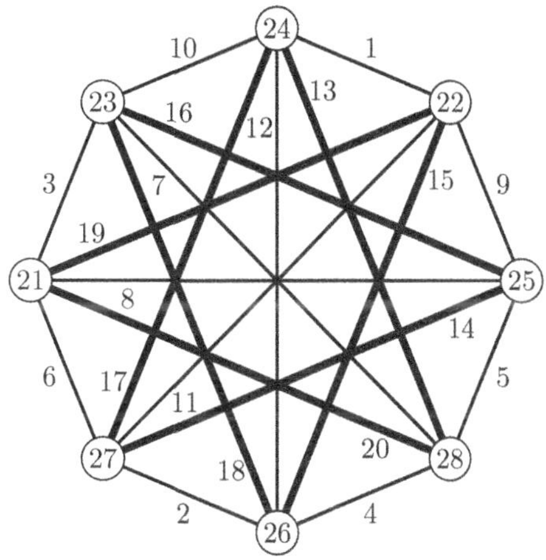

Fig. 3.10 A VMT labeling of 5-regular graph

Theorem 3.1.25 ([167]) *Let G be a $(2r + s)$-regular graph of order n such that it contains an s-regular factor G' which allows a VMT labeling with magic constant k' and vertex labels being consecutive integers starting at h. Then G has VMT labelings with magic constants*

$$k = \frac{1}{4}(s + 2r + 2)(n(s + 2r + 2) + 2) - \frac{1}{2}(n - 1) - t,$$

where $t \in \{h\} \cup \{n(s + 2i)/2 + 1 : i = 1, 2, \dots, r\}$.

The proof of Theorem 3.1.25 goes by induction. For $r = 0$ the claim follows immediately from the fact that a VMT labeling of an s-regular graph with consecutive vertex labels starting at h has a magic constant

$$k = \left(1 + \frac{s}{2}\right)\left(1 + n\left(1 + \frac{s}{2}\right)\right) - h - \frac{n-1}{2}.$$

The inductive step follows immediately from Theorems 3.1.24 and 3.1.23.

The drawback of the general approach of Theorem 3.1.25 is that the existence of VMT labelings of regular graphs is based on the existence of regular subgraphs. These do not exist in general (e.g., in some graphs with bridges). Moreover, the regular subgraphs are required to have certain VMT labelings. Again, these are not guaranteed in general (e.g., only a few families of even regular graphs on even order with a VMT labeling are known).

Using Theorem 3.1.25 we can find VMT labeling of large class of even regular graphs on an odd number of vertices.

Corollary 3.1.7 ([167]) *Let G be a $2r$-regular graph of odd order n which has a Hamiltonian cycle. Then G has a VMT labeling with the magic constants*

$$k = \frac{1}{4}(2r+2)(n(2r+2)+2) - \frac{1}{2}(n-1) - t,$$

where $t \in \{ni+1 : i = 0, 1, \ldots, r\}$.

Corollary 3.1.8 ([167]) *Let G be a $2r$-regular graph of odd order n, where $4r > n$. Then G has a VMT labeling.*

3.2 The Existence of Vertex-Magic Total Labelings

In [179] MacDougall, Miller, Slamin, and Wallis assert that a graph is likely to be VMT if and only if there is no much variation among the degrees of its vertices. This is reinforced in [181] with the suggestion that there might be a general principle to the effect that if a graph G contains a vertex whose degree is high relative to the degrees of all the other vertices of G, then G is not VMT.

Gray et al. [121] studied the vertex-magic properties of trees. The next theorem gives the proportion of leaves (vertices of degree 1) to internal vertices of the tree.

Theorem 3.2.1 ([121]) *Let T be a tree with n internal vertices and σn leaves. Then T does not admit a VMT labeling if*

$$\sigma > \frac{1+\sqrt{12n^2+4n+1}}{2n}.$$

A simple approximation of the result in Theorem 3.2.1 shows that a VMT labeling is impossible for more than $\sqrt{3}n+1$ leaves. This theorem does not provide a sufficient condition for the existence of a VMT labeling, however. The following result shows that there are also restrictions imposed by the degrees of the internal vertices.

Theorem 3.2.2 ([121]) *If Δ is the largest degree of any vertex in a tree T with p vertices, then T does not admit a VMT labeling whenever*

$$\Delta > \frac{\sqrt{32p+33}-7}{2}.$$

Theorems 3.2.1 and 3.2.2 still do not provide sufficient conditions since it is known, for example, that the tree with six vertices shown in [121], see Fig. 3.11, has no VMT labeling. If we consider the weights of the vertices in Fig. 3.11 we can see that the magic constant for vertex v is at least $1+2+3+4+5 = 15$ and the magic constant for leaf is at most $(11+10+\cdots+5+4)/4 = 15$. It follows that $k = 15$ and VMT labeling can only be achieved by the assignment of labels described. But

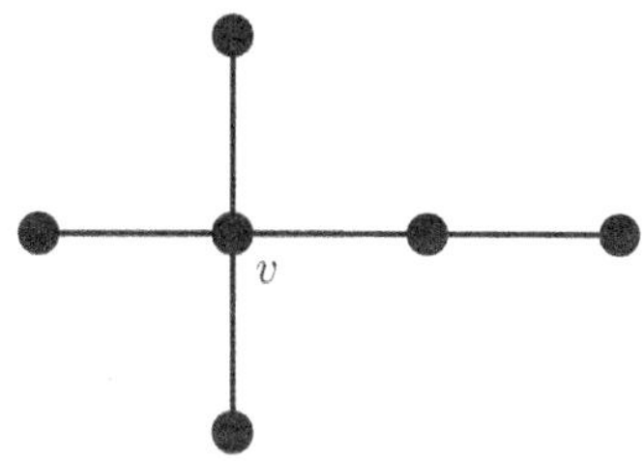

Fig. 3.11 A tree with no VMT labeling

this means that at least one of the edges incident with vertex v has label less than 4, which contradicts the assignment of labels to the leaf edges.

The comparison of maximum sum of weights on the leaves to minimum sum of weights on internal vertices gives the following analogue of Theorem 3.2.1.

Theorem 3.2.3 ([121]) *Let F be a forest of s components. If F has n internal vertices and σn leaves, then there is no VMT labeling whenever*

$$\sigma > \frac{2s - 1 + \sqrt{12n^2 + 4n(2s-1) - (4s^2 - 4s - 1)}}{2n}.$$

Theorem 3.2.2 implies that in any VMT tree with p vertices and $q = p - 1$ edges, the degree d of any vertex satisfies

$$d + \frac{7}{2} \leq \sqrt{\frac{32q^2 + 97q + 65}{4p}}. \tag{3.17}$$

Beardon [55] established the following result which holds for all graphs.

Theorem 3.2.4 ([55]) *Let G be a VMT (p, q) graph with C components. Then the degree d of any vertex of G satisfies*

$$d + 2 \leq \sqrt{\frac{7q^2 + (6C + 5)q + C^2 + 3C}{p}}. \tag{3.18}$$

In particular,

$$d + 2 \leq \sqrt{\frac{14q^2 + (16q + 4)}{p}}, \tag{3.19}$$

while if G is connected, then

$$d + 2 \leq \sqrt{\frac{7q^2 + 11q + 4}{p}}. \tag{3.20}$$

If G is a tree with p vertices, then $q = p - 1$, and it is clear that (3.20) is better than (3.17), for all sufficiently large p. A calculation shows this to be so whenever $p \geq 43$.

Beardon [55] proved that in any VMT graph, the degree of any vertex is bounded by a function of p and q.

Theorem 3.2.5 ([55]) *Suppose that a (p, q) graph G is VMT. Then the degree d of any vertex satisfies*

$$pd^2 + (5p - 2)d + p \leq p^2 + 2q^2 + 4pq + 2q.$$

Balbuena et al. [49] investigated the minimum degree of super VMT graph.

Theorem 3.2.6 ([49]) *The minimum degree of a super VMT graph G is at least two.*

Proof It follows from the definition of super VMT labeling that G cannot have more than one isolated vertex. If $v \in V(G)$ is the isolated vertex, then the weight of that vertex under VMT labeling λ satisfies $k = wt(v) = \lambda(v) \leq |V(G)|$. But the label of each edge is at least $|V(G)| + 1$. So, the weight of any vertex $u \in V(G)$ different from v satisfies $k = wt(u) \geq |V(G)| + 2$. This is a contradiction, thus the minimum degree is at least one. A VMT graph has no isolated edges, see [287]. This means that for every edge $uv \in E(G)$ we have $\sum_{w \in N(v)-u} \lambda(vw) > |V(G)|$ or $\sum_{w \in N(u)-v} \lambda(uw) > |V(G)|$ because G is super VMT. Thus, let us consider a vertex u of degree one and its neighbor $N(u) = \{v\}$. We have that $\lambda(u) + \lambda(uv) = \lambda(v) + \lambda(uv) + \sum_{w \in N(v)-u} \lambda(vw)$. This implies that $\lambda(u) - \lambda(v) = \sum_{w \in N(v)-u} \lambda(vw)$. But this is impossible because $0 < \lambda(u) - \lambda(v) < |V(G)|$ and $\sum_{w \in N(v)-u} \lambda(vw) > |V(G)|$. Therefore, the minimum degree is at least two.

□

As an immediate consequence of Theorem 3.2.6, we obtain the following corollary.

Corollary 3.2.1 ([49]) *Let G be a super VMT (p, q) graph. Then $q \geq p$.*

Corollary 3.2.1 implies that a tree is not super VMT. This is a result proved by MacDougall et al. [180].

The next result as a corollary of Theorem 3.2.6 shows that the only 2-regular super VMT graphs are the odd cycles or a disjoint union of cycles tC_s, for ts odd.

Corollary 3.2.2 ([49]) *If G is a 2-regular super VMT graph, then G has an odd number of vertices.*

It follows from (3.6) that the magic constant of super VMT labeling is $k = (p + 1)/2 + 3p + 1$ and it is not an integer for p even.

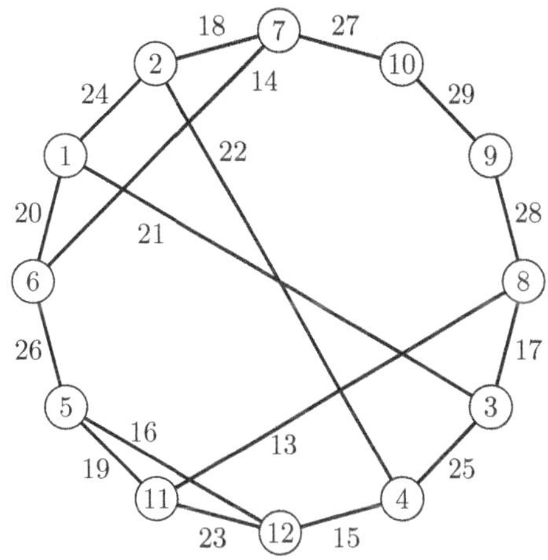

Fig. 3.12 A super VMT (12, 17) graph with minimum degree two

Most of the super VMT graphs have minimum degree at least three.

Theorem 3.2.7 ([49]) *Let G be a super VMT* (p, q) *graph.*

(i) If $2q \geq \sqrt{10p^2 - 6p + 1}$*, then the minimum degree of G is at least three.*
(ii) If $2q < \sqrt{10p^2 - 6p + 1}$*, then the maximum degree of G is at most six.*

According to Theorem 3.2.7, it is natural to ask for super VMT graphs with $2q < \sqrt{10p^2 - 6p + 1}$ having minimum degree two. Figure 3.12 proves the existence of super VMT graph with $2q \in \{3p-2, 3p\}$ and minimum degree two for $p = 12$ and $q = 17$.

Balbuena et al. [49] propose the following conjecture.

Conjecture 3.2.1 ([49]) Super VMT (p, q) graph such that $2q < \sqrt{10p^2 - 6p + 1}$ and minimum degree is 2 exists for all integer values of the magic constant $k = 2q + q(q+1)/p + (p+1)/2$.

The next theorem presents an upper bound for the maximum degree $\Delta(G)$ of a super VMT graph.

Theorem 3.2.8 ([49]) *Let G be a super VMT graph with size q and magic constant k. Then the maximum degree is at most*

$$\Delta(G) \leq -\frac{1}{2} + \sqrt{2(k - 2q) - \frac{7}{4}}.$$

The bounds for the degree of any vertex of a super VMT graph are given in the following theorem.

Theorem 3.2.9 ([49]) *Let G be a super VMT (p, q) graph with magic constant k. Then the degree d of any vertex of G satisfies*

$$p+q+\frac{1}{2}-\sqrt{\left(p+q+\frac{1}{2}\right)^2-2(k-p)} \le d \le -\frac{1}{2}+\sqrt{2(k-2q)-\frac{7}{4}}.$$

3.3 Vertex-Magic Total Labelings of Non-regular Graphs

3.3.1 Complete Bipartite Graphs

Let $K_{m,n}$ be the *complete bipartite graph* with $V(K_{m,n}) = \{u_i : 1 \le i \le m\} \cup \{v_j : 1 \le j \le n\}$ and $E(K_{m,n}) = \{u_i v_j : 1 \le i \le m \text{ and } 1 \le j \le n\}$. The VMT labelings for $K_{1,2}$ and $K_{2,2}$ exist and have been dealt with as a path and a cycle respectively. The first general result proved by MacDougall, Miller, Slamin, and Wallis shows that for a VMT labeling of a bipartite graph to exist, the parts must be nearly the same size.

Theorem 3.3.1 ([179]) *If $n > m + 1$, then $K_{m,n}$ has no VMT labeling.*

The proof of Theorem 3.3.1 can be found also in [287] and [122]. In contrast with the result of Theorem 3.3.1, the following theorem shows that a VMT labeling can always be found for $K_{m,m}$.

Theorem 3.3.2 ([179, 122]) *There is a VMT labeling for $K_{m,m}$, for every $m > 1$.*

The proof of Theorem 3.3.2 uses a magic square construction. For definition see Sect. 2.2. Let us demonstrate the idea of the proof by the following example. Consider a magic square of order 4, Fig. 3.13. All rows, all columns, the main diagonal, and the main back-diagonal have the same sum 34.

We decrease by 1 all the entries of the magic square, see Fig. 3.14.

If nonzero values of the first row determine the labels of vertices u_1, u_2, u_3 and nonzero values of the first column determine the labels of vertices v_1, v_2, v_3, then

Fig. 3.13 Magic square of order 4

1	15	8	10
12	6	13	3
14	4	11	5
7	9	2	16

Fig. 3.14 Square of order 4 after subtraction

0	14	7	9
11	5	12	2
13	3	10	4
6	8	1	15

Fig. 3.15 VMT labeling of $K_{3,3}$

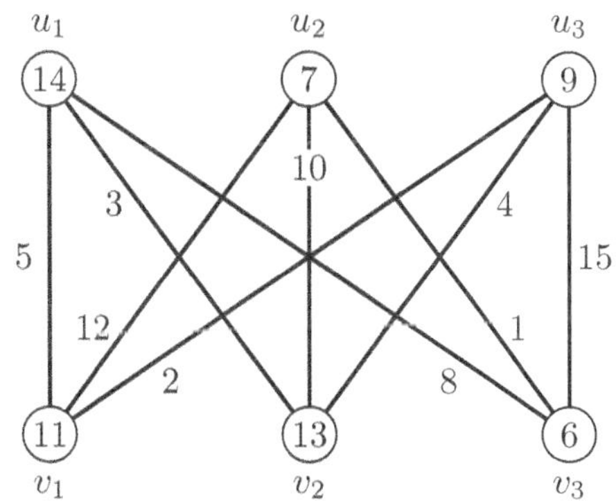

the square in Fig. 3.14 represents the matrix of corresponding VMT labeling of $K_{3,3}$, see Fig. 3.15.

A magic square exists for every order $n \geq 3$. The rows and columns of a magic square can be permuted so that value 1 will appear in the first row and the first column. The permutation rows and columns guarantee the same row sum and column sum.

Gray et al. [122] determined that all complete bipartite graphs $K_{m,m+1}$ have VMT labelings. They proved

Theorem 3.3.3 ([122]) *There exists a VMT labeling of $K_{2n-1,2n}$ with magic constant $4n^3 + 2n^2$.*

Theorem 3.3.4 ([122]) *There exists a VMT labeling of $K_{2n,2n+1}$ with magic constant $(n + 1)(2n + 1)^2$.*

3.3.2 Complete Multipartite Graphs

In the previous section a complete solution was given for the problem of the existence of VMT labeling for complete bipartite graphs. Cattell [79] studied a natural generalization of this problem for complete multipartite graphs. The graph G formed from the product $G = H_1 \oplus H_2$ is the graph with $V(G) = V(H_1) \cup V(H_2)$

and $E(G) = E(H_1) \cup E(H_2) \cup S$, where S is the set of edges connecting every vertex of H_1 to every vertex of H_2.

Theorem 3.3.5 ([79]) *If H is any graph such that $G = H \oplus \overline{K_n}$, then G can be VMT only if $|V(H)| \geq n - 1$.*

This theorem does not use the structure of H. As a consequence of the theorem it follows that $H \oplus \overline{K_n}$ is not VMT for $|V(H)| < n - 1$.

Since a complete multipartite graph $K_{m_1,m_2,\ldots,m_r,n}$ has the property of G in Theorem 3.3.5, the next corollary follows immediately.

Corollary 3.3.1 ([79]) *The complete multipartite graph $K_{m_1,m_2,\ldots,m_r,n}$ can be VMT only if $\sum_{i=1}^{r} m_i \geq n - 1$.*

In particular, for the tripartite graph $K_{t,m,n}$ to be VMT we need $m+t \geq n-1$. No stronger result can be obtained by these kinds of arguments since there are known labelings for cases where $m + t = n - 1$. A VMT labeling of $K_{1,1,3}$ is depicted in Fig. 3.16.

Cattell proposed the following

Open Problem 3.3.1 ([79]) Do all graphs satisfying Theorem 3.3.5 have VMT labelings?

There are certain families of tripartite graphs for which there have been found VMT labelings. No universal construction exists for VMT labelings of $K_{t,m,n}$. Cattell in [79] presented some constructions for two families of complete tripartite graphs which prove

Theorem 3.3.6 ([79]) *The tripartite graph $G = K_{1,n,n}$ has a VMT labeling with the magic constant $k = (n^3 + 6n^2 + 9n + 2)/2$ when n is odd.*

Unfortunately the same construction does not work for even n.

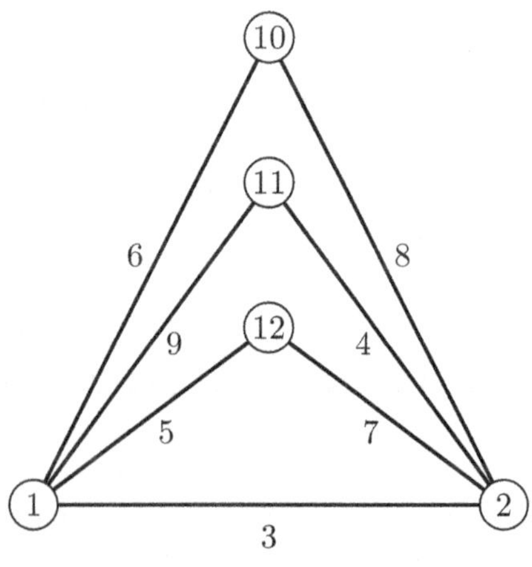

Fig. 3.16 VMT labeling of $K_{1,1,3}$

Theorem 3.3.7 ([79]) *The tripartite graph* $G = K_{2,n,n}$ *has a VMT labeling with the magic constant* $k = (n^3 + 10n^2 + 23n + 12)/2$ *whenever* $n \equiv 3 \pmod 4$.

3.3.3 Wheels and Related Graphs

We know that the range of feasible values for magic constant k shown in (3.6) is derived without a reference to the particular structure of any graph. It is certainly true that the presence of certain subgraphs in G or the presence of many vertices of high degree or of low degree in G can further restrict the admissible values of k.

Suppose W_n is the wheel whose n rim vertices form the cycle $(v_1, v_2, \dots, v_n)$. Thus $p = n + 1$ and $q = 2n$. Inequalities (3.6) yield only

$$\frac{13n^2 + 11n + 2}{2(n+1)} \le k \le \frac{17n^2 + 15n + 2}{2(n+1)} \tag{3.21}$$

but the permissible values of k will be determined by the degree of the hub vertex. The magic constant k is at least the vertex-weight of the hub vertex c,

$$k \ge wt(c) = 1 + 2 + \cdots + (n+1) = \frac{(n+1)(n+2)}{2}. \tag{3.22}$$

If we place the n largest labels on the rim edges (they are each counted twice) and the $2n$ next largest labels on the rim vertices and the spoke edges, then we have

$$wt(v_1) + wt(v_2) + \cdots + wt(v_n) \le \sum_{i=2}^{3n+1} i + \sum_{i=2n+2}^{3n+1} i = n(7n+6). \tag{3.23}$$

Since there are n rim vertices, then the magic constant k will be

$$k \le 7n + 6. \tag{3.24}$$

It is easy to see that

$$\frac{(n+1)(n+2)}{2} \le 7n + 6$$

only for $n \le 11$. Thus

Theorem 3.3.8 ([181]) *The wheel* W_n *has no VMT labeling for* $n > 11$.

MacDougall, Miller, and Wallis have found VMT labelings for W_n, for all n in the range $3 \le n \le 11$, see [181].

Recall that a fan F_n, $n \ge 2$, is a graph obtained by joining all vertices of path P_n to a further vertex called the center. Alternatively, a fan F_n can be constructed from a wheel W_n by removing one rim edge. A *friendship graph* f_n, $n \geq 1$, consists of n triangles with exactly one common vertex called the *center*. Alternatively, a friendship graph f_n can be constructed from a wheel W_{2n} by removing every second rim edge.

MacDougall, Miller, and Wallis proved

Theorem 3.3.9 ([181]) *For $n \geq 11$, the fan F_n has no VMT labeling.*

Theorem 3.3.10 ([181]) *For $n \geq 4$ the friendship graph f_n has no VMT labeling.*

In [181] MacDougall, Miller, and Wallis give tables with VMT labelings, for each fan F_n, for $n = 2, 3, \ldots, 10$, and for each friendship graph f_n, for $n = 1, 2, 3$.

Let G be the graph derived from a wheel by duplicating the hub vertex one or more times. Graph G is called a *t-fold wheel* if there are t hub vertices, each adjacent to all rim vertices, and not adjacent to each other. The next theorem is a generalization of Theorem 3.3.8.

Theorem 3.3.11 ([181, 287]) *Let G be a t-fold wheel with n rim vertices. Given t, there exists an integer N_t such that for all $n > N_t$, no VMT labeling of G exists.*

Rahim et al. [218] defined a *generalized helm* $H(n,t)$ and a *generalized web* $W(n,t)$ such that the $H(n,t)$ is a graph obtained from a wheel W_n by attaching a path on t vertices at each vertex of the n-cycle and the $W(n,t)$ is a graph obtained from a generalized helm $H(n,t)$ by joining the corresponding vertices of each path to form an n-cycle. Thus $W(n,t)$ has $(t+1)n+1$ vertices and $2(t+1)n$ edges. They proved in [218] that $H(n,1)$ has no VMT labeling for any $n \geq 3$, and that $W(n,t)$ has a VMT labeling for $n = 3$ or $n = 4$ and $t = 1$, but it is not VMT for $n \geq 17t + 12$ and $t \geq 0$.

Before presenting another result let us recall the corona product of two graphs. The *corona product* of G and H is the graph $G \odot H$ obtained by taking one copy of G, called the center graph along with $|V(G)|$ copies of H, called the outer graph, and making the ith vertex of G adjacent to every vertex of the ith copy of H, where $1 \leq i \leq |V(G)|$. This graph product was introduced by Frucht and Harary [105] in 1970.

Rahim and Slamin [217] proved that the disjoint union of coronas $C_{t_1} \odot K_1 \cup C_{t_2} \odot K_1 \cup \cdots \cup C_{t_n} \odot K_1$ has a VMT labeling with magic constant $6\sum_{k=1}^{n} t_k + 1$. In [216] they give the bounds for the number of vertices for Jahangir graphs, helms, webs, flower graphs, and sunflower graphs when the graphs considered are not VMT.

3.4 Disjoint Unions of Graphs

Wallis [288] has shown that if G is a regular graph of even degree that has a VMT labeling, then the graph consisting of an odd number of copies of G is VMT. He also proved that if G is a regular graph of odd degree, except K_1, that has a VMT labeling, then the graph consisting of any number of copies of G is VMT, see Theorem 3.1.13.

Gómez [116] described two new methods to obtain super VMT labelings of graphs. The first method provides a super VMT labeling for the graph mG defined as the disjoint union of m copies of G, for a large number of values of m. The next lemma gives a relationship between the magic constant of a graph G that admits a super VMT labeling and the magic constant of a graph mG.

Lemma 3.4.1 ([116]) *Let $k(G)$ be the magic constant of a Δ-regular (p, q) graph G. The magic constant of the graph mG is given by $k(mG) = mk(G) - (m-1)(\Delta+1)/2$.*

Observe that, since we have assumed that G admits a super VMT labeling, $k(G)$ is a positive integer. Therefore, if $(m-1)(\Delta+1)/2$ is not an integer, then mG does not admit a super VMT labeling. The following theorem gives a positive answer for super VMT labeling of the disjoint union of m copies of G.

Theorem 3.4.1 ([116]) *Let m be a positive integer. If the graph G is Δ-regular graph that admits a super VMT labeling and $(m-1)(\Delta+1)/2$ is an integer, then the graph mG has a super VMT labeling.*

Corollary 3.4.1 ([116]) *Let n and m be two positive integers. If n and m are odd or $n = 4l$, $l = 2, 3, \ldots$, then the graph mK_n has a super VMT labeling.*

It was shown in [116] that a super VMT labeling of kK_n exists for n odd and any k, for $4 < n \equiv 0 \pmod 4$ and any k, and for $n = 4$ and k even. Gómez and Kovář [117] proved the following lemmas and theorem.

Lemma 3.4.2 ([117]) *If k is even and n is odd, then kK_n does not admit a super VMT labeling.*

Lemma 3.4.3 ([117]) *Let l be a nonnegative integer. If k is odd, then the graph kK_{4l+2} does not admit a super VMT labeling.*

Theorem 3.4.2 ([117]) *The graph $2mK_{4l+2}$ admits a super VMT labeling for any positive integers m and l.*

There remains one unsolved case, namely finding a super VMT labeling of kK_n for $n = 4$ and odd $k \geq 3$. In [117] Gómez and Kovář showed that for $3K_4$ the super VMT labeling exists and they suggest the following conjecture.

Conjecture 3.4.1 ([117]) If k is an odd integer, $k > 1$, then kK_4 admits a super VMT labeling.

The second method described in [116], starting from a graph G_0 which admits a super VMT labeling, provides a large number of super VMT labelings for the graphs G_0' obtained by means of the addition of various sets of edges to G_0. The method can be applied to the disjoint union of m cycles of length s, mC_s, for odd m and s, due to the fact that these graphs admit a super VMT labeling. For details see [116].

McQuillan and McQuillan [185] investigated the existence of VMT labelings of disjoint union of m copies of cycle C_3.

Theorem 3.4.3 ([185]) *Let m be an odd positive integer. There exists a VMT labeling of mC_3 with a magic constant of $(21m+3-6j)/2$ for $0 \le j \le m$.*

Moreover, for every even integer $m \ge 6$, mC_3 has VMT labelings with at least $2m-2$ different magic constants.

Corollary 3.4.2 ([185]) *Let $m = 2t$ be a positive even integer. If $t \ge 3$, then there exists a VMT labeling of mC_3 with the magic constant of $21t-1-3j$, for $0 \le j \le m-2$. Furthermore, there exists a VMT labeling of mC_3 with the magic constant of $21t-2-3j$, for $0 \le j \le m-2$.*

They also showed that $mC_3, m = 2t$, has a VMT labeling with the magic constant $21t+1$ and, by duality, it also has another VMT labeling with magic constant $15t+2$. This means

Theorem 3.4.4 ([185]) *Let $m = 2t$ be an even integer, and assume that $t \ge 3$ is odd. Then there is a VMT labeling of mC_3 with the magic constant of $k = 21t+1$ and another VMT labeling with the magic constant of $15t+2$.*

McQuillan in [184] described a technique for constructing VMT labelings of 2-regular graphs.

Slamin et al. [259] constructed VMT labelings for the disjoint union of two copies of the generalized Petersen graph $P(n,k)$ and Silaban et al. [253] extended this result to any number of copies of $P(n,k)$.

Chapter 4
Edge-Magic Total Labelings

4.1 Basic Ideas

Let G be a (p, q) graph. A bijective function f from the set $V(G) \cup E(G)$ to the set of integers $\{1, 2, \ldots, p+q\}$ is called an *edge-magic total* (EMT) *labeling* of G if there exists a constant k, called the *magic sum* of f, such that

$$f(u) + f(uv) + f(v) = k,$$

for any edge uv of G. Figure 4.1 illustrates an EMT labeling of the wheel W_6 with magic sum $k = 32$.

Originally the EMT labeling was introduced and studied by Kotzig and Rosa [163, 164], who called it *magic valuations*. Interest in these labelings has been rekindled due to Ringel and Lladó's paper [221] in 1996.

In the computation of the edge-weights of a (p, q) graph G with an EMT labeling f, each edge label is used once and each label of vertex $v \in V(G)$ is used $\deg(v)$ times. Thus the following equation holds

$$\sum_{v \in V(G)} \deg(v) \cdot f(v) + \sum_{e \in E(G)} f(e) = kq \tag{4.1}$$

and

$$\sum_{v \in V(G)} f(v) + \sum_{e \in E(G)} f(e) + \sum_{v \in V(G)} (\deg(v) - 1) \cdot f(v) = kq. \tag{4.2}$$

If

$$\sum_{v \in V(G)} f(v) + \sum_{e \in E(G)} f(e) = \frac{(p+q+1)(p+q)}{2} \tag{4.3}$$

M. Bača et al., *Magic and Antimagic Graphs*, Developments in Mathematics 60,
https://doi.org/10.1007/978-3-030-24582-5_4

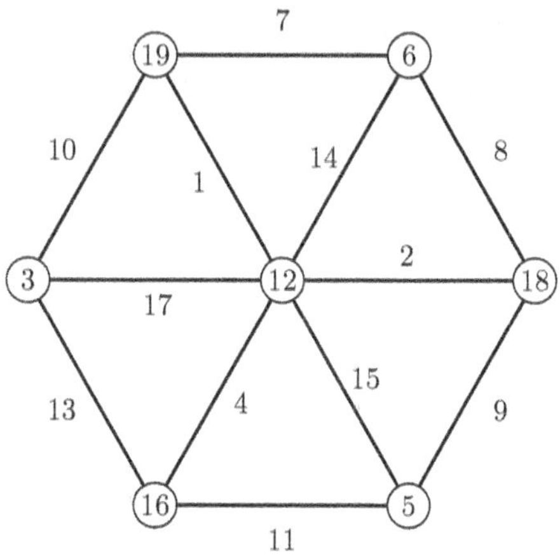

Fig. 4.1 An EMT labeling of the wheel W_6

then the Eq. (4.2) gives

$$\frac{(p+q+1)(p+q)}{2} + \sum_{v \in V(G)} (\deg(v) - 1) \cdot f(v) = kq. \tag{4.4}$$

If q is even, $\deg(v)$ is odd for every vertex of G and $p+q \equiv 2 \pmod 4$, then (4.4) is impossible. We have

Theorem 4.1.1 ([221]) *Let G be a (p, q) graph such that the degree of every vertex is odd, q is even, and $p + q \equiv 2 \pmod 4$. Then G is not EMT.*

It has been conjectured in [164], and also in [221], that

Conjecture 4.1.1 ([164, 221]) Every tree is EMT.

However, proving or disproving this conjecture seems to be a difficult problem.

An edge-magic total labeling f of a (p, q) graph G is called *super edge-magic total* (super EMT) if $f(V(G)) = \{1, 2, \ldots, p\}$. If f is a super EMT labeling of G, then there exists an integer s such that $s + p + q = k$ and

$$S = \{f(u) + f(v) : uv \in E(G)\} = \{s, s+1, \ldots, s+q-1\}. \tag{4.5}$$

On the other hand, there exists exactly one extension of a vertex labeling f : $V(G) \rightarrow \{1, 2, \ldots, p\}$ satisfying (4.5) to a super EMT labeling of G such that $f(uv) = s + p + q - f(u) - f(v)$ for any edge $uv \in E(G)$, see Lemma 1 in [98].

A graph G is called EMT (super EMT) if there exists an EMT labeling (super EMT labeling, respectively). The concept of super EMT graphs was introduced by Enomoto et al. [96]. Wallis [287] and later Marr and Wallis [182] call this labeling *V-super edge-magic*. In [182] is noted that some authors may call such

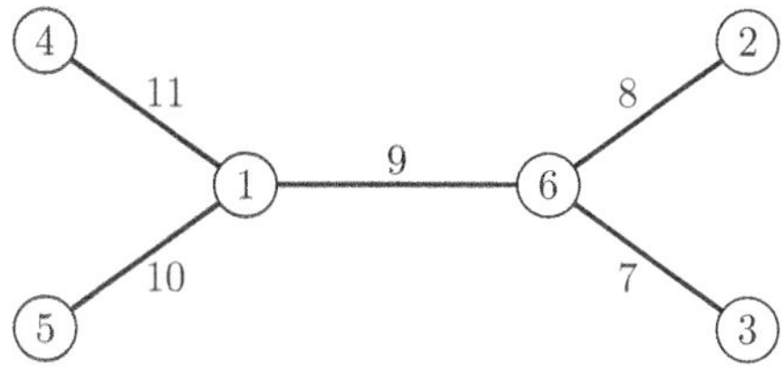

Fig. 4.2 A super EMT labeling of the double star $S(3, 3)$

an edge-magic labeling *strong* and the resulting graph a *strongly edge-magic graph*. Figure 4.2 shows a super EMT labeling of the double star on 6 vertices with magic sum $k = 16$.

Enomoto et al. strengthened the Conjecture 4.1.1 as follows.

Conjecture 4.1.2 ([96]) Every tree is super EMT.

Using a computer this conjecture has been verified by Lee and Shan [172] for all trees with up to 17 vertices. There are several results on super EMT labelings for special families of trees. For example: Hussain et al. [133] described super EMT labelings for certain classes of banana trees. Salman et al. [227] constructed super EMT labelings for subdivision of stars. Several results on super EMT labelings of w-trees, subdivision of w-trees and for extended w-trees can be found in [14, 145, 146, 147]. It seems that proving Conjecture 4.1.2 is difficult problem.

Next lemma gives interesting necessary condition for a graph to admit super EMT labeling.

Lemma 4.1.1 ([96]) *If a (p, q) graph G is super EMT, then $q \leq 2p - 3$.*

The condition $q \leq 2p - 3$ in Lemma 4.1.1 is not a sufficient condition for the graph G to be super EMT. A counterexample is an even cycle with p vertices and $q = p$ edges which satisfies the condition. These graphs are not super EMT.

In [98] it is noticed that if $q = 2p - 3$, then the vertices labeled with the following pairs of integers $(1, 2)$, $(1, 3)$, $(p, p - 2)$ and $(p, p - 1)$ have to be adjacent since there is a unique way of expressing $3, 4, 2p - 2$ and $2p - 1$ as sum of two distinct elements in the set $\{1, 2, \ldots, p\}$. As a corollary to Lemma 4.1.1 in [98] is proved the following result.

Corollary 4.1.1 ([98]) *Every super EMT (p, q) graph contains at least two vertices of degree less than* 4.

Proof Assume, to the contrary, that $p - 1$ vertices of G have degree at least 4. Then graph G has at least $2p - 2$ edges and by Lemma 4.1.1 we have

$$2(2p - 2) \leq 2q \leq 2(2p - 3) = 4p - 6$$

which is a contradiction. □

4.2 Edge-Magic Total and Super Edge-Magic Total Labelings of Regular Graphs

In the computation of the edge-weights of an r-regular super EMT (p, q) graph each edge label is used once and each vertex label is used r times. Thus we have

$$r\sum_{i=1}^{p} i + \sum_{j=p+1}^{p+q} j = kq \tag{4.6}$$

and (4.6) gives

$$\frac{(p+1)pr}{2} + \frac{(2p+q+1)q}{2} = kq. \tag{4.7}$$

Since $pr = 2q$, from (4.7) we obtain the magic sum of any r-regular super EMT graph

$$k = \frac{4p+q+3}{2}. \tag{4.8}$$

The next lemma shows that every regular graph with super EMT labeling has odd size.

Lemma 4.2.1 ([98]) *If G is an r-regular super EMT (p, q) graph, where $r > 0$, then q is odd and the magic sum of any super EMT labeling of G is $k = (4p + q + 3)/2$.*

From Eq. (4.8) it follows that q is odd. If f is a super EMT labeling of G with q edges, then the set S in (4.5) consists of q consecutive integers and $\lfloor q/2 \rfloor$ or $\lceil q/2 \rceil$ of the elements in S are odd. Clearly every odd element in S is sum of an even and an odd vertex label. Thus we have

Lemma 4.2.2 ([98]) *Let G be a super EMT graph of size q and f be a super EMT labeling of G. Then there are exactly $\lfloor q/2 \rfloor$ or $\lceil q/2 \rceil$ edges between V_e and V_o, where $V_e = \{v \in V(G) : f(v) \text{ is even}\}$ and $V_o = \{v \in V(G) : f(v) \text{ is odd}\}$.*

Fukuchi proved the following lemma.

Lemma 4.2.3 ([107]) *Let r be an odd integer. Let p be an integer, and let G be an r-regular graph of order p.*

(i) If $p \equiv 4 \pmod 8$, then G is not EMT.
(ii) If $p \equiv 0 \pmod 4$, then G is not super EMT.

Proof Suppose that f is an EMT labeling of an r-regular graph G of order p with magic sum k. Then G has $q = rp/2$ edges and the Eq. (4.4) gives

$$\frac{1}{2}\left(p+\frac{rp}{2}+1\right)\left(p+\frac{rp}{2}\right)+(r-1)\sum_{v\in V(G)} f(v)=\frac{krp}{2}. \tag{4.9}$$

If $p \equiv 4 \pmod 8$, then both $krp/2$ and $(r-1)\sum_{v\in V(G)} f(v)$ are even, but the expression

$$\frac{1}{2}\left(p+\frac{rp}{2}+1\right)\left(p+\frac{rp}{2}\right)$$

is odd, which is a contradiction.

Next we suppose that f is a super EMT labeling of an r-regular graph of order p, $p \equiv 0 \pmod 4$. If $p = 4m$ and

$$\sum_{v\in V(G)} f(v)=\sum_{i=1}^{p} i=\frac{p(p+1)}{2},$$

then from Eq. (4.9) it follows that

$$m(r+2)(4m+2mr+1)+2m(r-1)(4m+1)=2kmr. \tag{4.10}$$

Consequently,

$$(r+2)(2m(r+2)+1)+2(r-1)(4m+1)=2kr. \tag{4.11}$$

We can see that the right side of the Eq. (4.11) is even and the expression $2(r-1)(4m+1)$ is also even, but the expression $(r+2)(2m(r+2)+1)$ is odd, which is a contradiction. □

Ichishima et al. in [135] present constructions for generating large classes of super EMT 2-regular graphs from previously known super EMT 2-regular graphs.

4.2.1 Cycles

The cycle C_n of order n is regular of degree 2. In [163], it is proved that cycle C_n has an EMT labeling, for all $n \geq 3$, see also [56, 112, 223].

In [182] all EMT labelings of cycles up to C_6 are listed. Thus for C_4 the possibilities for magic sums are $k = 12, 13, 14$, and 15.

For $k = 12$ there is the unique solution with cyclic vertex labels $S = (1, 3, 2, 6)$ that gives an EMT labeling. For $k = 13$ there are two solutions with cyclic vertex labels $S = (1, 5, 2, 8)$ and $S = (1, 4, 6, 5)$ that give EMT labelings. The other two cases $k = 14$ and $k = 15$ are duals of the cases $k = 12$ and $k = 13$.

For C_5 the feasible magic sums are $k = 14, 15, 16$ and their duals. The unique solution for $k = 14$ has cyclic vertex labels $S = (1, 4, 2, 5, 3)$. There are no solutions for $k = 15$. For $k = 16$ there are two solutions with cyclic vertex labels $S = (1, 5, 9, 3, 7)$ and also $S = (1, 7, 3, 4, 10)$.

For C_6 the possible magic sums are $k = 17, 18, 19$ and their duals. For $k = 17$ there are three solutions with cyclic vertex labels $S = (1, 5, 2, 3, 6, 7)$, $S = (1, 6, 7, 2, 3, 5)$ and $S = (1, 5, 4, 3, 2, 9)$. There is one solution for $k = 18$ with cyclic vertex labels $S = (1, 8, 4, 2, 5, 10)$, and six solutions for $k = 19$ with cyclic vertex labels $S = (1, 6, 11, 3, 7, 8)$, $S = (1, 7, 3, 12, 5, 8)$, $S = (1, 8, 7, 3, 5, 12)$, $S = (1, 8, 9, 4, 3, 11)$, $S = (2, 7, 11, 3, 4, 9)$, and $S = (3, 4, 5, 6, 11, 7)$.

In the case of C_7, the possible magic sums run from 19 to 26. Godbold and Slater [112] found that all these magic sums can be realized. There are 118 labelings up to isomorphism. The corresponding numbers of EMT labelings for C_8, C_9, and C_{10} are 282, 1540, and 7092, see [112].

In [96] is proved that the cycle C_n is super EMT if and only if n is odd. The fact that C_n is not super EMT for n even follows from Lemma 4.2.1.

4.2.2 Complete Graphs

In 2001 Valdés presented the following result.

Theorem 4.2.1 ([286]) *Suppose the complete graph K_n has an EMT labeling with magic sum k. The number t of vertices that receive even labels satisfies the following conditions:*

(i) If $n \equiv 0$ or 3 $\pmod 4$ and k is even, then $t = \frac{1}{2}(n - 1 \pm \sqrt{n+1})$.
(ii) If $n \equiv 1$ or 2 $\pmod 4$ and k is even, then $t = \frac{1}{2}(n - 1 \pm \sqrt{n-1})$.
(iii) If $n \equiv 0$ or 3 $\pmod 4$ and k is odd, then $t = \frac{1}{2}(n + 1 \pm \sqrt{n+1})$.
(iv) If $n \equiv 1$ or 2 $\pmod 4$ and k is odd, then $t = \frac{1}{2}(n + 1 \pm \sqrt{n+3})$.

Kotzig and Rosa [164] proved the following.

Theorem 4.2.2 ([164]) *The complete graph K_n has an EMT labeling if and only if $n = 1, 2, 3, 5$ or 6.*

Wallis et al. [289] constructed EMT labeling of K_n, $n = 1, 2, 3, 5$ and 6, for all possible values of the magic sum k. Thus, K_2 is trivially EMT with labels 1, 2, and 3 and with magic sum $k = 6$.

For K_3 the feasible magic sums are $k = 9, 10, 11$, and 12. If $k = 9$, then cyclic vertex labels $S = (1, 2, 3)$ give an EMT labeling. If $k = 10$, then $S = (1, 3, 5)$, if $k = 11$, then $S = (2, 4, 6)$, and if $k = 12$, then $S = (4, 5, 6)$.

From Lemma 4.2.3 it follows that K_4 is not EMT.

For K_5 the possible magic sums are $k = 18, 21, 24, 27, 30$. According to Theorem 4.2.1 there exist no solutions when k is odd. So only magic sums 18, 24, and 30 are listed. If $k = 18$, then cyclic vertex labels $S = (1, 2, 3, 5, 9)$ give an EMT labeling. If $k = 24$ there are two solutions with cyclic vertex labels $S = (1, 8, 9, 10, 12)$ and $S = (4, 6, 7, 8, 15)$. If $k = 30$, then $S = (7, 11, 13, 14, 15)$.

For K_6 the feasible magic sums are $k = 21, 25, 29, 33, 37, 41$, and 45. There are no solutions when $k = 21, 33$, and 45. If $k = 25$, then cyclic vertex labels $S = (1, 3, 4, 5, 9, 14)$ give an EMT labeling. If $k = 29$, then $S = (2, 6, 7, 8, 10, 18)$. If $k = 37$, then $S = (4, 12, 14, 15, 16, 20)$. If $k = 41$, then $S = (8, 11, 17, 18, 19, 21)$.

In [96] Enomoto et al. proved that the complete graph K_n is super EMT if and only if $n = 1, 2$, or 3.

4.2.3 Generalized Petersen Graphs

It follows from Lemma 4.1.1 that if an r-regular graph is super EMT, then $r \leq 3$. Since the generalized Petersen graph $P(n, m)$ forms an interesting class of 3-regular graphs, it is desirable to determine for which values of parameters n and m it is super EMT graph. $P(n, m)$ has $2n$ vertices and $3n$ edges and as a corollary to Lemma 4.2.1 we get the following result.

Corollary 4.2.1 ([294]) *If $P(n, m)$ is super EMT, then n is odd and the magic sum of any super EMT labeling of $P(n, m)$ is $k = (11n + 3)/2$.*

In [284], Tsuchiya and Yokomura constructed a super EMT labeling of generalized Petersen graph $P(n, m)$ in the case where n is odd and $m = 1$. Fukuchi considered the case where n is odd and $m = 2$, and proved the following theorem.

Theorem 4.2.3 ([107]) *The generalized Petersen graph $P(n, 2)$ is super EMT for odd $n \geq 3$.*

Xirong et al. [294] proved that

Theorem 4.2.4 ([294]) *The generalized Petersen graph $P(n, 3)$ is super EMT for odd $n \geq 5$.*

Bača, Baskoro, Simanjuntak, and Sugeng obtain the following result for $m = (n - 1)/2$.

Theorem 4.2.5 ([27]) *For n odd, $n \geq 3$, every generalized Petersen graph $P(n, (n - 1)/2)$ has a super EMT labeling.*

Constructions that will produce super EMT labeling of $P(n, m)$, for n odd and $4 \leq m \leq (n - 3)/2$, have not been found yet. Nevertheless, we suggest the following.

Conjecture 4.2.1 ([27]) There is a super EMT labeling for the generalized Petersen graph $P(n, m)$, for every n odd, $n \geq 9$, and $4 \leq m \leq (n - 3)/2$.

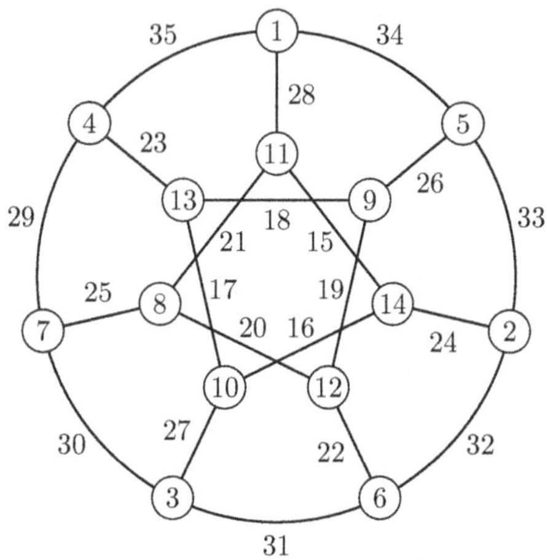

Fig. 4.3 A super EMT labeling of $P(7, 2)$ with magic sum $k = 40$

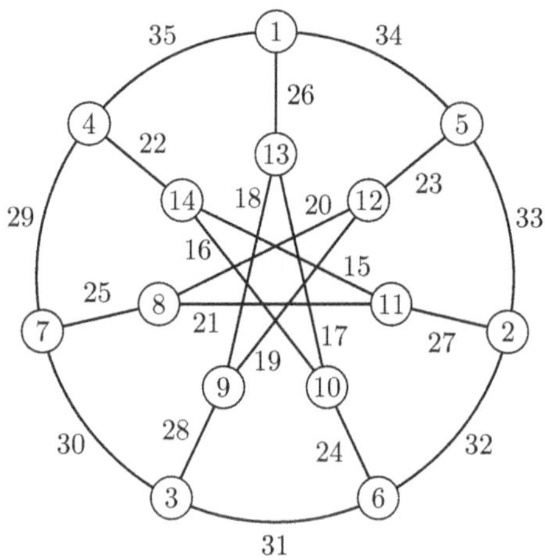

Fig. 4.4 A super EMT labeling of $P(7, 3)$ with magic sum $k = 40$

Applying the construction described in the proof of Theorem 4.2.3, see [107], we obtain the super EMT labeling of $P(7, 2)$ shown in Fig. 4.3. Figure 4.4 illustrates a super EMT labeling of $P(7, 3)$ constructed by algorithm given in the proof of Theorem 4.2.4, see [294].

4.3 Labelings of Certain Families of Connected Graphs

We will use results from the previous section to study the (super) EMT properties of certain graphs.

4.3.1 Wheels

From Theorem 4.1.1 it follows that wheel W_n is not EMT if $n \equiv 3 \pmod 4$. For other wheels Enomoto, Lladó, Nakamigawa, and Ringel conjectured that

Conjecture 4.3.1 ([96]) W_n is EMT if $n \not\equiv 3 \pmod 4$.

This conjecture was proved for $n \equiv 0, 1 \pmod 4$ by Phillips et al. in [210], see also [108]. Slamin et al. proved the following.

Theorem 4.3.1 ([257]) *For $n \equiv 6 \pmod 8$, every wheel W_n has an EMT labeling with the magic sum $k = 5n + 2$.*

There remains one case to be settled. It is $n \equiv 2 \pmod 8$. For this case in [257] EMT labelings for W_{10}, W_{18} and W_{26} are constructed, but authors have been unable to generalize these labelings for every $n \equiv 2 \pmod 8$. As an example, Fig. 4.5 shows an EMT labeling for W_{10} with $k = 52$.

Thus we propose the following open problem.

Open Problem 4.3.1 ([257]) For wheel W_n, $n = 2 \pmod 8$, determine if there is an EMT labeling.

4.3.2 Fans and Friendship Graphs

The fan is always EMT as it is shown in the next theorem.

Theorem 4.3.2 ([98, 257]) *The fan F_n is EMT for every positive integer $n \geq 2$.*

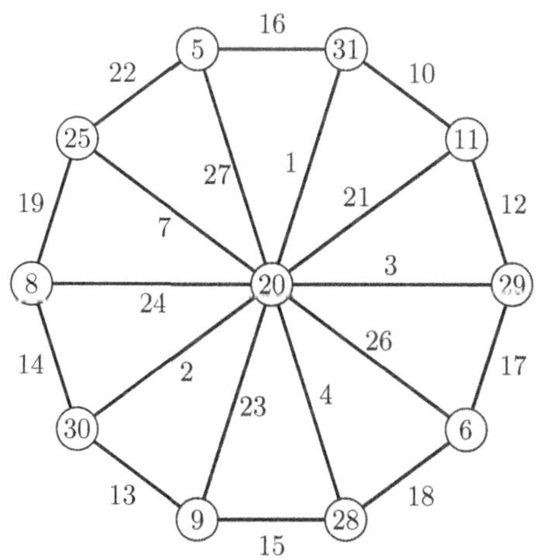

Fig. 4.5 An EMT labeling of the wheel W_{10}

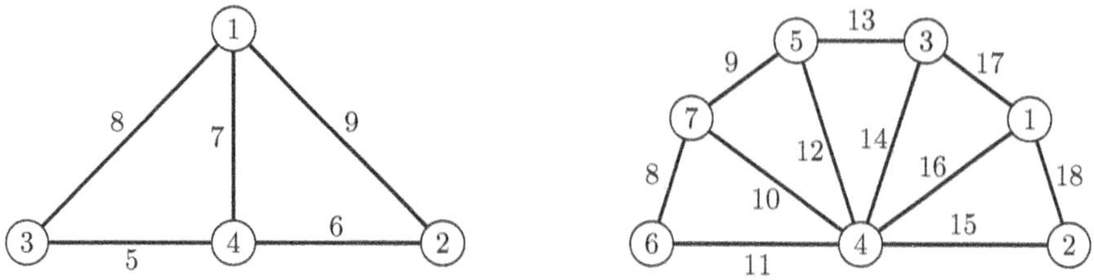

Fig. 4.6 Super EMT labelings of fans F_3 and F_6

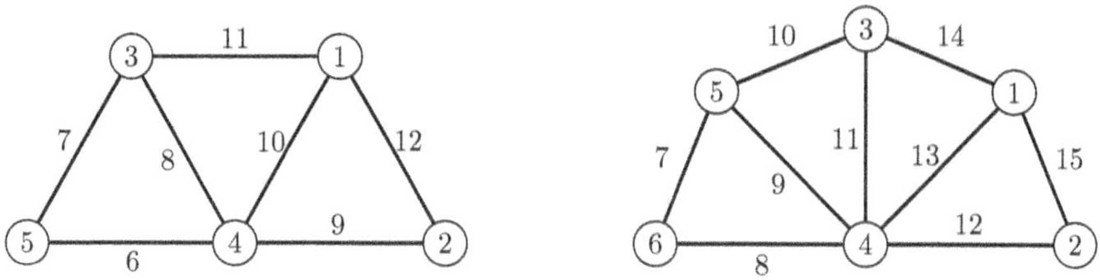

Fig. 4.7 Super EMT labelings of fans F_4 and F_5

In [98] is constructed an EMT labeling for F_n with the magic sum $k = 3n + 3$ and in [257] is described an EMT labeling for F_n with the magic sum $k = 4n + 2$.

The next result is interesting because it analyzes some (p, q) graphs for which $q = 2p - 3$.

Theorem 4.3.3 ([98]) *The fan F_n is super EMT if and only if $1 \leq n \leq 6$.*

Proof It is easy to see that the fans $F_1 \cong K_2$ and $F_2 \cong K_3$ are both super EMT. The super EMT labelings for F_n, $3 \leq n \leq 6$, are shown in Figs. 4.6 and 4.7.

Assume, to the contrary, that F_n with $n + 1$ vertices and $2n - 1$ edges admits a super EMT labeling g for every integer $n \geq 7$. Since $g(v_i) = i$, $1 \leq i \leq n + 1$, from (4.5) it follows that the set S consists of $2n - 1$ consecutive integers, namely, $S = \{3, 4, \dots, 2n + 1\}$. The vertices $v_1, v_2, v_3, v_4, v_{n-2}, v_{n-1}, v_n, v_{n+1}$ are all distinct because $n \geq 7$. The elements 3, 4, $2n$, and $2n + 1$ from S can be expressed uniquely as sums of two distinct vertex labels, namely, $3 = 1 + 2$, $4 = 1 + 3$, $2n = (n - 1) + (n + 1)$, and $2n + 1 = n + (n + 1)$. Therefore v_1v_2, v_1v_3, $v_{n-1}v_{n+1}$, and v_nv_{n+1} are four necessary edges in F_n.

On the other hand, each of the elements 5 and $2n - 1$ from S can be expressed in exactly two ways as sums of two distinct vertex labels, namely, $5 = 1 + 4 = 2 + 3$ and $2n - 1 = (n - 2) + (n + 1) = (n - 1) + n$. Thus, there are four mutually exclusive possibilities for edges in F_n: either v_1v_4, $v_{n-2}v_{n+1}$ or v_1v_4, $v_{n-1}v_n$ or v_2v_3, $v_{n-2}v_{n+1}$ or v_2v_3, $v_{n-1}v_n$.

If we arbitrarily add any of these four pairs of edges to the four necessary edges in F_n, we obtain a forbidden subgraph of F_n, namely, either $2K_{1,3}$ or $K_{1,3} \cup K_3$ or $2K_3$, which is a contradiction. □

The friendship graph f_n was defined in Sect. 3.3.3. Let $V(f_n) = \{u_i, v_i : 1 \leq i \leq n\} \cup \{c\}$ and $E(f_n) = \{u_i v_i : 1 \leq i \leq n\} \cup \{cu_i : 1 \leq i \leq n\} \cup \{cv_i : 1 \leq i \leq n\}$. The next auxiliary lemma shows that the center of a super EMT friendship graph has to admit an even value.

Lemma 4.3.1 ([257]) *If friendship graph f_n is super EMT, then the value of the center is even and $n \not\equiv 2 \pmod 4$.*

Proof Assume that friendship graph f_n is super EMT. Then there exists a bijection $g : V(f_n) \to \{1, 2, \ldots, 2n+1\}$ and the set S from (4.5) consists of $3n$ consecutive integers, namely, $S = \{g(u) + g(v) : uv \in E(f_n)\} = \{s, s+1, \ldots, s+3n-1\}$.

Let $l = g(c)$ be the value of the center, $1 \leq l \leq 2n+1$, and $\{g(u_i)+g(c), g(v_i)+g(c) : 1 \leq i \leq n\} = S_1 \cup S_2$, where each of the sets $S_1 = \{l+1, l+2, \ldots, 2l-1\}$ and $S_2 = \{2l+1, 2l+2, \ldots, l+2n+1\}$ consists of consecutive integers.

$S_3 = \{g(u_i) + g(v_i) : 1 \leq i \leq n\} \cup \{2l\}$ is defined as the set of elements, where each element is a sum of two distinct elements either from the set $V_1 = \{1, 2, \ldots, l-1\} - \{l-j\}$ or from the set $V_2 = \{l+1, l+2, \ldots, 2n+1\} - \{l+j\}$ (but not from both) with the restriction that the values less than $l+1$ (respectively greater than $l+2n+1$) are obtained as sums of two distinct elements in the set V_1 (respectively V_2), and the value $2l = (l-j) + (l+j)$ for $1 \leq j \leq l-1$. Thus $S_1 \cup S_2 \cup S_3 \cup \{2l\} = S$.

We can see that the number of elements in the set V_1 must be even, which implies that l is even. The sum of all values in the set S_1 is $3l(l-1)/2$, in the set S_2 is $(2n-l+1)(3l+2n+2)/2$ and in the set S_3 is $l(l-1)/2 + (2n+l+2)(2n-l+1)/2 - 2l$.

The sum of all values in the sets S_i, $i = 1, 2, 3$, plus the value $2l$ is equal to the sum of all values in the set S. Thus the following equation holds:

$$\frac{3l(l-1)}{2} + \frac{(2n-l+1)(3l+2n+2)}{2} + \frac{l(l-1)}{2} + \frac{(2n+l+2)(2n-l+1)}{2} - 2l + 2l = 3ns + \frac{3n(3n-1)}{2}. \tag{4.12}$$

Then we have

$$l(l-1) + (2n-l+1)(n+l+1) = \frac{3n(2s+3n-1)}{4}. \tag{4.13}$$

From (4.13) we get the following.

(i) If n is even, then $n \equiv 0 \pmod 4$.
(ii) If n is odd, then $2s+3n-1 \equiv 0 \pmod 4$, which means that for $n \equiv 1 \pmod 4$ the value s is odd and for $n \equiv 3 \pmod 4$ the value s is even.

□

With the previous lemma in hand we are ready to give a characterization for super EMT labeling of f_n.

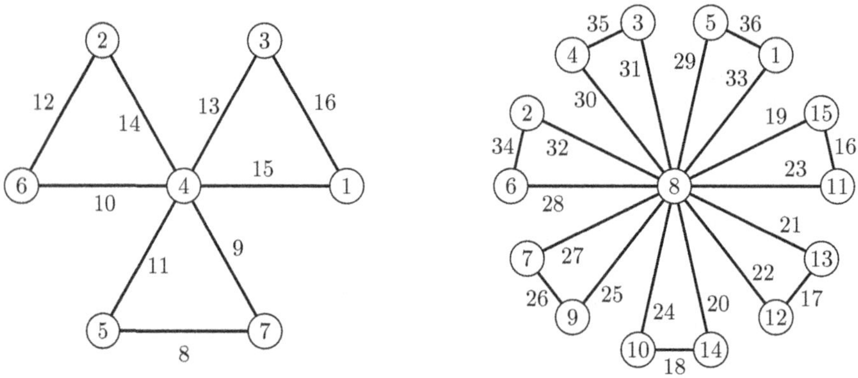

Fig. 4.8 Super EMT labelings of f_3 and f_7

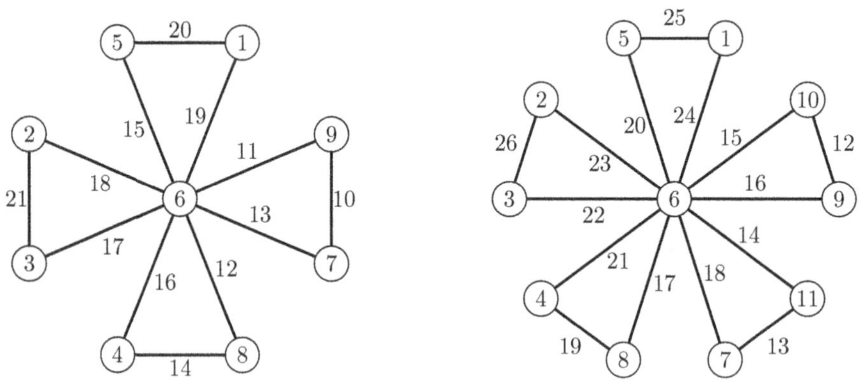

Fig. 4.9 Super EMT labelings of f_4 and f_5

Theorem 4.3.4 ([257]) *The friendship graph f_n is super EMT if and only if $3 \le n \le 5$ and $n = 7$.*

Proof First, we show that f_n admits super EMT labeling for $3 \le n \le 5$ and $n = 7$. These labelings are given in Figs. 4.8 and 4.9.

For the converse, we consider the set $V_1 = \{1, 2, \ldots, l-1\} - \{l-j\}$ from the Lemma 4.3.1, where

$$1 \le j \le l - 1. \tag{4.14}$$

The sums, corresponding to distinct pairs in a matching of the set V_1, constitute the set of consecutive integers $\{s, s+1, \ldots, s+l/2-2\}$ and hence

$$(1 + 2 + \cdots + l - 1) - (l - j) = s + (s + 1) + \cdots + (s + \frac{l}{2} - 2) \tag{4.15}$$

or, equivalently,

$$\frac{(l-1)(l-2)}{2} - 1 + j = \frac{(l-2)s}{2} + \frac{(l-2)(l-4)}{8}. \tag{4.16}$$

Combining (4.14) and (4.16) we have

$$1 \leq \frac{8+6l-3l^2}{8} + \frac{(l-2)s}{2} \leq l-1 \tag{4.17}$$

and hence

$$\frac{3l}{4} \leq s \leq \frac{3l}{4} + 2. \tag{4.18}$$

If we consider l even, $2 \leq l \leq 2n$, and $n \not\equiv 2 \pmod 4$, then the following table gives all possible integer values of parameters s and l for $3 \leq n \leq 11$, which are the solutions of (4.13).

n	3	4	4	4	4	5	7	9	11
s	4	3	4	5	6	5	6	7	8
l	4	2	4	6	8	6	8	10	12

It is easy to see that the Condition (4.18) can be realized only for $3 \leq n \leq 5$ and $n = 7$. □

4.3.3 Ladders and Generalized Prisms

The *ladder* L_n can be defined as the Cartesian product $L_n \cong P_n \square P_2$ of a path on n vertices with the path on 2 vertices. In [98] and [284] the authors listed a super EMT labeling of L_n with the magic sum $k = (11n + 1)/2$. Thus we have

Theorem 4.3.5 ([98]) *The ladder $L_n \cong P_n \square P_2$ is super EMT, where n is odd.*

From Lemma 4.2.1 it follows that ladder $L_2 \cong P_2 \square P_2 \cong C_4$ is not super EMT. However for L_4 and L_6 the corresponding super EMT labelings are given in [98], see Figs. 4.10 and 4.11. They suspect that a super EMT labeling might be found for larger even values of n. Thus the following conjecture may hold.

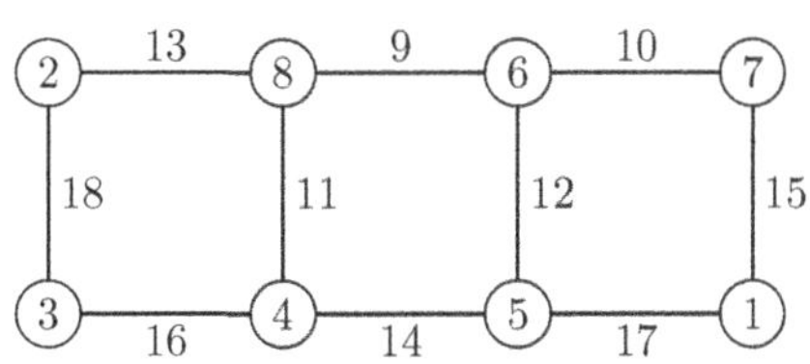

Fig. 4.10 A super EMT labeling of L_4 with the magic sum $k = 23$

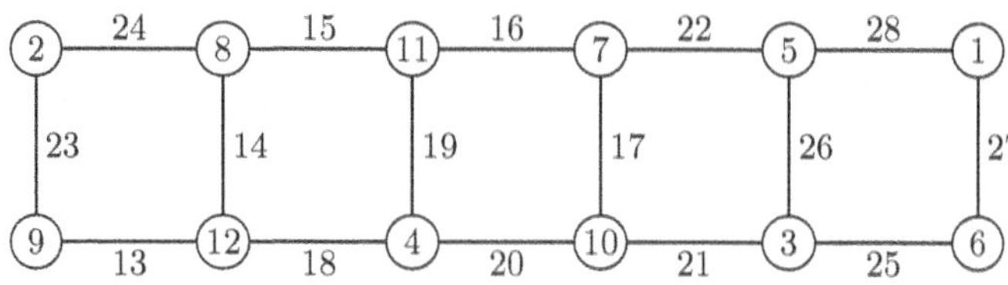

Fig. 4.11 A super EMT labeling of L_6 with the magic sum $k = 34$

Conjecture 4.3.2 ([270]) The ladder $L_n \cong P_n \Box P_2$ is super EMT if n is even.

The *generalized prism* $C_m \Box P_n$ is sometimes super EMT. The construction of a super EMT labeling of generalized prism for m odd is given in [98] and [284].

Theorem 4.3.6 ([98]) *The generalized prism $C_m \Box P_n$ is super EMT if m is odd and $n \geq 2$.*

Figueroa-Centeno et al. in [98] notice that for m even the graph $C_m \Box P_2$ is not super EMT. López and Muntaner-Batle in [176] remark that as far as they know, for $n > 2$ and m even, the super edge-magicness of $C_m \Box P_n$ is unknown. In view of this, we suggest the following open problem.

Open Problem 4.3.2 ([176]) For the generalized prism $C_m \Box P_n$, $n > 2$ and m even, determine if there is a super EMT labeling.

4.3.4 Paths

Let P_n be the path with $V(P_n) = \{u_i : 1 \leq i \leq n\}$ and $E(P_n) = \{u_i u_{i+1} : 1 \leq i \leq n-1\}$. Assume that a bijection $f : V(P_n) \cup E(P_n) \to \{1, 2, \ldots, 2n-1\}$ is super EMT with a magic sum k. Then we have

$$2\sum_{i=1}^{n} f(u_i) - f(u_1) - f(u_n) + \sum_{i=1}^{n-1} f(u_i u_{i+1}) = (n-1)k,$$

where

$$\sum_{i=1}^{n} f(u_i) = \frac{(n+1)n}{2} \text{ and } \sum_{i=1}^{n-1} f(u_i u_{i+1}) = \frac{3n(n-1)}{2}$$

and, consequently,

$$(n-1)k = \frac{5n^2 - n}{2} - (f(u_1) + f(u_n)). \tag{4.19}$$

Theorem 4.3.7 ([38]) *The path P_n, $n \geq 2$, has a super EMT labeling if and only if one of the following conditions is satisfied.*

(i) $k = 5n/2 + 1$, *for n even.*
(ii) $k = (5n + 1)/2$ *or* $k = (5n + 3)/2$, *for n odd.*

Proof In Eq. (4.19) the number $2(n-1)$ is a factor of $5n^2 - n - 2(f(u_1) + f(u_n))$ if there exists a real number s, such that $2(n-1)\,(5n/2 + s) = 5n^2 - n - 2(f(u_1) + f(u_n))$, whence it follows that

$$f(u_1) + f(u_n) = 2n + s(1-n). \tag{4.20}$$

Considering the extreme values of the vertex labels u_1 and u_n, we get

$$3 \leq f(u_1) + f(u_n) \leq 2n - 1, \tag{4.21}$$

and Eq. (4.20) gives that $0 < s < 2$.

So, the magic sum $k = 5n/2 + s$ is an integer if and only if one of the followings holds.

(i) n is even and $s = 1$.
(ii) n is odd and $s = 1/2$ or $s = 3/2$.

Let us consider three cases.

Case A If n is even and $s = 1$, then, from Eq. (4.20), we have that $f(u_1) + f(u_n) = n + 1$. Define a vertex labeling f_1 and an edge labeling f_2 as follows.

$$f_1(u_i) = \begin{cases} \frac{i+1}{2} & \text{if } i \text{ is odd} \\ \left\lceil \frac{n+i}{2} \right\rceil & \text{if } i \text{ is even} \end{cases}$$

$$f_2(u_i u_{i+1}) = 2n - i \quad \text{if } 1 \leq i \leq n-1.$$

The labelings f_1 and f_2 combine to a desired super EMT labeling with $k = 5n/2+1$, see Fig. 4.12 for P_6.

Case B If n is odd and $s = 1/2$, then $f(u_1) + f(u_n) = (3n + 1)/2$. The desired super EMT labeling with $k = (5n + 1)/2$ is obtained by combining the following

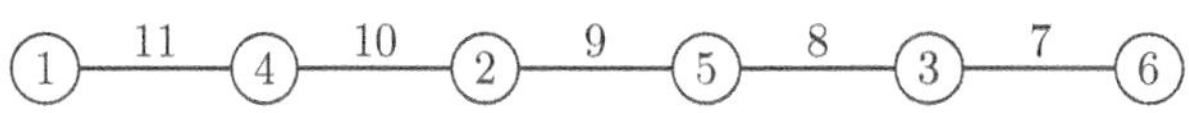

Fig. 4.12 Super EMT labeling of P_6 with magic sum $k = 16$

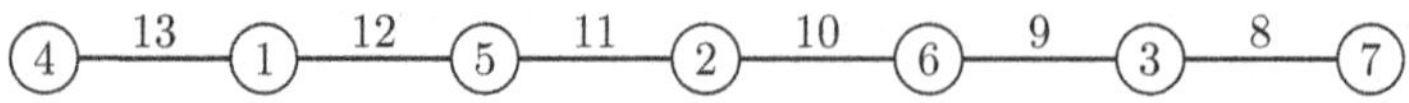

Fig. 4.13 Super EMT labeling of P_7 with magic sum $k = 18$

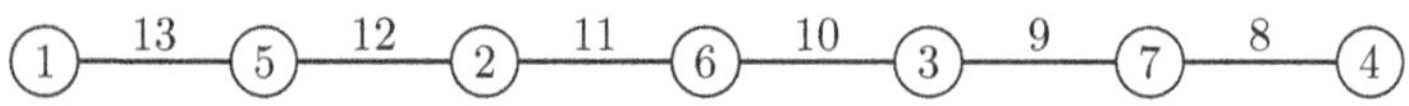

Fig. 4.14 Super EMT labeling of P_7 with magic sum $k = 19$

labelings f_3 and f_4, see Fig. 4.13 for P_7.

$$f_3(u_i) = \begin{cases} \frac{n+i}{2} & \text{if } i \text{ is odd} \\ \frac{i}{2} & \text{if } i \text{ is even} \end{cases}$$

$$f_4(u_i u_{i+1}) = 2n - i \quad \text{if } 1 \le i \le n-1.$$

Case C If n is odd and $s = 3/2$, then $f(u_1) + f(u_n) = (n+3)/2$. To establish a required super EMT labeling with $k = (5n+3)/2$, it suffices to present

$$f_5(u_i) = f_1(u_i)$$

$$f_6(u_i u_{i+1}) = f_2(u_i u_{i+1}).$$

□

Figure 4.14 illustrates the desired super EMT labeling of P_7 with $s = 3/2$.

4.3.5 *Path-Like Trees*

In this section we study the embedding of paths in the 2-dimensional grid and consider a set of transformations which keep the edge-magic character of the paths.

Let P_n, $n \ge 4$, be the path with $V(P_n) = \{w_i : 1 \le i \le n\}$ and $E(P_n) = \{w_i w_{i+1} : 1 \le i \le n-1\}$. We embed the path P_n as a subgraph of the 2-dimensional grid. Consider the ordered set of subpaths $S_1, S_2, \dots, S_l$, which are maximal straight segments in the embedding and have the property that the end of S_j is the beginning of S_{j+1}, for any $j = 1, 2, \dots, l-1$.

Figure 4.15 shows such an embedding of the path P_{24} with a vertex labeling. It is an easy exercise to check that the set of edge-weights, under the vertex labeling, consists of consecutive integers $\{14, 15, \dots, 36\}$. If the edge labeling with values $25, 26, \dots, 47$ is combined with the vertex labeling, then we are able to obtain the resulting super EMT labeling with the magic sum $k = 61$.

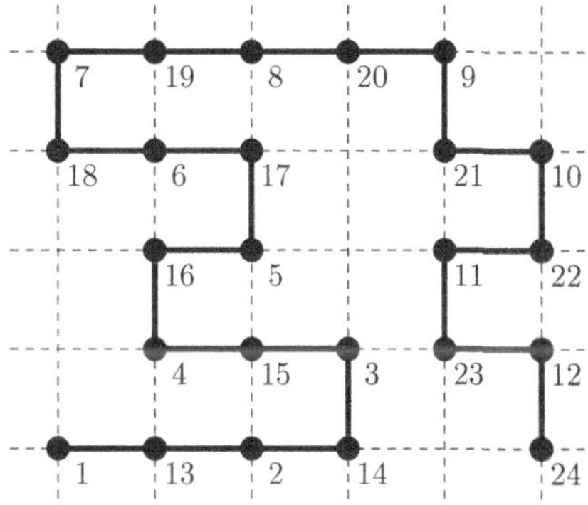

Fig. 4.15 A vertex labeling of the path P_{24}

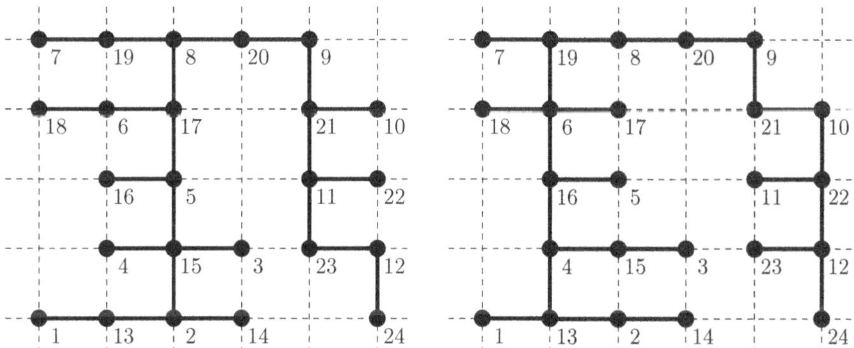

Fig. 4.16 Two examples of path-like trees with vertex labelings

In Fig. 4.15 the ordered set of subpaths is $S_1 \cong P_4$, $S_2 \cong P_2$, $S_3 \cong P_3$, $S_4 \cong P_2$, $S_5 \cong P_2$, $S_6 \cong P_2$, $S_7 \cong P_3$, $S_8 \cong P_2$, $S_9 \cong P_5$, $S_{10} \cong P_2$, $S_{11} \cong P_2$, $S_{12} \cong P_2$, $S_{13} \cong P_2$, $S_{14} \cong P_2$, $S_{15} \cong P_2$, and $S_{16} \cong P_2$.

Suppose that $S_j \cong P_2$, for some j, $1 < j < l$, $V(S_j) = \{u_0, v_0\}$, $u_0 \in V(S_{j-1}) \cap V(S_j)$, $v_0 \in V(S_j) \cap V(S_{j+1})$, and that some vertex u of S_{j-1} is at distance 1 in the grid to a vertex v of S_{j+1}. The distance of u_0 and u in S_{j-1} is equal to the distance of v_0 and v in S_{j+1}. An *elementary transformation* of the path refers to replacing the edge u_0v_0 by a new edge uv.

We say that a tree T of order n is a *path-like tree* if it can be obtained from some embedding of P_n in the 2-dimensional grid by a set of elementary transformations. This concept of path-like tree was introduced by Barrientos in [52].

In Fig. 4.16, we exhibit two different examples of path-like trees, obtained from the embedding of P_{24} in Fig. 4.15. The edge-weights of these vertex labelings form the set of consecutive integers and after completing the edge labeling we obtain super EMT labelings with magic sum $k = 61$.

Now, we present the following theorem.

Theorem 4.3.8 ([38]) *Every path-like tree admits a super EMT labeling.*

Proof Let T be a path-like tree of order n. Consider the embedding of P_n, from which the path-like tree T can be obtained. Denote the vertices of P_n successively as $w_1, w_2, \ldots, w_n$, starting from one pendant vertex of P_n.

Consider the vertex labeling f_1 of the path P_n, used in the proof of Theorem 4.3.7, where $f_1(w_i) + f_1(w_{i+1}) = \lceil n/2 \rceil + i + 1$ for $1 \leq i \leq n-1$.

Let $T_0 = P_n, T_1, T_2, \ldots, T_t = T$ be the series of trees, obtained by successively applying the appropriate elementary transformations of P_n, to obtain T. We will show that the edge-weights under the vertex labeling f_1 successively attain the values $\lceil n/2 \rceil + 2, \lceil n/2 \rceil + 3, \ldots, \lceil 3n/2 \rceil$ on the edges of each of the trees in the series $T_0, T_1, \ldots, T_t = T$.

Suppose that the tree T_r, $0 \leq r < t$, from the series of trees, has the property that the edge-weights, under the vertex labeling f_1, successively attain the values $\lceil n/2 \rceil + 2, \lceil n/2 \rceil + 3, \ldots, \lceil 3n/2 \rceil$, and that T_{r+1} is obtained from T_r by replacing the edge u_0v_0 by a new edge uv.

Without loss of generality, we may assume that $S_j \cong P_2$, $V(S_j) = \{u_0, v_0\}$ and $u_0 = w_i$, $v_0 = w_{i+1}$. Let S_{j-1} and S_{j+1} be maximal straight segments in T_r, such that $\{u_0, u\} \subseteq V(S_{j-1})$, $\{v_0, v\} \subseteq V(S_{j+1})$, the vertex u is at distance one in the grid to the vertex v, and the distance between u_0 and u in S_{j-1} is equal to the distance between v_0 and v in S_{j+1}, say, $d(u_0, u) = d(v_0, v) = m$.

Let $u = w_{i-m}$ and $v = w_{i+m+1}$. We know that

$$f_1(u_0) + f_1(v_0) = f_1(w_i) + f_1(w_{i+1}) = \left\lceil \frac{n}{2} \right\rceil + i + 1.$$

If m is even and i is odd, or if m is odd and i is even, then

$$f_1(u) = f_1(w_{i-m}) = \frac{i-m+1}{2}$$

$$f_1(v) = f_1(w_{i+m+1}) = \left\lceil \frac{n+i+m+1}{2} \right\rceil.$$

If m and i are even, or if m and i are both odd, then

$$f_1(u) = f_1(w_{i-m}) = \left\lceil \frac{n+i-m}{2} \right\rceil$$

$$f_1(v) = f_1(w_{i+m+1}) = \frac{i+m+2}{2}.$$

In both these cases,

$$f_1(u) + f_1(v) = \left\lceil \frac{n}{2} \right\rceil + i + 1 = f_1(u_0) + f_1(v_0),$$

so that the edge-weights of the tree T_{r+1} successively assume the values $\lceil n/2 \rceil + 2, \lceil n/2 \rceil + 3, \ldots, \lceil 3n/2 \rceil$.

Similarly, if we consider the vertex labeling f_3 from Theorem 4.3.7, then

$$f_3(u_0) + f_3(v_0) = f_3(w_i) + f_3(w_{i+1}) = \frac{n+1}{2} + i$$

$$f_3(w_{i-m}) = \frac{n+i-m}{2}$$

$$f_3(w_{i+m+1}) = \frac{i+m+1}{2}$$

if m and i do not have the same parity, and

$$f_3(w_{i-m}) = \frac{i-m}{2}$$

$$f_3(w_{i+m+1}) = \frac{n+i+m+1}{2}$$

if m and i have the same parity.

Thus,

$$f_3(u) + f_3(v) = \frac{n+1}{2} + i = f_3(u_0) + f_3(v_0)$$

and the edge-weights of the tree T_{r+1}, under the vertex labeling f_3, successively attain the values $(n+1)/2 + 1, (n+1)/2 + 2, \ldots, (3n-1)/2$.

For each tree T_r, $0 \leq r \leq t$, we are able to complete an edge labeling with the values $n+1, n+2, \ldots, 2n-1$, in such a way that this edge labeling and the vertex labeling f_1 (respectively, f_3) combine to a super EMT labeling. This completes the proof. □

Note that Fukuchi [106] shows how to recursively create super EMT trees from certain kinds of existing super EMT trees. Ngurah et al. [204] provide a method for constructing new (super) EMT graphs from existing ones. One of their results is that if G has an EMT labeling and G has order p and size p or $p-1$, then $G \odot nK_1$ has an EMT labeling.

4.4 Labelings of Certain Families of Disconnected Graphs

In this section we consider when the disjoint union of multiple copies of a (super) EMT graph admits a (super) EMT labeling. The next theorem allows us to generate infinite classes of disconnected (super) EMT graphs.

Theorem 4.4.1 ([100]) *If G is a (super) EMT bipartite or tripartite graph, and m is odd, then mG is (super) EMT.*

Since all cycles C_n, $n \geq 3$, are EMT (see [112, 223, 56]), from the previous theorem it follows immediately that

Corollary 4.4.1 ([100]) *If m is odd and $n > 1$, then the 2-regular graph mC_{2n} is EMT.*

Figueroa-Centeno et al. have shown in [99] that the 2-regular graph mC_n has super EMT labeling if and only if m and n are odd. Therefore, mC_n is EMT for m odd and every $n \geq 3$. Ahmad et al. [7] constructed super EMT labelings for disjoint union of banana trees.

4.4.1 Disjoint Union of Stars

Figueroa-Centeno et al. [100] proved that disjoint union of stars $K_{1,m} \cup K_{1,n}$ is EMT if and only if mn is even. For super EMT labeling of disjoint union of stars they proved only the sufficient condition and conjectured the necessary condition.

Ivančo and Lučkaničová [141] give a characterization of super EMT labeling of $K_{1,m} \cup K_{1,n}$ in the next theorem.

Theorem 4.4.2 ([141]) *$K_{1,m} \cup K_{1,n}$ is a super EMT graph if and only if either m is a multiple of $n + 1$ or n is a multiple of $m + 1$.*

Proof Let $V(K_{1,m} \cup K_{1,n}) = \{v_{i,j} :$ either $i = 1$ and $j = 0, 1, \dots, m$ or $i = 2$ and $j = 0, 1, \dots, n\}$ be the vertex set and $E(K_{1,m} \cup K_{1,n}) = \{v_{i,0}v_{i,j} : i \in \{1, 2\}, j \geq 1\}$ be the edge set of disjoint union of stars. Assume that $K_{1,m} \cup K_{1,n}$ is a super EMT graph and f is a corresponding super EMT labeling. Then there exists an integer s such that $s + 2(m + n + 1) = k$ and (4.5) gives that

$$S = \{f(u)+f(v) : uv \in E(K_{1,m} \cup K_{1,n})\} = \{s, s+1, \dots, s+m+n-1\}. \quad (4.22)$$

The sum of the elements of S is

$$\sum_{uv \in E(K_{1,m} \cup K_{1,n})} (f(u) + f(v)) = (m + n)s + \frac{(m + n)(m + n - 1)}{2}. \quad (4.23)$$

Let $s_1 = f(v_{1,0})$ and $s_2 = f(v_{2,0})$. In the computation of the Eq. (4.23), the labels s_1 and s_2 are used m and n times, respectively, and the labels of the remaining vertices are used once each. The sum of all the vertex labels used to calculate the

Eq. (4.23) is equal to

$$(m-1)f(v_{1,0})+(n-1)f(v_{2,0})+\sum_{k=1}^{m+n+2}k$$
$$=(m-1)s_1+(n-1)s_2+\frac{(m+n+3)(m+n+2)}{2}. \tag{4.24}$$

Thus, from (4.23) and (4.24) we have

$$(m+n)s=3(m+n+1)+(m-1)s_1+(n-1)s_2. \tag{4.25}$$

Clearly, $s_1+s_2\notin\{s,s+1,\ldots,s+m+n-1\}$ because exactly one endpoint of any edge belongs to $\{v_{1,0},v_{2,0}\}$. Without loss of generality, we may assume that $s_1+s_2<s$ (if $s_1+s_2>s+m+n-1$, then we consider the dual labeling g given by $g(v_{i,j})=m+n+3-f(v_{i,j})$).

If $1\notin\{s_1,s_2\}$, then

$$s>s_1+s_2>\min_{1\le j\le m}f(v_{1,j})+s_2\ge 1+s_2\ge s$$

or

$$s>s_1+s_2>s_1+\min_{1\le j\le n}f(v_{2,j})\ge s_1+1\ge s,$$

a contradiction.

Suppose $s_1=2$ and $s_2=1$. From Eq. (4.25), we have that

$$(m+n)(s-4)=m. \tag{4.26}$$

This implies that m is a multiple of $m+n$, a contradiction.

Suppose $s_1>2$ and $s_2=1$. We can say that $s=s_1+2$ because if

$$\min_{1\le j\le n}f(v_{2,j})=2,$$

then

$$\min_{1\le j\le n}f(v_{2,j})+s_2<s_1+s_2<s,$$

thus the vertex labeled by 2 must belong to $K_{1,m}$. It follows from Eq. (4.25) that

$$(s_1-2)(n+1)=m,$$

which means that $m>n$ and m is a multiple of $n+1$.

On the other hand, assume that $m = t(n+1)$. Consider the vertex labeling f_1 described by Ivančo and Lučkaničová in [141];

$$f_1(v_{i,j}) = \begin{cases} 2+t & \text{if } i=1 \text{ and } j=0 \\ \lceil \frac{j}{t} \rceil + j & \text{if } i=1 \text{ and } j=1,2,\ldots,m \\ 1 & \text{if } i=2 \text{ and } j=0 \\ 1+(j+1)(t+1) & \text{if } i=2 \text{ and } j=1,2,\ldots,n. \end{cases}$$

It is not difficult to check that the vertex labeling f_1 satisfies (4.22) for $s = t+4$. Then there exists exactly one extension of the vertex labeling f_1 by edge labeling to a super EMT labeling of the disjoint union of stars $K_{1,m} \cup K_{1,n}$ with magic sum $k = s + 2(m+n+1)$. □

In [8] Ahmad et al. also study the super edge-magicness of disjoint union of stars.

4.4.2 Disjoint Union of Paths

Let $F \cong \bigcup_{j=1}^{m} P_n^j$ be a disjoint union of m paths each on n vertices, $m > 1, n \geq 4$, with $V(F) = \{u_i^j : 1 \leq i \leq n,\ 1 \leq j \leq m\}$ and $E(F) = \{u_i^j u_{i+1}^j : 1 \leq i \leq n-1,\ 1 \leq j \leq m\}$. Now, we use the vertex labeling f_1 and the edge labeling f_2 from Theorem 4.3.7 for constructing a labeling g_1 of F in the following way.

$$g_1(u_i^j) = \begin{cases} m(f_1(u_i)-1)+j & \text{if } i \text{ is odd, } 1 \leq j \leq m \\ m(f_1(u_i)-1)+\frac{m-j}{2}+1 & \text{if } i \text{ is even, } j \text{ is odd} \\ mf_1(u_i)+\frac{2-j}{2} & \text{if } i \text{ is even, } j \text{ is even} \end{cases}$$

$$g_1(u_i^j u_{i+1}^j) = \begin{cases} mf_2(u_i u_{i+1})+\frac{1-j}{2} & \text{if } j \text{ is odd, } 1 \leq i \leq n-1 \\ mf_2(u_i u_{i+1})+\frac{1-j-m}{2} & \text{if } j \text{ is even, } 1 \leq i \leq n-1. \end{cases}$$

Theorem 4.4.3 ([40]) *If m is odd, $m \geq 3$ and $n \geq 4$, then g_1 is a super EMT labeling for $F \cong \bigcup_{j=1}^{m} P_n^j$ with magic sum $k = m(2n + \lceil n/2 \rceil) + (3-m)/2$.*

Proof It is not difficult to check that if $f_1(u_i) \in \{1,2,\ldots,n\}$, then $g_1(V(F)) = \bigcup_{j=1}^{m} \bigcup_{i=1}^{n} \{mf_1(u_i)+1-j\} = \{1,2,\ldots,mn\}$ and if $f_2(u_i u_{i+1}) \in \{n+1, n+2, \ldots, 2n-1\}$, then $g_1(E(F)) = \bigcup_{j=1}^{m} \bigcup_{i=1}^{n} \{mf_2(u_i u_{i+1})+1-j\} = \{mn+1, mn+2, \ldots, 2mn-m\}$. Moreover, we can see that for every $1 \leq i \leq n-1$ and $1 \leq j \leq m$ the edge-weight $w_{g_1}(u_i^j u_{i+1}^j) = g_1(u_i^j) + g_1(u_i^j u_{i+1}^j) + g_1(u_{i+1}^j) = m(f_1(u_i) + f_2(u_i u_{i+1}) + f_1(u_{i+1})) + (3-3m)/2$. Since $f_1(u_i) + f_2(u_i u_{i+1}) +$

$f_1(u_{i+1}) = 2n + \lceil n/2 \rceil + 1$ for all $1 \leq i \leq n-1$, it follows that the labeling g_1 is a super EMT labeling with $k = m(2n + \lceil n/2 \rceil) + (3-m)/2$. □

Let us remark that Theorem 4.4.3 follows from Theorem 4.4.1. We described only one convenient total labeling g_1 which will be useful for finding a super EMT labeling for an union of path-like trees.

Notice that for the case when m is even, we know very few results. Two of them are the following.

Theorem 4.4.4 ([100]) *For every positive integer $n \equiv 1, 5$ or $7 \pmod{12}$, the 2-regular graph $G \cong 2C_n$ is EMT.*

Theorem 4.4.5 ([100]) *The forest $F \cong 2P_n$, $n > 1$, is super EMT if and only if $n \neq 2$ or 3.*

Why does the previous theorem exclude $n = 2$ and $n = 3$? In [100] it is proved that a disjoint union of two stars $K_{1,2} \cup K_{1,n}$ is super EMT if and only if n is a multiple of 3. It proves that the forest $2P_3$ is not super EMT. Moreover Kotzig and Rosa in [163] proved that the forest mP_2 is EMT if and only if m is odd.

The EMT analogue to Theorem 4.4.5 is as follows;

Theorem 4.4.6 ([100]) *The forest $F \cong 2P_n$, $n > 1$, is EMT if and only if $n \neq 2$.*

When we have a disjoint union of non-isomorphic paths, only few results are known. The next theorems generalize the result found in [99] that the forest $P_2 \cup P_m$ is super EMT for every integer $m \geq 3$.

Theorem 4.4.7 ([100]) *For every two integers $m \geq 4$ and $n \geq 1$, the forest $K_{1,n} \cup P_m$ is super EMT.*

Theorem 4.4.8 ([100]) *The forest $F \cong K_{1,n} \cup 2mP_2$, where m and n are positive integers, is super EMT. Furthermore, if $n + 2m$ and $2m + 1$ are relatively prime, then only the magic sums $2n + 9m + 4$ and $3n + 9m + 3$ are attained by the super EMT labelings of F.*

Sudarsana et al. [264] showed that $P_n \cup P_{n+1}$ is super EMT with the magic sums $5n + 2$ and $5n + 4$ for every n odd. Later they completed this result by proving the following theorems.

Theorem 4.4.9 ([265]) *Let n be odd. Graph $P_n \cup P_{n+1}$ is super EMT with the magic sum k if and only if $k = 5n + 2, 5n + 3$ or $5n + 4$.*

Theorem 4.4.10 ([265]) *For $n \geq 2$, the graph $P_n \cup P_{n+2}$ has a super EMT labeling with the magic sum $k = 5n + 6$.*

Theorem 4.4.11 ([265]) *For $n \geq 2$, the graph $P_n \cup P_{n+3}$ is super EMT with the magic sum $k = 5n + 7, 5n + 8$ or $5n + 9$.*

4.4.3 Disjoint Union of Path-Like Trees

In this part we show that a forest, in which every component is a path-like tree, has a super EMT labeling. Suppose $F \cong \bigcup_{j=1}^{m} T_j$ is the disjoint union of m path-like trees each of order n, $n \geq 4$, and they may be non-isomorphic.

Consider an embedding of a disjoint union of m paths $P_n^1 \cup P_n^2 \cup \cdots \cup P_n^m$ in the 2-dimensional grid where P_n^j is a path with vertices $V(P_n^j) = \{w_i^j : 1 \leq i \leq n\}$ and edges $E(P_n^j) = \{w_i^j w_{i+1}^j : 1 \leq i \leq n-1\}$ from which the path-like tree T_j can be obtained, for $j = 1, 2, \ldots, m$. Let $P_n^j = T_j^0, T_j^1, \ldots, T_j^{s_j} = T_j$ be the series of trees obtained by successively applying the appropriate elementary transformations of P_n^j to obtain T_j, for $j = 1, 2, \ldots, m$. Note that we allow a different series of trees $T_j^0, T_j^1, \ldots, T_j^{s_j}$ for different s_j, i.e., the forest F may be a disjoint union of different path-like trees $T_1, T_2, \ldots, T_m$, each of order n. Now, in light of Theorem 4.4.3 we are able to prove the following theorem.

Theorem 4.4.12 ([40]) *Let T_j, $1 \leq j \leq m$, be a path-like tree of order n. If m is odd, $m \geq 3$ and $n \geq 4$, then a forest $F \cong \bigcup_{j=1}^{m} T_j$ admits a super EMT labeling with magic sum $k = m(2n + \lceil n/2 \rceil) + (3 - m)/2$.*

Proof Consider an embedding of a disjoint union of m paths $\bigcup_{j=1}^{m} P_n^j$, and label the vertices of $\bigcup_{j=1}^{m} P_n^j$ using the labeling g_1 as described in Theorem 4.4.3. In order to prove the result, it suffices to show that if $u_0 v_0 = w_i^j w_{i+1}^j$, then

$$g_1(w_i^j) + g_1(w_{i+1}^j) = g_1(w_{i-t}^j) + g_1(w_{i+1+t}^j)$$

whenever w_{i-t}^j and $w_{i+1+t}^j \in V(\bigcup_{j=1}^{m} P_n^j)$.

In accordance with the parity of i, j, and t, there are 2^3 cases to consider. However, the cases are similar to each other, and hence, we will only show one case to illustrate how the proof of all cases works. Let i and j be odd, t be even. Then

$$\begin{aligned}
g_1(w_i^j) + g_1(w_{i+1}^j) &= m(f_1(w_i) - 1) + j + m(f_1(w_{i+1}) - 1) + \frac{m-j}{2} + 1 \\
&= m(f_1(w_i) + f_1(w_{i+1})) + \frac{2 - 3m + j}{2} \\
&= m(f_1(w_{i-t}) + f_1(w_{i+1+t})) + \frac{2 - 3m + j}{2} \\
&= m(f_1(w_{i-t}) - 1) + j + m(f_1(w_{i+1+t}) - 1) + \frac{m-j}{2} + 1 \\
&= g_1(w_{i-t}^j) + g_1(w_{i+1+t}^j).
\end{aligned}$$

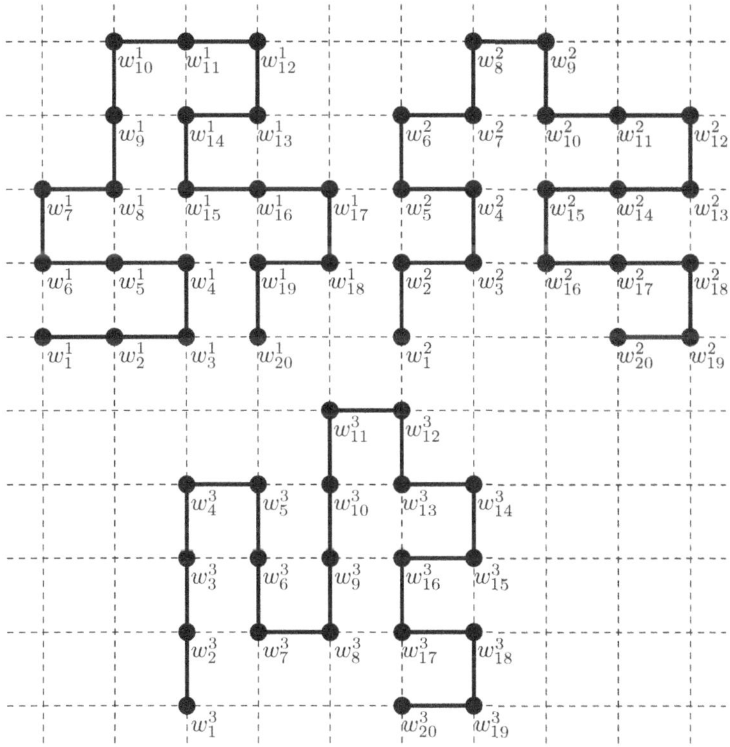

Fig. 4.17 Union of paths $\bigcup_{j=1}^{3} P_{20}^{j}$

Thus the elementary transformation keeps the super edge-magic character of the forest and the resulting labeling of the $F \cong \bigcup_{j=1}^{m} T_j$ is a super EMT with $k = m(2n + \lceil n/2 \rceil) + (3 - m)/2$. $\square$

Figure 4.17 shows an embedding of a disjoint union of three paths $P_{20}^1 \cup P_{20}^2 \cup P_{20}^3$ in the 2-dimensional grid. After a sequence of elementary transformations on every path we obtain the disjoint union of three path-like trees $\bigcup_{j=1}^{3} T_j$. Figure 4.18 depicts a vertex labeling of the union of three path-like trees, where the edge-weights form the sequence of consecutive integers, namely $33, 34, \ldots, 89$. It is easy to complete an edge labeling with values $61, 62, \ldots, 117$ such that we obtain a super EMT labeling with magic sum $k = 150$.

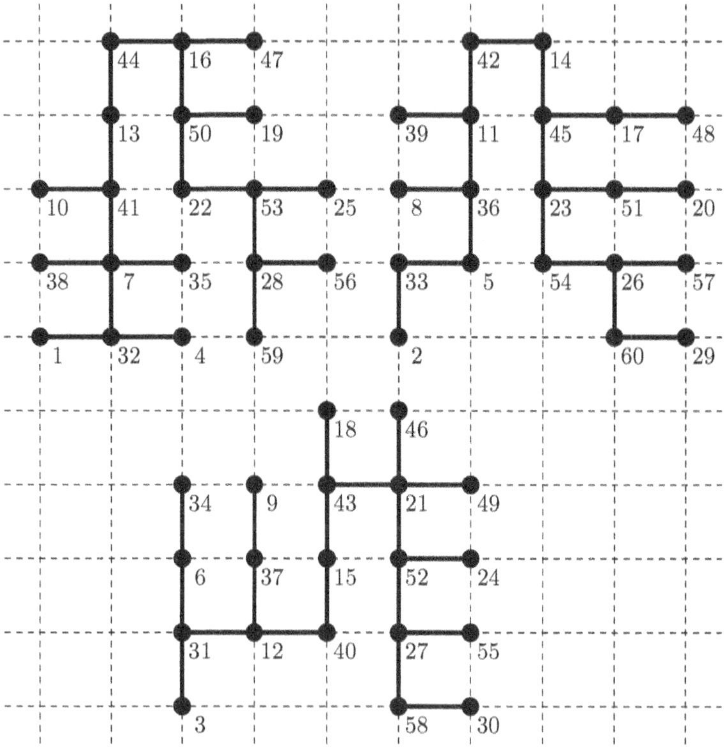

Fig. 4.18 Vertex labeling of the forest $F \cong \bigcup_{j=1}^{3} T_j$

4.5 Strong Super Edge-Magic Labeling

In this section, we introduce the concept of a strong super EMT labeling as a particular class of super EMT labelings and we use such labelings in order to show that the number of super EMT labelings of an odd union of path-like trees $(\bigcup_{j=1}^{m} T_j)$, all of them of the same order, grows at least exponentially with m.

We consider a path P_n to be a particular case of a linear forest and for a linear forest G we introduce the concept of strong super EMT labeling which is defined as follows.

Let G be a (p, q) linear forest, and assume that $f : V(G) \cup E(G) \to \{1, 2, \ldots, p+q\}$ is a super EMT labeling of G with the additional property that if $uv \in E(G)$, $u'v' \notin E(G)$ and $d_G(u, u') = d_G(v, v') < \infty$, then we have that $f(u) + f(v) = f(u') + f(v')$. From now on, we will call this property *strong*. Then, we call f a *strong super EMT labeling* of G, and we call G a strong super EMT linear forest.

For instance, for the path P_n the vertex labeling f_1 and the edge labeling f_2 from Theorem 4.3.7 combine in fact a strong super EMT labeling. Thus for n odd

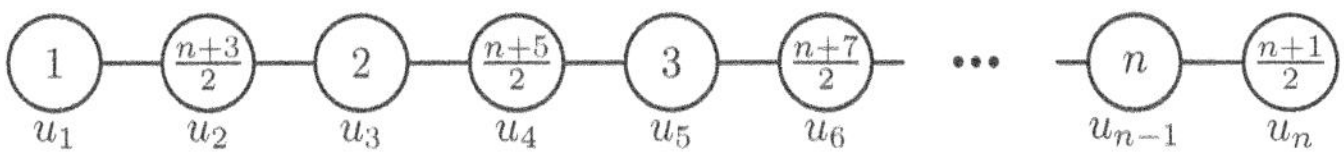

Fig. 4.19 Example of a strong super EMT labeling of P_n

Fig. 4.19 illustrates the vertex labeling f_1 of P_n where $f_1(u_3) + f_1(u_4) = f_1(u_2) + f_1(u_5) = f_1(u_1) + f_1(u_6) = (n + 9)/2$.

A *graceful labeling* of a (p, q) graph G is an injection $\phi : V(G) \to \{1, 2, \dots, q+1\}$ such that, when each edge uv is assigned the label $|\phi(u)-\phi(v)|$, the resulting edge labels (or weights) are distinct from the set $\{1, 2, \dots, q\}$. Note that, when originally defined by Rosa in [224], graceful labeling was called β-valuation and used the injection $\phi : V(G) \to \{0, 1, \dots, q\}$.

When the graceful labeling ϕ has the property that there exists an integer λ such that for each edge uv either $\phi(u) \leq \lambda < \phi(v)$ or $\phi(v) \leq \lambda < \phi(u)$, ϕ is called an α-*labeling*. More information and properties of graceful and α-labeling are given in Chap. 7.

Let $1 \leq d < n$ and let P_n be a path with $V(P_n) = \{u_i : 1 \leq i \leq n\}$ and $E(P_n) = \{u_iu_{i+1} : 1 \leq i \leq n-1\}$. Let f be an α-labeling of P_n. Then f will be called an α_d-labeling of P_n if $\min\{f(u_1), f(u_n)\} = d$.

The next lemma gives a relationship between an α_d-labeling of P_n for n odd and $d = 1, 2, 3$, and a super EMT labeling of cycle C_n.

Lemma 4.5.1 ([39]) *Let P_n be a path on n vertices, $n \geq 3$ odd. If P_n admits an α_d-labeling for $d = 1, 2, 3$, then the cycle C_n admits a super EMT labeling.*

Proof Abrham and Kotzig [4] proved that if f is an α-labeling of the path P_n, $n = 2t + 1$, and $\min\{f(u_1), f(u_n)\} \leq t$, then $f(u_1) + f(u_n) = t + 2 = \lambda + 1$. In this case the vertices with the values $> \lambda$ and the vertices with the values $\leq \lambda$ necessarily alternate.

Let f be an α_d-labeling of P_{2t+1}, for $d = 1, 2, 3$, satisfying $f(u_1) < f(u_n)$ and $f(u_1) \leq t$. Consider the following labeling of the vertices of P_{2t+1}.

$$g(u_i) = \begin{cases} f(u_i) & \text{if } i \text{ is even} \\ t + 2 - f(u_i) & \text{if } i \text{ is odd.} \end{cases}$$

Since $f(u_1) \leq t = \lambda - 1$, then the labels assigned by g to the vertices u_i, i odd, are $1, 2, \dots, \lambda - 1, \lambda$, and those assigned to the vertices u_i, i even, are $\lambda + 1, \lambda + 2, \dots, n$. Thus, g is an injection from $V(P_n)$ onto $\{1, 2, \dots, n\}$. Furthermore $\{|f(u_i) - f(u_{i+1})| : i = 1, 2, \dots, n-1\} = \{1, 2, \dots, n-1\}$ since f is an α-labeling and $\{g(u_i)+g(u_{i+1}) = t+2+|f(u_i)-f(u_{i+1})| : i = 1, 2, \dots, n-1\} = \{t + 3, t + 4, \dots, 3t + 2\}$.

If $d = 1$, then $f(u_1) = 1$, $f(u_n) = t + 1$ and $g(u_1) = t + 1$, $g(u_n) = 1$.
If $d = 2$, then $f(u_1) = 2$, $f(u_n) = t$ and $g(u_1) = t$, $g(u_n) = 2$.
If $d = 3$, then $f(u_1) = 3$, $f(u_n) = t - 1$ and $g(u_1) = t - 1$, $g(u_n) = 3$.

For each above case $g(u_1) + g(u_n) = t + 2$ and the vertex labeling g can be extended to a super EMT labeling of cycle C_n with magic sum $5t + 4$. □

Let $\mathcal{N}_d(n)$ denotes the number of α_d-labelings of P_n. The next lemma gives an exponential lower bound for the number of super EMT labelings of the cycle C_n, where n is odd.

Lemma 4.5.2 ([39]) *Let C_n be a cycle on n vertices, $n \geq 11$ odd. The number of super EMT labelings of the cycle C_n is at least* $5 \cdot 2^{\lfloor n/3 \rfloor}/4 + 1$.

Proof Abrham and Kotzig [4] proved that $\mathcal{N}_1(n) = 1$ for every $n \geq 2$, $\mathcal{N}_2(n) \geq 2^{\lfloor n/3 \rfloor}/4$ for every $n \geq 6$, and $\mathcal{N}_3(n) \geq 2^{\lfloor n/3 \rfloor}$ for every $n \geq 10$.

With respect to Lemma 4.5.1 and Abrham and Kotzig's result, we have that for every $n \geq 11$ odd, the number of super EMT labelings of the cycle C_n is at least

$$\mathcal{N}_1(n) + \mathcal{N}_2(n) + \mathcal{N}_3(n) \geq \frac{5}{4} 2^{\lfloor \frac{n}{3} \rfloor} + 1.$$ □

Let us state the next lemma, which follows immediately from the definition of the strong super EMT labeling and the properties of the set (4.5).

Lemma 4.5.3 ([39]) *A (p, q) linear forest G is strong super EMT if and only if there exists a bijective function $f : V(G) \to \{1, 2, \ldots, p\}$, such that the following conditions hold.*

(i) The set $S = \{f(u) + f(v) : uv \in E(G)\}$ consists of q consecutive integers.
(ii) If $uv \in E(G)$ and $d_G(u, u') = d_G(v, v') < \infty$ for two vertices $u', v' \in V(G)$, and $u'v' \notin E(G)$, then $f(u) + f(v) = f(u') + f(v')$.

In such a case, f can be extended to a strong super EMT labeling of G with magic sum $k = p + q + s$, where $s = \min(S)$ and

$$S = \{f(u) + f(v) : uv \in E(G)\} = \{k - (p + 1), k - (p + 2), \ldots, k - (p + q)\}.$$

Thus, due to Lemma 4.5.3, it is sufficient to exhibit the vertex labels of a strong super EMT labeling.

It is clear that if mP_n is strong super EMT for m odd, then the forest $F \cong \bigcup_{j=1}^{m} T_j$, where each T_j is a path-like tree of order n, is super EMT.

We show that the number of non-isomorphic strong super EMT labelings of the graph mP_n, for m odd and any n, grows very fast with m. This allows us to generate an exponential number of non-isomorphic super EMT labelings of the forest $F \cong \bigcup_{j=1}^{m} T_j$. We will use a technique introduced in [102], see also [134] and [177], that involves products of digraphs. In the next lines we will describe this operation on digraphs.

Let D be a digraph, $out(v)$ be the outdegree and $in(v)$ be the indegree of vertex v and let $\Gamma = \{F_1, F_2, \ldots, F_s\}$ be a family of digraphs that meet the following conditions.

(i) $V(F_i) = V$ for every $i \in \{1, 2, \ldots, s\}$.
(ii) $out(v) = in(v) = 1$ for every $v \in V(F_i)$.
(iii) F_i may contain loops. In other words, each F_i of the same order is either a cycle or union of cycles (possibly loops) such that each component has been oriented cyclically.
(iv) Each vertex of F_i takes the name of a super EMT labeling of F_i.

Consider any function $h : E(D) \to \Gamma$. Then the product $D \bigotimes_h \Gamma$ is a digraph with vertex set $V(D \bigotimes_h \Gamma) = V(D) \times V$ and

$$((a, b), (c, d)) \in E(D \bigotimes_h \Gamma) \to [(a, c) \in E(D) \wedge (b, d) \in E(h(a, c))].$$

Notice that the adjacency matrix of $D \bigotimes_h \Gamma$, denoted by $A(D \bigotimes_h \Gamma)$, is obtained by multiplying every 0 entry of $A(D)$, where $A(D)$ denotes the adjacency matrix of D, by the $|V| \times |V|$ null square matrix, and every 1 entry of $A(D)$ by $A(h(a, c))$, where $A(h(a, c))$ denotes the adjacency matrix of $h(a, c)$. Notice that when the function h is a constant, we have the classical Kronecker matrix product.

In [102] Figueroa-Centeno et al. call a digraph D super EMT if the underlying graph of the digraph D, $und(D)$, is super EMT, and they proved the following two results.

Theorem 4.5.1 ([102]) *Let D be a super EMT digraph for which each vertex takes the name of its label. Let $\Gamma = \{F_1, F_2, \ldots, F_s\}$ be a family of all super EMT 1-regular labeled digraphs (each F_i, $1 \leq i \leq s$, is either a cycle or union of cycles such that each component has been oriented cyclically) of the same odd order each, where each vertex of F_i, $1 \leq i \leq s$, takes the name of its label. Consider any function $h : E(D) \to \Gamma$. Then the digraph $D \bigotimes_h \Gamma$ is super EMT.*

Theorem 4.5.2 ([102]) *Let $\overrightarrow{T}$ be any oriented tree. Let $\Gamma = \{F_1, F_2, \ldots, F_s\}$ be a family of 1-regular digraphs of order m each. Consider any function $h : E(\overrightarrow{T}) \to \Gamma$. Then $und(\overrightarrow{T} \bigotimes_h \Gamma) = mT$.*

Next, we will describe an algorithm that will allow us to create strong super EMT labelings for the graph mP_n; $m = 2\mu + 1$; $\mu, n \in \mathbb{N}$. Then, we will illustrate the algorithm with a specific example.

Input

1. Oriented path $\overrightarrow{P_n}$ with:

- Vertex set $V(\overrightarrow{P_n}) = \{x_i\}_{i=1}^{n}$ and $E(\overrightarrow{P_n}) = \{(x_i x_{i+1})\}_{i=1}^{n-1}$
- Consider a function $f : V(\overrightarrow{P_n}) \to \{1, 2, \ldots, n\}$ defined by the rule

$$f(x_i) = \begin{cases} \frac{i+1}{2} & \text{if } i \text{ is odd} \\ \lceil \frac{n}{2} \rceil + \frac{i}{2} & \text{if } i \text{ is even.} \end{cases}$$

Observation The labeling f, which is a strong super EMT labeling of $\overrightarrow{P_n}$, could be substituted by any strong super EMT labeling of the oriented path $\overrightarrow{P_n}$.

2. The set $\Gamma_m = \{F_1, F_1', F_2, F_2', \ldots, F_{s/2}, F_{s/2}'\}$ is the family of all 1-regular digraphs where each digraph of order $m = 2\mu + 1$ is labeled in a super edge-magic way, and each vertex takes the name of its label. Each couple (F_j, F_j') comes from the same underlying 2-regular graph, but it has been oriented in an opposite way. That is to say, if a component is oriented clockwise in F_j, then the corresponding component is oriented counter clockwise in F_j', and vice versa.
3. Let (F, F') be a fixed couple from the family $\Gamma_m = \{(F_j, F_j') : j = 1, 2, \ldots, s/2\}$. Define a function $h : E(\overrightarrow{P_n}) \to \{F, F'\}$ with

$$h(x_{i-1}x_i) = \begin{cases} F & \text{if } i \text{ is even} \\ F' & \text{if } i \text{ is odd.} \end{cases}$$

Observation

$$\begin{cases} f(x_{i-1}) = x \\ f(x_i) = x' \end{cases} \to h(x_{i-1}x_i) = \begin{cases} F & \text{if } x + x' + \lceil \frac{n}{2} \rceil \text{ is even} \\ F' & \text{if } x + x' + \lceil \frac{n}{2} \rceil \text{ is odd.} \end{cases}$$

Algorithm

Step 1. Rename each vertex of $\overrightarrow{P}_n$ with the name of its label, creating a new graph $\overrightarrow{P^l}_n$.

Step 2. Compute $\overrightarrow{P^l}_n \otimes_h \Gamma_m = \overrightarrow{Q}$.

Step 3. Take $und(\overrightarrow{Q}) = Q$.

Step 4. Let $(a_i, b_i) \in V(Q)$. Relabel the vertex (a_i, b_i) with z_i where z_i is computed using the formula

$$z_i = m(a_i - 1) + b_i$$

creating the new graph Q^l.

Output $Q^l = (2\mu + 1)P_n$ labeled in a strong super edge-magic way.

Proof By Theorems 4.5.1 and 4.5.2, it is known that Q^l is super EMT and that $Q^l \cong (2K + 1)P_n$. It only remains to be shown that the obtained labeling preserves the "strong property."

Let (x, y), (x', y'), and (x'', y'') be the three vertices of Q such that

$$\left.\begin{array}{l} \{(x, y), (x', y')\} \\ \{(x', y'), (x'', y'')\} \end{array}\right\} \in E(Q).$$

If $x + x'$ is odd, then $x' + x''$ is even and vice versa. Since F and F' have opposite orientations, we obtain that $y = y''$.

Let (a_i, b_i), $1 \leq i \leq n$ be the labels of the n "consecutive" vertices of a component of Q. By the previous observation we have that $b_i = b_j$ if $|i - j|$ is even. Hence $b_i + b_{i+1} = b_{i-r} + b_{i+r+1}$, for every $i \in \{1, 2, \ldots, n - 2\}$, $r \leq \min\{i - 1, n - i - 1, 1\}$.

Now, following the notation introduced in the algorithm, we denote by z_i the vertex of Q^l that corresponds to the vertex (a_i, b_i) in Q. We want to show that

$$z_i + z_{i+1} = z_{i-r} + z_{i+1+r}.$$

Notice that

$$\begin{aligned} z_i + z_{i+1} &= m(a_i + a_{i+1} - 2) + (b_i + b_{i+1}) \\ z_{i-r} + z_{i+1+r} &= m(a_{i-r} + a_{i+1+r} - 2) + (b_{i-r} + b_{i+1+r}) \end{aligned}$$

- $b_i + b_{i+1} = b_{i-r} + b_{i+1+r}$ by the previous argument
- $a_i + a_{i+1} = a_{i-r} + a_{i+1+r}$ since these are the same labels that we used in the oriented path $\overrightarrow{P}^l_n$, and the labeling of $\overrightarrow{P}^l_n$ is a strong super EMT labeling.

Therefore,

$$z_i + z_{i+1} = z_{i-r} + z_{i+1+r}.$$

□

In the following example we use the previous algorithm in order to obtain a strong super EMT labeling of $5P_6$. Let $\overrightarrow{P}_6$ be the following digraph, where each vertex of $\overrightarrow{P_6}$ takes the name of the strong super EMT labeling described in the algorithm, see Fig. 4.20. Figure 4.21 depicts the adjacency matrix of the digraph $\overrightarrow{P_6}$.

Let $\Gamma_5 = \{F_1, F'_1, F_2, F'_2, F_3, F'_3\}$ be the family of all super EMT 1-regular digraphs of order 5 with each component oriented cyclically and each vertex of each digraph taking the name of the label assigned by the super EMT labeling. Figures 4.22, 4.23, 4.24, 4.25, 4.26, 4.27, 4.28, 4.29, 4.30, 4.31, 4.32, and 4.33 illustrate the digraphs and their corresponding adjacency matrices.

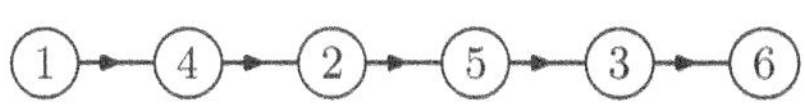

Fig. 4.20 Example of a vertex labeling of the digraph $\overrightarrow{P_6}$

Fig. 4.21 Adjacency matrix of the digraph $\overrightarrow{P_6}$

$A(\overrightarrow{P_6}) =$

	1	2	3	4	5	6
1	0	0	0	1	0	0
2	0	0	0	0	1	0
3	0	0	0	0	0	1
4	0	1	0	0	0	0
5	0	0	1	0	0	0
6	0	0	0	0	0	0

Fig. 4.22 Digraph F_1

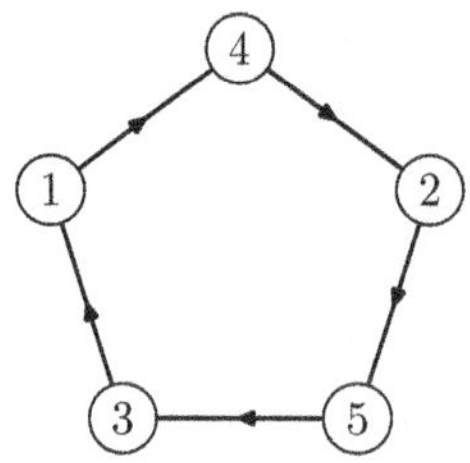

Fig. 4.23 Adjacency matrix of digraph F_1

$A(F_1) =$

	1	2	3	4	5
1	0	0	0	1	0
2	0	0	0	0	1
3	1	0	0	0	0
4	0	1	0	0	0
5	0	0	1	0	0

Fig. 4.24 Digraph F_1'

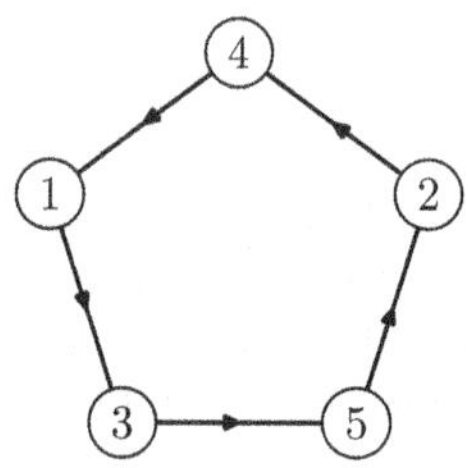

Fig. 4.25 Adjacency matrix of digraph F_1'

$A(F_1') =$

	1	2	3	4	5
1	0	0	1	0	0
2	0	0	0	1	0
3	0	0	0	0	1
4	1	0	0	0	0
5	0	1	0	0	0

Fig. 4.26 Digraph F_2

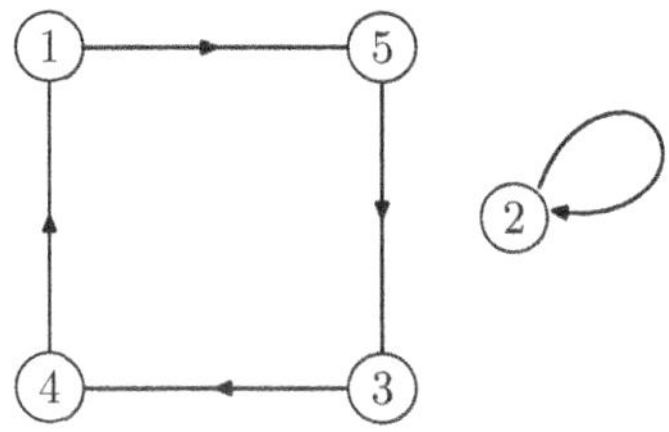

Fig. 4.27 Adjacency matrix of digraph F_2

$A(F_2) =$

	1	2	3	4	5
1	0	0	0	0	1
2	0	1	0	0	0
3	0	0	0	1	0
4	1	0	0	0	0
5	0	0	1	0	0

Fig. 4.28 Digraph F_2'

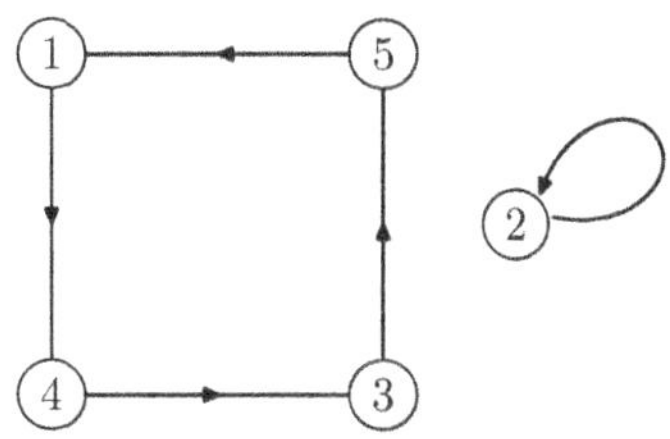

Fig. 4.29 Adjacency matrix of digraph F_2'

$A(F_2') =$

	1	2	3	4	5
1	0	0	0	1	0
2	0	1	0	0	0
3	0	0	0	0	1
4	0	0	1	0	0
5	1	0	0	0	0

Fig. 4.30 Digraph F_3

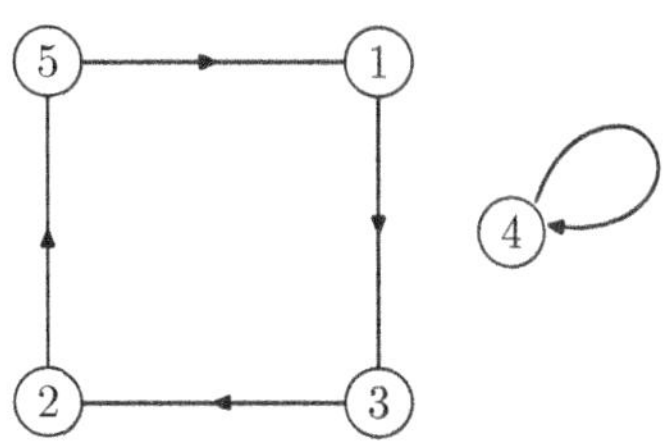

Fig. 4.31 Adjacency matrix of digraph F_3

$A(F_3) =$

	1	2	3	4	5
1	0	0	1	0	0
2	0	0	0	0	1
3	0	1	0	0	0
4	0	0	0	1	0
5	1	0	0	0	0

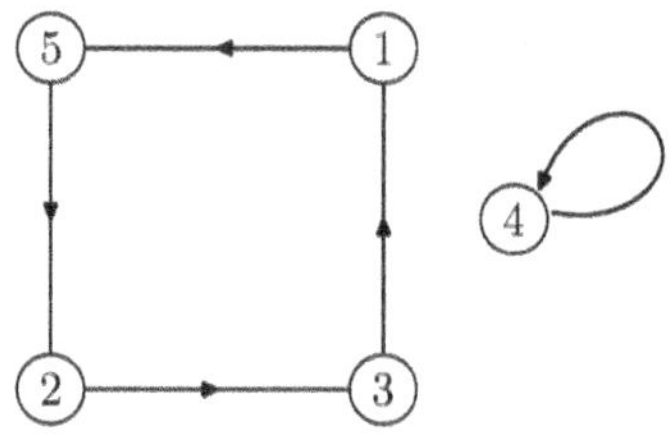

Fig. 4.32 Digraph F_3'

Fig. 4.33 Adjacency matrix of digraph F_3'

$A(F_3') =$

	1	2	3	4	5
1	0	0	0	0	1
2	0	0	1	0	0
3	1	0	0	0	0
4	0	0	0	1	0
5	0	1	0	0	0

Now, define the function $h : E(\overrightarrow{P_6}) \to \{F_1, F_1'\}$ such that for every edge $xx' \in E(\overrightarrow{P}_6)$, we assign

$$h(x, x') = \begin{cases} F_1 & \text{if } x + x' \equiv 1 \pmod 2 \\ F_1' & \text{if } x + x' \equiv 0 \pmod 2). \end{cases}$$

The adjacency matrix of $\overrightarrow{P_6} \bigotimes_h \{F_1, F_1'\}$ is obtained by multiplying every 0 entry of $A(\overrightarrow{P_6})$ by the 5×5 null square matrix and every 1 entry of $A(\overrightarrow{P_6})$ by $A(F_1)$ or $A(F_1')$, see Fig. 4.34. The underlying graph of $\overrightarrow{P_6} \bigotimes_h \{F_1, F_1'\}$ is isomorphic to $5P_6$. The adjacency matrix $A(\overrightarrow{P_6} \bigotimes_h \{F_1, F_1'\})$ describes the corresponding vertex labeling of a strong super EMT labeling of $5P_6$, see Fig. 4.35.

Let m be an odd positive integer, $m \geq 3$, and denote by $\mathcal{N}(m)$ the number of non-isomorphic strong super EMT labelings of the graph mP_n, $n \geq 4$. The next theorem gives an exponential lower bound for $\mathcal{N}(m)$.

Theorem 4.5.3 ([39]) *Let $m \geq 5$ be an odd integer. Then*

$$\mathcal{N}(m) \geq \frac{5}{2} 2^{\lfloor \frac{m}{3} \rfloor} + 1.$$

$A(\overrightarrow{P_6} \otimes_h \{F_1, F_1'\}) =$

	$1 \dots 5$	$6 \dots 10$	$11 \dots 15$	$16 \dots 20$	$21 \dots 25$	$26 \dots 30$
$1 \dots 5$	0	0	0	$A(F_1)$	0	0
$6 \dots 10$	0	0	0	0	$A(F_1)$	0
$11 \dots 15$	0	0	0	0	0	$A(F_1)$
$16 \dots 20$	0	$A(F_1')$	0	0	0	0
$21 \dots 25$	0	0	$A(F_1')$	0	0	0
$26 \dots 30$	0	0	0	0	0	0

Fig. 4.34 Adjacency matrix of $\overrightarrow{P_6} \bigotimes_h \{F_1, F_1'\}$

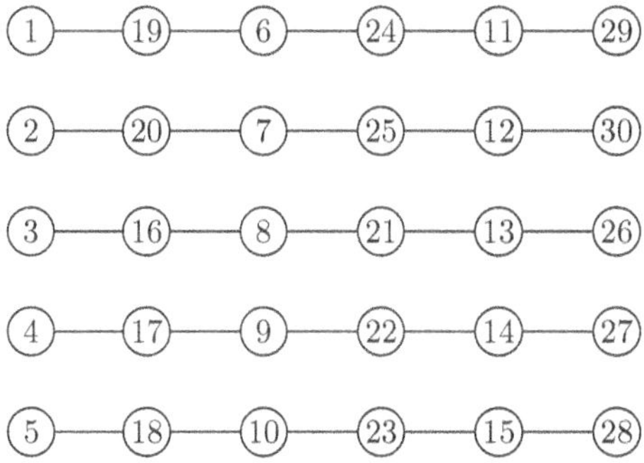

Fig. 4.35 Corresponding vertex labeling of a strong super EMT labeling of $5P_6$

Proof In [102] it was shown that if $h : E(D) \to \Gamma$ and $h' : E(D) \to \Gamma$ are two different functions, then the labelings obtained from the product are non-isomorphic.

Consider a strong super EMT labeling of P_n, $n \geq 4$, and $\Gamma_m = \{F_1, F_1', F_2, F_2', \dots, F_{s/2}, F_{s/2}'\}$ as a family of super EMT 1-regular digraphs of order m. The family Γ_m consists of $s/2$ couples of the same super EMT labeled 1-regular digraphs (F_j, F_j'), $j = 1, 2, \dots, s/2$, but with opposite orientations. With respect to the previous algorithm for generating strong super EMT labelings of the union of paths

$m P_n$, there are s different functions $h_j : E(D) \to \{F_j, F_j'\}$, for $j = 1, 2, \ldots, s/2$ (each couple (F_j, F_j') has two possible orientations), and also s non-isomorphic strong super EMT labelings of $m P_n$.

It remains to investigate how many different couples (F_j, F_j') contain the family Γ_m. Let us distinguish the following four cases, according to the order m.

Case A: $m = 5$ We have three couples (F_1, F_1'), (F_2, F_2'), and (F_3, F_3'), see Figs. 4.22, 4.23, 4.24, 4.25, 4.26, 4.27, 4.28, 4.29, 4.30, 4.31, and 4.32. With respect to the two possible orientations of each couple we can see that in this case the lower bound is tight.

Case B: $m = 7$ There are at least 14 couples. They are described in [102]. Let us rewrite them in Table 4.1.

Case C: $m = 9$ There are at least 39 couples. Table 4.2 shows these super EMT 2−regular graphs of order 9, where each component has been oriented cyclically.

Case D: $m \geq 11$ If we consider only super EMT cyclically oriented cycles of order m, as elements of the family Γ_m, then from Lemma 4.5.2 it follows that, for $m \geq 11$, there exist at least $5 \cdot 2^{\lfloor m/3 \rfloor}/4 + 1$ couples, where each couple comes from the same super EMT labeled cycle but with opposite orientations.

Since in the last three cases each couple of oriented cycles has two possible orientations, there are at least $5 \cdot 2^{\lfloor m/3 \rfloor}/2 + 1$ non-isomorphic strong super EMT labelings of the graph $m P_n$. □

According to the previous two cases, for $m = 7$ and 9, we can observe that there exist also super EMT 1-regular disconnected graphs of order m. It means that, in reality, the lower bound of Theorem 4.5.3 is bigger.

Table 4.1 Super EMT 2-regular graphs of order 7

Graph	Vertex labeling	Number of possible orientations
$C_5 \cup C_1 \cup C_1$	$1-4-7-2-6 \cup 3 \cup 5$	2
$C_6 \cup C_1$	$1-6-3-2-4-7 \cup 5$	2
$C_6 \cup C_1$	$1-4-6-5-2-7 \cup 3$	2
$C_3 \cup C_3 \cup C_1$	$1-5-6 \cup 2-3-7 \cup 4$	4
C_7	$1-5-2-6-3-7-4$	2
C_7	$1-6-5-3-7-2-4$	2
C_7	$1-7-3-6-5-2-4$	2
C_7	$1-4-3-7-2-6-5$	2
C_7	$1-7-2-3-4-6-5$	2
C_7	$1-6-4-7-2-3-5$	2
C_7	$1-6-2-3-7-4-5$	2
C_7	$1-5-2-3-6-4-7$	2
C_7	$1-6-5-4-2-3-7$	2

Table 4.2 Super EMT 2-regular graphs of order 9

Graph	Vertex labeling	Number of possible orientations
$C_3 \cup C_3 \cup C_3$	$1-5-9 \cup 2-6-7 \cup 3-4-8$	8
$C_3 \cup C_3 \cup C_3$	$1-6-8 \cup 2-4-9 \cup 3-5-7$	8
$C_5 \cup C_4$	$1-5-2-6-8 \cup 3-7-4-9$	4
$C_5 \cup C_4$	$2-9-5-8-4 \cup 1-6-3-7$	4
$C_5 \cup C_3 \cup C_1$	$2-9-5-4-8 \cup 1-6-7 \cup 3$	4
$C_5 \cup C_3 \cup C_1$	$1-7-6-4-8 \cup 2-5-9 \cup 3$	4
$C_5 \cup C_3 \cup C_1$	$1-5-6-2-8 \cup 3-4-9 \cup 7$	4
$C_5 \cup C_3 \cup C_1$	$2-6-4-3-9 \cup 1-5-8 \cup 7$	4
$C_6 \cup C_1 \cup C_1 \cup C_1$	$1-6-2-9-4-8 \cup 3 \cup 5 \cup 7$	2
$C_7 \cup C_1 \cup C_1$	$1-7-4-9-5-2-8 \cup 3 \cup 6$	2
$C_7 \cup C_1 \cup C_1$	$1-5-8-2-9-3-6 \cup 4 \cup 7$	2
$C_7 \cup C_1 \cup C_1$	$1-6-8-4-9-2-7 \cup 3 \cup 5$	2
$C_7 \cup C_1 \cup C_1$	$1-8-3-9-4-2-6 \cup 5 \cup 7$	2
$C_7 \cup C_1 \cup C_1$	$1-9-2-5-7-6-8 \cup 3 \cup 4$	2
$C_7 \cup C_1 \cup C_1$	$1-8-5-3-4-2-9 \cup 6 \cup 7$	2
$C_8 \cup C_1$	$1-9-4-7-5-2-6-8 \cup 3$	2
$C_8 \cup C_1$	$1-5-2-7-6-8-3-9 \cup 4$	2
$C_8 \cup C_1$	$1-7-2-4-9-3-8-6 \cup 5$	2
$C_8 \cup C_1$	$1-9-5-8-3-4-2-7 \cup 6$	2
$C_8 \cup C_1$	$1-6-3-5-8-4-2-9 \cup 7$	2
C_9	$1-6-2-7-3-8-4-9-5$	2
C_9	$1-5-7-2-6-8-3-4-9$	2
C_9	$1-5-9-2-6-7-3-4-8$	2
C_9	$1-5-7-3-4-9-2-6-8$	2
C_9	$1-6-7-3-5-9-2-4-8$	2
C_9	$1-6-7-2-4-8-3-5-9$	2
C_9	$1-6-8-3-5-7-2-4-9$	2

Theorem 4.5.4 ([39]) *Let P_n be strong super EMT for $n \geq 4$. If m is odd, $m \geq 5$, then a forest $F \cong \bigcup_{j=1}^{m} T_j$, where each T_j is a path-like tree of order n, admits at least $5 \cdot 2^{\lfloor m/3 \rfloor}/2 + 1$ non-isomorphic super EMT labelings.*

Proof From Theorem 4.5.3 it follows that if P_n admits a strong super EMT labeling, then there are at least $5 \cdot 2^{\lfloor m/3 \rfloor}/2 + 1$ non-isomorphic strong super EMT labelings of the disjoint union of paths $\bigcup_{j=1}^{m} P_j$ for m odd.

Consider an embedding of the disjoint union of paths $P_1, P_2, \ldots, P_m$ in the 2-dimensional grid. Let $P_j = T_j^0, T_j^1, \ldots, T_j^{s_j} = T_j$ be the series of trees obtained by successively applying the appropriate elementary transformations of P_j to obtain T_j, for $j = 1, 2, \ldots, m$, which keeps the super edge-magic character of the path P_j.

There are different series of trees $T_j^0, T_j^1, \ldots, T_j^{s_j}$ for different s_j, i.e., the forest F is a disjoint union of different path-like trees $T_1, T_2, \ldots, T_m$, each of order n. For each strong super EMT labeling of $\bigcup_{j=1}^{m} P_j$, m odd, there exists a super EMT labeling of the forest $F \cong \bigcup_{j=1}^{m} T_j$.

Thus, the forest $F \cong \bigcup_{j=1}^{m} T_j$ admits at least $5 \cdot 2^{\lfloor m/3 \rfloor}/2 + 1$ non-isomorphic super EMT labelings. □

Ahmad et al.[9] studied the super edge-magicness of odd union of non-necessarily isomorphic acyclic graphs. They found exponential lower bounds for the number of super EMT labelings of these unions.

Figure 4.36 depicts vertex labelings for two different disjoint unions of five path-like trees obtained by applying the appropriate elementary transformations on $5P_6$ which can be completed by edge labelings to super EMT labelings.

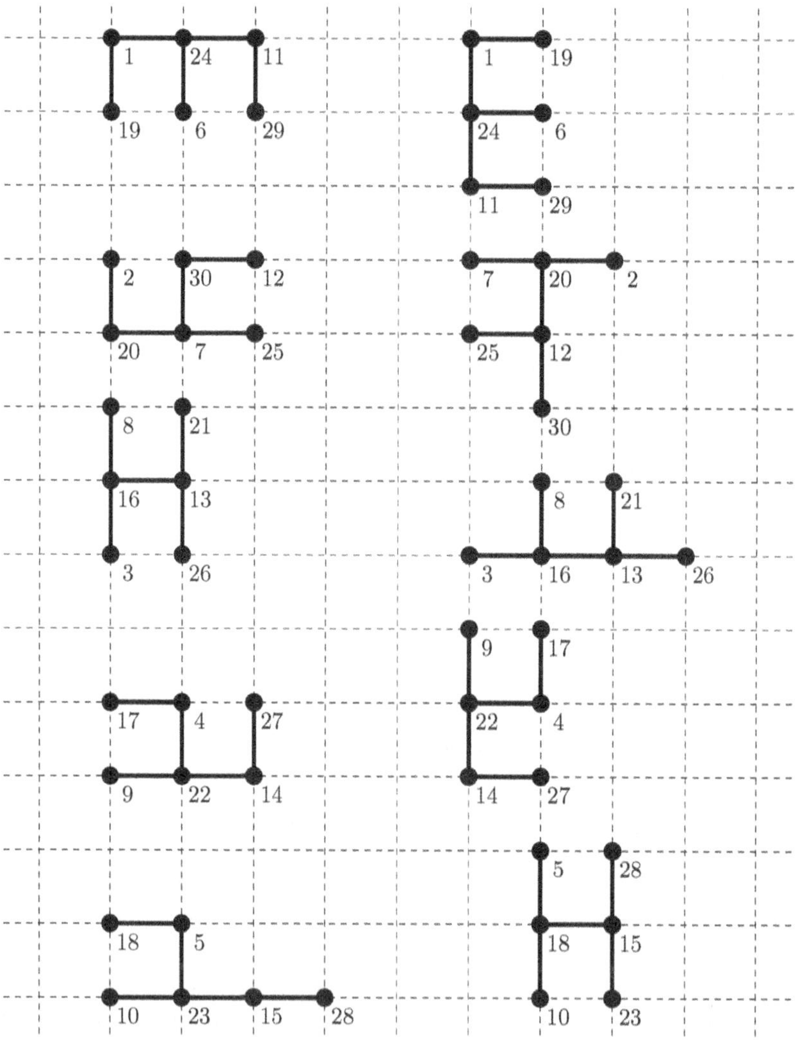

Fig. 4.36 Vertex labeling of super EMT labelings of $\bigcup_{j=1}^{5} T_j$

At this point we do not know anything in general about the existence of strong super EMT labelings for the graph $G \cong (2m)P_n$, except for the fact that $(2m)P_2$ is not super EMT. It is very interesting to know the super EMT properties of an even union of path-like trees. In [39] Bača, Lin, and Muntaner-Batle posed the following open problem.

Open Problem 4.5.1 ([39]) Let $G \cong (2m)P_n$, $n \neq 2$, $m \geq 1$. Is G a strong super EMT?

If the answer to Open Problem 4.5.1 is yes, then it leads to the following.

Open Problem 4.5.2 ([39]) Let $G \cong (2m)P_n$, $n \neq 2$, $m \geq 1$. How many non-isomorphic strong super EMT labelings does G admit?

Open Problem 4.5.3 ([39]) Let $G \cong \bigcup_{j=1}^{2m} T_j$ be a disjoint union of an even number of path-like trees, all of them of the same order, and such that $T_j \neq P_2$ for $j = 1, 2, \ldots, 2m$. Is G a super EMT graph?

4.6 Relationships Super Edge-Magic Total Labelings with Other Labelings

In this section we exhibit the relationships between super EMT labelings and other well-studied classes of labelings, namely sequential, harmonious, and cordial labelings.

The definition of sequential labeling was introduced by Grace [118]. A *sequential labeling* of a (p, q) graph G is an injective function $f : V(G) \to \{0, 1, \ldots, q-1\}$, with the label q allowed if G is a tree, such that the induced edge labeling given by $f(uv) = f(u) + f(v)$ has the property that

$$\{f(uv) : uv \in E(G)\} = \{m, m+1, \ldots, m+q-1\},$$

for some integer m. Graph G is said to be *sequential* if such a labeling exists.

Next theorem gives a connection between super EMT labeling and sequential labeling.

Theorem 4.6.1 ([98]) *If a (p, q) graph G that is a tree or where $q \geq p$ is super EMT, then G is sequential.*

Proof Suppose that $f : V(G) \cup E(G) \to \{1, 2, \ldots, p+q\}$ is a super EMT labeling of a (p, q) graph G with magic sum k. Then from (4.5) it follows that $S = \{f(u) + f(v) : uv \in E(G)\} = \{s, s+1, \ldots, s+q-1\}$, where $s = k - (p+q)$.

Define the injective vertex function $g : V(G) \to \{0, 1, \ldots, p-1\}$ such that

$$g(v) = f(v) - 1,$$

for each vertex $v \in V(G)$.

Thus,

$$\{g(u) + g(v) = (f(u) - 1) + (f(v) - 1) : uv \in E(G)\}$$
$$= \{t, t + 1, \ldots, t + q - 1\},$$

where $t = k - (p + q) - 2$. This implies that the injection g is a sequential labeling of G. □

Harmonious graphs naturally arose in the study of modular version of error-correcting codes and channel assignment problems. Graham and Sloane [119] defined a (p, q) graph G to be *harmonious* if there is an injective function $f : V(G) \rightarrow \mathbb{Z}_q$, where $\mathbb{Z}_q$ is the group of integers modulo q, such that the induced function $f^* : E(G) \rightarrow \mathbb{Z}_q$, defined by

$$f^*(uv) = f(u) + f(v)$$

for each edge $uv \in E(G)$, is a bijection.

The function f is called a *harmonious labeling* and the image of f denoted by $Im(f)$ is called the corresponding set of vertex labels. When G is a tree or in general for a graph G with $p = q + 1$, exactly one label may be used on two vertices.

Grace [118] showed that sequential (p, q) graphs with $q \geq p$ are harmonious. According to Grace's result and Theorem 4.6.1 we have

Theorem 4.6.2 ([98]) *If a (p, q) graph G with $q \geq p$ is super EMT, then G is harmonious.*

Theorem 4.6.2 can be extended to trees if we reduce the edge labels in f^* modulo $p - 1$. Thus we have

Theorem 4.6.3 ([98]) *If a tree T of order p is super EMT, then T is harmonious.*

This theorem implies that Conjecture 4.1.2 is at least as hard as the following conjecture of Graham and Sloane.

Conjecture 4.6.1 ([119]) All trees are harmonious.

The relationships between super EMT, harmonious, and sequential labelings of certain 2-regular graphs are investigated in [101].

Cahit [76] has introduced cordial labeling as a variation of both graceful and harmonious labelings. A *cordial labeling* of G is a function $f : V(G) \rightarrow \mathbb{Z}_2$ with an induced edge labeling

$$f(uv) \equiv f(u) - f(v) \pmod 2$$

such that if $v_f(i)$ and $e_f(i)$ are the number of vertices v and edges e satisfying that $f(v) = i$ and $f(e) = i$ for all $i \in \mathbb{Z}_2$, respectively, then

$$|v_f(0) - v_f(1)| \leq 1$$

and

$$|e_f(0) - e_f(1)| \leq 1.$$

A graph that admits a cordial labeling is said to be *cordial*. Cahit [77] proved that every tree is cordial, the complete graph K_n is cordial if and only if $n \leq 3$, the complete bipartite graph $K_{m,n}$ is cordial for all m and n, the friendship graph f_n is cordial if and only if $n \not\equiv 2 \pmod 4$, all fans are cordial, the wheel W_n is cordial if and only if $n \not\equiv 3 \pmod 4$. The relationship between super EMT labeling and cordial labeling gives the next theorem.

Theorem 4.6.4 ([98]) *If a graph G is super EMT, then G is cordial.*

Proof Assume that a graph G admits a super EMT labeling f. Define the function $g : V(G) \cup E(G) \to \mathbb{Z}_2$ in the following way.

$$\begin{aligned} g(v) &\equiv f(v) \pmod 2, && \text{for every vertex } v \in V(G) \\ g(uv) &\equiv g(u) - g(v) \pmod 2, && \text{for every edge } e \in E(G). \end{aligned}$$

Clearly,

$$g(uv) \equiv g(u) - g(v) \equiv g(u) + g(v) \equiv f(u) + f(v) \pmod 2.$$

Since f is super EMT labeling, then $f(V(G))$ and $S = \{f(u)+f(v) : uv \in E(G)\}$ are sets of consecutive integers and this implies that

$$|v_g(0) - v_g(1)| \leq 1$$

and

$$|e_g(0) - e_g(1)| \leq 1. \qquad \square$$

Chapter 5
Vertex-Antimagic Total Labelings

5.1 Vertex-Antimagic Edge Labeling

Let us recall that for an edge labeling $g : E(G) \to \{1, 2, \dots, q\}$ of a (p, q) graph G, the associated vertex-weight of a vertex $v \in V(G)$ is

$$w_g(v) = \sum_{u \in N(v)} g(vu).$$

By an (a, d)*-vertex-antimagic edge* (VAE) *labeling* of a graph G we mean a one-to-one mapping from $E(G)$ onto $\{1, 2, \dots, q\}$ such that the set of all vertex-weights in G is $\{a, a+d, \dots, a+(p-1)d\}$, where $a > 0$ and $d \geq 0$ are two fixed integers.

The (a, d)-VAE labeling was originally defined by Bodendiek and Walther [68] who called it (a, d)-antimagic labeling. This labeling is a special case of the more general vertex-antimagic labeling introduced by Hartsfield and Ringel in [125]. For more details on vertex-antimagic labeling see [19, 43, 191, 192, 193, 208, 209, 226, 290, 291]. Bodendiek and Walther in [70] and [71] proved that the Herschel graph is not (a, d)-VAE and obtained both positive and negative results about (a, d)-VAE labelings for various cases of graphs called *parachutes* $P_{\alpha,\beta}$. Note that $P_{\alpha,\beta}$ is the graph obtained from the wheel $W_{\alpha+\beta}$ by deleting β consecutive spokes.

Furthermore, (a, d)-VAE labelings for some classes of graphs (for example, paths, cycles, and complete graphs) are described in [69, 72] and [154].

Characterization of all (a, d)-VAE graphs of the prism $C_n \square P_2$ when n is even is given in [32] from which we get the following results.

Theorem 5.1.1 ([32]) *Let the prism* $C_n \square P_2$ *be* (a, d)*-VAE.*

(i) If n *is even, then either* $d = 1$ *and* $a = (7n+4)/2$ *or* $d = 3$ *and* $a = (3n+6)/2$.
(ii) If n *is odd, then either* $d = 2$ *and* $a = (5n+5)/2$ *or* $d = 4$ *and* $a = (n+7)/2$.

M. Bača et al., *Magic and Antimagic Graphs*, Developments in Mathematics 60,
https://doi.org/10.1007/978-3-030-24582-5_5

Theorem 5.1.2 ([32]) *If n is even, $n \geq 4$, then the prism $C_n \square P_2$ has a $((7n+4)/2, 1)$-VAE labeling and a $((3n+6)/2, 3)$-VAE labeling.*

Theorem 5.1.3 ([32]) *If n is odd, $n \geq 3$, then the prism $C_n \square P_2$ has a $((5n+5)/2, 2)$-VAE labeling.*

In [32] the authors also show that the prism $C_5 \square P_2$ is not $(6, 4)$-VAE and the prisms $C_n \square P_2$, for $n = 7, 9, 11$ admit $(6, 4)$-VAE labelings. This prompted the authors to propose the following conjecture.

Conjecture 5.1.1 ([32]) If n is odd, $n \geq 7$, then the prism $C_n \square P_2$ is $((n+7)/2, 4)$-VAE.

In [22] it is proved that (a, d)-VAE labelings of antiprisms do not exist for all values of (a, d) other than $(6n+3, 2)$, $(4n+4, 4)$, and $(2n+5, 6)$. Moreover, $(6n+3, 2)$-VAE and $(4n+4, 4)$-VAE labelings of antiprism A_n, for $n \geq 3$, $n \not\equiv 2 \pmod 4$, are given. Miller et al. in [188] have proved that every antiprism A_n is $(6n+3, 2)$-VAE and $(4n+4, 4)$-VAE. They also considered $(2n+5, 6)$-VAE labelings of antiprism and showed that A_3 does not have $(11, 6)$-VAE labeling but there exist $(2n+5, 6)$-VAE labelings for A_4 and A_7. They posed the following conjecture.

Conjecture 5.1.2 ([188]) For $n \geq 4$, the antiprism A_n has a $(2n+5, 6)$-VAE labeling.

Nicholas et al. [205] obtained some results about (a, d)-VAE labelings for special trees (caterpillars), unicyclic graphs, and complete bipartite graphs. They suggest the following conjecture.

Conjecture 5.1.3 ([205]) For n odd, $n \geq 3$, $K_{n,n+2}$ is $((n+1)(n^2-1)/2, n+1)$-VAE.

Next we describe a technique that allows us to construct several (a, r)-VAE labelings for any $2r$-regular graph G of odd order provided the graph is Hamiltonian or has a 2-regular factor that is $(b, 1)$-VAE. A similar technique allows us to construct a super (a, d)-VAT labeling for any $2r$-regular Hamiltonian graph of odd order with differences $d = 1, 2, \ldots, r$ and $d = 2r + 2$.

Let H be a subgraph of a graph G. By $G - H$ we understand the maximal factor of G containing no edge of H.

Theorem 5.1.4 ([6]) *Let G be a $2r$-regular Hamiltonian graph of odd order p. Then G is (a, r)-VAE for $a = pr^2 - (p-1)r/2 + r$.*

Proof Let G be a $2r$-regular Hamiltonian graph of order p, p odd. By F_1 we denote a Hamiltonian cycle in G. As the graph $G - F_1$ is even regular, then, using Theorem 3.1.23, we construct a 2-factorization of $G - F_1$. We denote the 2-factors by F_j, $j = 2, 3, \ldots, r$. Thus $V(G) = V(F_j)$ for all j, $j = 1, 2, \ldots, r$ and $E(G) = \bigcup_{j=1}^{r} E(F_j)$. We denote the vertices of G by $v_1, v_2, \ldots, v_p$ in such a way that $v_1 v_2 \ldots v_p v_1$ is the Hamiltonian cycle F_1. Each factor F_j is a collection of cycles. We order and orient the cycles arbitrarily. By the symbol $e_j^{out}(v_i)$ we denote the unique outgoing arc from the vertex v_i in the factor F_j and by the symbol $e_j^{in}(v_i)$ we denote the unique incoming arc to v_i in the factor F_j. Note that each edge is denoted by two symbols.

Let α be the following edge labeling of G. First we label the edges of F_1.

$$\alpha(v_i v_{i+1}) = \begin{cases} 1 + \frac{(i-1)r}{2} & \text{if } i = 1, 3, \ldots, p-2 \\ 1 + \frac{(p-1+i)r}{2} & \text{if } i = 2, 4, \ldots, p-1 \end{cases}$$

$$\alpha(v_p v_1) = 1 + \frac{(p-1)r}{2}.$$

Notice that, since p is odd, this gives at the same time an (a, r)-VAE labeling of F_1. We label the edges of F_2 as

$$\alpha(e_2^{out}(v_i)) = \frac{(3p-3)r}{2} + 4 - \alpha(v_{i-1} v_i) - \alpha(v_i v_{i+1}),$$

for every $i = 1, 2, \ldots, p$ and

$$\alpha(e_j^{out}(v_i)) = (p-1)r + 2j - 1 - \alpha(e_{j-1}^{in}(v_i)),$$

for every $j = 3, 4, \ldots, r$ and $i = 1, 2, \ldots, p$. It is easy to see that the set of the edge labels in each factor F_j is

$$\{\alpha(e) \colon e \in E(F_j)\} = \{j, r+j, 2r+j, \ldots, (p-1)r+j\}. \tag{5.1}$$

Thus, α is a bijection $E(G) \to \{1, 2, \ldots, pr\}$. The vertex-weight of vertex v_i, $i = 1, 2, \ldots, p$, in α is

$$\begin{aligned} w_\alpha(v_i) = &(\alpha(v_{i-1} v_i) + \alpha(v_i v_{i+1})) + \left(\alpha(e_2^{out}(v_i)) + \alpha(e_2^{in}(v_i))\right) \\ &+ \left(\alpha(e_3^{out}(v_i)) + \alpha(e_3^{in}(v_i))\right) + \cdots + \left(\alpha(e_r^{out}(v_i)) + \alpha(e_r^{in}(v_i))\right) \\ = &\alpha(v_{i-1} v_i) + \alpha(v_i v_{i+1}) \\ &+ \left(\frac{(3p-3)r}{2} + 4 - \alpha(v_{i-1} v_i) - \alpha(v_i v_{i+1}) + \alpha(e_2^{in}(v_i))\right) \\ &+ \left((p-1)r + 5 - \alpha(e_2^{in}(v_i)) + \alpha(e_3^{in}(v_i))\right) \\ &+ \left((p-1)r + 7 - \alpha(e_3^{in}(v_i)) + \alpha(e_4^{in}(v_i))\right) + \cdots \end{aligned}$$

$$+\Big((p-1)r+2r-1-\alpha(e_{r-1}^{in}(v_i))+\alpha(e_r^{in}(v_i))\Big)$$
$$=pr^2-\frac{(p-1)r}{2}+\alpha(e_r^{in}(v_i)).$$

Using (5.1), we observe that

$$\Big\{\alpha(e_r^{in}(v_i))\colon i=1,2,\ldots,p\Big\}=\big\{\alpha(e_r^{out}(v_i))\colon i=1,2,\ldots,p\big\}=\{r,2r,\ldots,pr\}.$$

Thus, the vertex-weights $w_\alpha(v_i)$ form the set

$$\left\{pr^2-\frac{(p-1)r}{2}+r,\, pr^2-\frac{(p-1)r}{2}+2r,\ldots,pr^2-\frac{(p-1)r}{2}+pr\right\},$$

which is an arithmetic progression with the difference r. We conclude that the labeling α is a $(pr^2-(p-1)r/2+r, r)$-VAE labeling of G. □

In the previous theorem we used the fact that every Hamiltonian cycle on an odd number of vertices admits a $(b, 1)$-VAE labeling. In fact the graph G does not need to be Hamiltonian, it can contain any 2-regular factor with a $(b, 1)$-VAE labeling. Such graphs exist, see Fig. 5.1, where integers in italic font denote vertex-weights. We claim the following

Theorem 5.1.5 ([6]) *Let G be a $2r$-regular graph of odd order that contains a 2-regular factor that admits a $(b, 1)$-VAE labeling. Then G is (a, r)-VAE for some integer a.*

The proof is almost identical to the proof of Theorem 5.1.4. On the other hand, we cannot lift the requirement of G being of odd order. To make use of the construction above, G has to be of odd order since there do not exist any $(a, 1)$-VAE Hamiltonian cycle F_1 for p even. The following theorem shows that there exists no even regular $(a, 1)$-VAE graph on an even number of vertices.

Theorem 5.1.6 ([6]) *If G is an even regular (a, d)-VAE graph on an even number of vertices, then d is even.*

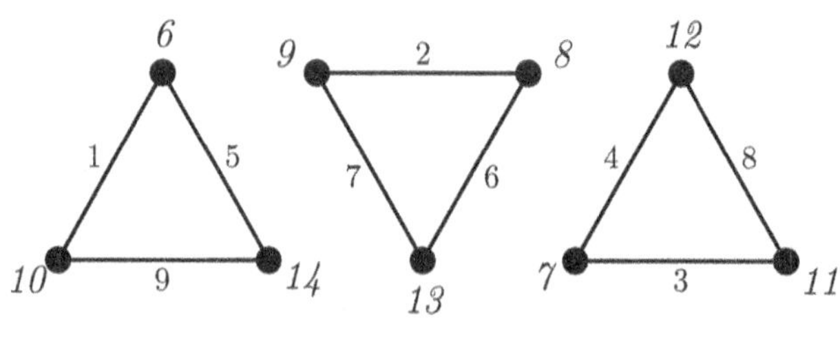

Fig. 5.1 A (6, 1)-VAE labeling of $3C_3$

Proof Suppose α is an (a, d)-VAE labeling of $2r$-regular graph G on $p = 2t$ vertices. We proceed by double counting the sum of all vertex-weights. First, each of the edge labels $1, 2, \ldots, 2tr$ is counted twice. Thus

$$\sum_{e \in E(G)} \alpha(e) = 2(1 + 2 + \cdots + 2tr) = 2tr(1 + 2tr). \tag{5.2}$$

Second, since the vertex-weights form an arithmetic progression with difference d, we also have

$$\sum_{v \in V(G)} w_{\alpha}(v) = a + (a + d) + \cdots + (a + (2t - 1)d) = t(2a + (2t - 1)d). \tag{5.3}$$

Comparing (5.2) and (5.3) it is obvious that d has to be even in order to keep parity. □

5.2 Vertex-Antimagic Total and Super Vertex-Antimagic Total Labelings

In this section we will focus on the (a, d)-*vertex-antimagic total labelings*. A total labeling on a (p, q) graph G is a bijection λ from $V(G) \cup E(G)$ onto the integers $1, 2, \ldots, p + q$ with the property that the set of vertex-weights is $W = \{wt_{\lambda}(v) : v \in V(G)\} = \{a, a + d, \ldots, a + (p - 1)d\}$, where $a > 0$ and $d \geq 0$ are two fixed integers and for such total labeling λ the associated vertex-weight of a vertex $v \in V(G)$ is

$$wt_{\lambda}(v) = \lambda(v) + \sum_{u \in N(v)} \lambda(vu).$$

For short, we call a vertex-antimagic total labeling a *VAT labeling*. If $d = 0$, then we have vertex-magic total (VMT) labeling. As an example, a $(14, 4)$-VAT labeling of K_4 is depicted in Fig. 5.2, where the vertex-weights are 14, 18, 22, and 26.

The (a, d)-VAT labeling was introduced in [28] as a natural extension of the notion of VMT labeling, defined in [179] and [181].

Assume that a (p, q) graph G has an (a, d)-VAT labeling $\lambda : V(G) \cup E(G) \to \{1, 2, \ldots, p + q\}$. Let S_v be the sum of the vertex labels and S_e the sum of the edge labels. If we let $wt_{\lambda}(v_i) = a + id$, then summing the weights over all vertices adds each vertex label once and each edge label twice, so we get

$$S_v + 2S_e = \sum_{i=0}^{p-1} (a + id) = pa + \frac{p(p - 1)d}{2}. \tag{5.4}$$

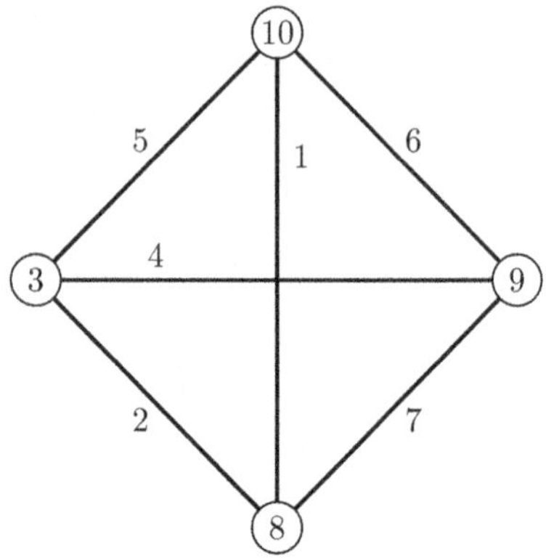

Fig. 5.2 (14, 4)-VAT labeling of K_4

The edge labels could conceivably receive the q smallest labels or, at the other extreme, the q largest labels, or anything between. Consequently, we have

$$\sum_{i=1}^{q} i \leq S_e \leq \sum_{i=p+1}^{p+q} i. \tag{5.5}$$

A corresponding result holds for S_v. Combining (5.4) and (5.5) results in the inequalities

$$2(1+2+\cdots+q)+((q+1)+(q+2)+\cdots+(p+q)) \leq pa+\frac{p(p-1)d}{2}$$
$$\leq (1+2+\cdots+q)+2((p+1)+(p+2)+\cdots+(p+q)),$$

which restrict the feasible values for a and d. For particular graphs, however, we can often exploit the structure to get considerably stronger restrictions. We note that if δ is the smallest degree in G, then the minimum possible vertex-weight is at least $1+2+\cdots+(\delta+1)$. Consequently,

$$a \geq \frac{(\delta+1)(\delta+2)}{2}. \tag{5.6}$$

Similarly, if Δ is the largest degree, then the maximum vertex-weight is no more than the sum of the $\Delta+1$ largest labels. Thus

$$a+(p-1)d \leq \sum_{i=p+q-\Delta}^{p+q} i = \frac{(2p+2q-\Delta)(\Delta+1)}{2}. \tag{5.7}$$

Combining Inequalities (5.6) and (5.7) gives the following upper bound on the values of d.

$$d \leq \frac{(2p+2q-\Delta)(\Delta+1)-(\delta+1)(\delta+2)}{2(p-1)}. \tag{5.8}$$

Given one VAT labeling on a graph, it may be possible to construct other VAT labelings from it.

Theorem 5.2.1 ([28]) *The dual of an (a, d)-VAT labeling for a graph G is an (a', d)-VAT labeling for some a' if and only if G is regular.*

Proof Suppose λ is an (a, d)-VAT labeling for G and let $wt_\lambda(v)$ be the weight of vertex v under the labeling λ. Then $\{wt_\lambda(v) : v \in V(G)\} = \{a, a+d, \ldots, a+(p-1)d\}$ is the set of vertex-weights of G. For any vertex $v \in V(G)$ with respect to (3.7) we have

$$\begin{aligned} wt_{\lambda'}(v) &= \lambda'(v) + \sum_{u\in N(v)} \lambda'(vu) = p+q+1-\lambda(v) + \sum_{u\in N(v)} (p+q+1-\lambda(vu)) \\ &= (r_v+1)(p+q+1) - wt_\lambda(v), \end{aligned}$$

where r_v is the number of edges incident to the given vertex v (degree of the vertex v). Clearly, the set $W' = \{wt_{\lambda'}(v) : v \in V(G)\}$ consists of an arithmetic progression with difference $d' = d$ if and only if r_v is constant for every v, that is, if and only if G is regular. □

Corollary 5.2.1 ([28]) *Let G be a regular graph of degree r. Then G has an (a, d)-VAT labeling if and only if G has an (a', d)-VAT labeling where $a' = (r+1)(p+q+1) - a - (p-1)d$.*

Proof Let G be a regular graph of degree r and λ be an (a, d)-VAT labeling for G. If λ' is the dual labeling of λ, then for every vertex $v \in V(G)$ we have $wt_{\lambda'}(v) = (r+1)(p+q+1) - wt_\lambda(v)$, where $wt_\lambda(v)$ is the weight of the vertex v under the labeling λ. We have $wt_\lambda(v) = a + (p-1)d$ as the maximum vertex-weight under the labeling λ if and only if $wt_{\lambda'}(v) = (r+1)(p+q+1) - a - (p-1)d$ is the minimum vertex-weight under the labeling λ'. □

Can a VAT labeling on a graph G be used to derive a VAT labeling for a subgraph of G? This seems to be a difficult question in general. The following theorem provides one case in which it is possible.

Theorem 5.2.2 ([28]) *Let G be a regular graph of degree r labeled in such a way that some edge e receives the label 1. Then G has an (a, d)-VAT labeling if and only if $G - \{e\}$ has an (a', d)-VAT labeling with $a' = a - r - 1$.*

Proof Assume that G is an r-regular graph and λ is the (a, d)-VAT labeling on G. Define a new mapping ρ by

$$\begin{aligned} \rho(v) &= \lambda(v) - 1, && \text{for any vertex } v \in V(G) \\ \rho(uv) &= \lambda(uv) - 1, && \text{for any edge } uv \in E(G). \end{aligned}$$

Clearly, the map ρ is one-to-one and the label 0 is assigned to the edge e by ρ. Then we have

$$\begin{aligned} wt_\rho(u) &= \rho(u) + \sum_{uv \in E(G)} \rho(uv) = \lambda(u) - 1 + \sum_{uv \in E(G)} (\lambda(uv) - 1) \\ &= \lambda(u) + \sum_{uv \in E(G)} \lambda(uv) - r - 1 = wt_\lambda(u) - r - 1, \end{aligned}$$

where the above summations are taken over all vertices adjacent to u. Clearly, the minimum value of $wt_\rho(u)$ occurs when $wt_\lambda(u) = a$.

If we delete the edge e from G, we obtain a graph $G - \{e\}$ and the restriction of the mapping ρ to $G - \{e\}$ is an $(a - r - 1, d)$-VAT labeling.

The proof of the converse is as follows. Let λ be the VAT labeling for $G - \{e\}$. Define a new mapping ρ in G by

$$\begin{aligned} \rho(e) &= 1, \\ \rho(u) &= \lambda(u) + 1, && \text{for all } u \in V(G) \\ \rho(uv) &= \lambda(uv) + 1, && \text{for all } uv \neq e \in E(G). \end{aligned}$$

Then it is easy to check that ρ is the appropriate VAT labeling for G. □

We show that it is possible in some cases to derive a VAT labeling from some other appropriate labeling of the graph. There is a relationship between supermagic labeling and VAT labeling. Stewart [263] showed that the complete graph K_n is supermagic if and only if either $n > 5$ and $n \not\equiv 0 \pmod 4$, or $n = 2$, see Theorem 2.6.6. For K_n we have $p = n$ and $q = n(n-1)/2$. Let $f : E(K_n) \to \{1, 2, \ldots, q\}$ be the supermagic labeling of K_n. Thus the sum of all edge labels is equal to

$$\sum_{e \in E(K_n)} f(e) = \frac{(n^2 - n + 2)(n^2 - n)}{8}$$

and, since each label is used by two vertices, the magic constant at each vertex is

$$k = \frac{(n^2 - n + 2)(n - 1)}{4}.$$

If we now label the vertices in G with $\{q+1, q+2, \dots, p+q\}$, then these labels together with the edge labels from f combine to give an (a, d)-VAT labeling where

$$a = k + q + 1 = \frac{n^3 + n + 2}{4} \text{ and } d = 1.$$

A similar argument applies for any graph G that has a supermagic labeling and so, more generally, we have

Theorem 5.2.3 ([28]) *Every supermagic graph G has an $(a, 1)$-VAT labeling.*

We know that both K_n and $K_{n,n}$ have supermagic labelings; see [263] or Theorem 2.6.6. Consequently, we have the following two corollaries.

Corollary 5.2.2 ([28]) *If $n > 5$ and $n \not\equiv 0 \pmod 4$ or $n = 2$, then the complete graph K_n has an $(a, 1)$-VAT labeling.*

Corollary 5.2.3 ([28]) *There is an $(a, 1)$-VAT labeling for $K_{n,n}$ for all $n \geq 3$.*

Sugeng, Miller, Lin and Bača suggest the following problem for further investigation.

Open Problem 5.2.1 ([272]) For the complete graph K_n and complete bipartite graph $K_{n,n}$, determine if there is an (a, d)-VAT labeling for every feasible value of $d > 1$.

Another corollary concerning quartic graphs is given bellow.

Corollary 5.2.4 ([33]) *If $n = 4k$ or $n = 4k + 2$, $k \geq 1$, then the quartic graphs R_n have an $(a, 1)$-VAT labeling.*

Readers interested in quartic graphs are directed to [33].

The next theorem gives an example of how one may construct a VAT labeling from a VMT labeling.

Theorem 5.2.4 ([28]) *Let G be a graph with a total labeling whose vertex labels constitute an arithmetic progression with difference d. Then G has a VMT labeling with magic constant k if and only if G has an $(a', 2d)$-VAT labeling where $a' = k + (1 - p)d$.*

Proof Let λ be a VMT labeling of G and k the magic constant for λ. Suppose that, under the labeling λ, the vertex labels of G constitute an arithmetic progression with difference d. In other words,

$$\{\lambda(v_i) : v_i \in V(G)\} = \{s + (i - 1)d : i = 1, 2, \dots, p\} = \{s, s + d, \dots, s + (p - 1)d\},$$

$s \in \mathbb{Z}^+$. Then, under the edge labeling λ_E induced by λ, the weights of the vertices constitute the arithmetic progression

$$\begin{aligned}\{wt_{\lambda_E}(v_i) : v_i \in V(G)\} &= \{wt_\lambda(v_i) - \lambda(v_i) : v_i \in V(G)\}\\ &= \{k - s - (i-1)d : i = 1, 2, \dots, p\}\\ &= \{k - s, k - s - d, \dots, k - s - (p-1)d\}.\end{aligned}$$

Define a new mapping ρ by

$$\begin{aligned}\rho(e) &= \lambda(e), && \text{for all } e \in E(G)\\ \rho(v_i) &= s + (p - i)d, && \text{for all } v_i \in V(G).\end{aligned}$$

It can be seen that the weights of the vertices, under the new mapping ρ, constitute the set

$$\begin{aligned}W &= \{wt_\rho(v_i) : v_i \in V(G)\} = \{k + (p + 1 - 2i)d : i = 1, 2, \dots, p\}\\ &= \{k + (p-1)d, k + (p-3)d, \dots, k + (1-p)d\},\end{aligned}$$

i.e., the weights of the vertices constitute an arithmetic progression with difference $2d$ and the minimum value of the weight $k + (1 - p)d$. Hence ρ is a VAT labeling on G.

The proof of the converse is similar. □

An (a, d)-VAT labeling λ is called a *super* (a, d)-VAT if it has the property that the vertex labels are the smallest possible labels.

Assume that a (p, q) graph G has a super (a, d)-VAT labeling $\lambda : V(G) \cup E(G) \to \{1, 2, \dots, p+q\}$ with the set of the vertex-weights $W = \{a, a+d, \dots, a+(p-1)d\}$. If δ is the minimum degree of G, then the minimum possible vertex-weight is at least $1 + (p + 1) + (p + 2) + \dots + (p + \delta)$. Thus,

$$a \geq 1 + p\delta + \frac{\delta(\delta + 1)}{2}. \tag{5.9}$$

On the other hand, if Δ is the maximum degree of G, then the maximum possible vertex-weight is no more than the sum of p, the maximum vertex label, and the Δ largest edge labels $p + q - \Delta + 1, p + q - \Delta + 2, \dots, p + q$. Consequently,

$$a + (p - 1)d \leq p + \sum_{i=0}^{\Delta-1} (p + q - i). \tag{5.10}$$

Combining (5.9) and (5.10) we get that

$$d \leq 1 + \frac{(2p + 2q - \Delta + 1)\Delta - (2p + \delta + 1)\delta}{2(p - 1)}. \tag{5.11}$$

Thus we have obtained an upper bound on the feasible value of the difference d.

Summing the vertex-weights over all the vertices in G is equal to summing all the values of the vertex labels and edge labels, where each vertex label is used once and each edge label is used twice.

Thus we get

$$\sum_{v \in V(G)} \lambda(v) + 2 \sum_{e \in E(G)} \lambda(e) = \sum_{v \in V(G)} wt_\lambda(v). \tag{5.12}$$

Investigation of the parities of the equation sides leads to the following results.

Theorem 5.2.5 ([48]) *Let G be a (p, q) graph.*

(i) If $p = 2 \pmod 4$, then G is not super (a, d) VAT for every even d.
(ii) If $p \equiv 0 \pmod 4$ and $q \equiv 2 \pmod 4$, then G is not super (a, d)-VAT for every odd d.
(iii) If $p \equiv 0 \pmod 8$ and $q \equiv 2 \pmod 4$, then G is not super (a, d)-VAT for every d.

Proof Suppose λ is a super (a, d)-VAT labeling for a (p, q) graph G. Consequently,

$$\sum_{e \in E(G)} \lambda(e) = \frac{q(2p + q + 1)}{2}, \quad \sum_{v \in V(G)} \lambda(v) = \frac{p(p + 1)}{2}$$

and

$$\sum_{v \in V(G)} wt_\lambda(v) = ap + \frac{p(p - 1)d}{2}.$$

Then from (5.12) we have the following equation:

$$(2p + q + 1)q = \frac{p}{2}\left((p - 1)(d - 1) + 2(a - 1)\right). \tag{5.13}$$

Case A If $p \equiv 2 \pmod 4$ and d is even, then the right-hand side of (5.13) is odd. On the other hand, the left-hand side is always even, which is a contradiction.

Case B If $q \equiv 2 \pmod 4$, then $2p+q+1$ is odd and $(2p+q+1)q \equiv 2 \pmod 4$. If $p \equiv 0 \pmod 4$ and d is odd, then $p\left((p - 1)(d - 1) + 2(a - 1)\right)/2 \equiv 0 \pmod 4$. This contradicts Eq. (5.13).

Case C If $p \equiv 0 \pmod 8$ and $q \equiv 2 \pmod 4$, then the left-hand side of (5.13) is congruent to 2 modulo 4 but the right-hand side of (5.13) is congruent to 0 modulo 4. This leads to a contradiction. □

In Theorem 3.2.6 it is proved that the minimum degree of a super VMT graph is at least 2. It is trivial that any disconnected graph with empty edge set is super $(a, 1)$-VAT. Thus for the super $(a, 1)$-VAT graphs with nonempty edge set, we obtain

Lemma 5.2.1 ([13]) *The minimum degree of a super $(a, 1)$-VAT (p, q) graph G with $q \geq 1$ is at least one.*

Proof Suppose that a graph G with at least one edge admits a super $(a, 1)$-VAT labeling. If G has an isolated vertex, then its vertex-weight is at most p. However, the label of each edge is at least $p + 1$. Thus the vertex-weight of any non-isolated vertex in G is at least $p + 2$. This is a contradiction, so the minimum degree is greater than zero. □

The sum of all the vertex labels and all the edge labels under a super (a, d)-VAT labeling λ is

$$\sum_{v \in V(G)} \lambda(v) + 2 \sum_{e \in E(G)} \lambda(e) = \frac{p(p+1)}{2} + 2pq + q(q+1) \tag{5.14}$$

and the sum of the vertex-weights over all the vertices is

$$\sum_{v \in V(G)} wt_\lambda(v) = ap + \frac{pd(p-1)}{2}. \tag{5.15}$$

Thus from (5.12) by using (5.14) and (5.15) we obtain the minimum vertex-weight

$$a = \frac{1}{2}(p + 1 - (p-1)d) + 2q + \frac{q(q+1)}{p}. \tag{5.16}$$

Now, let us consider a graph G with a minimum degree at least one.

Lemma 5.2.2 ([13]) *Let G be a super $(a, 1)$-VAT (p, q) graph. Then*

$$\frac{\sqrt{1+8q}+1}{2} \leq p \leq \frac{2q - 1 + \sqrt{8q^2+1}}{2}.$$

Proof Let G be a super $(a, 1)$-VAT (p, q) graph. Its minimum vertex-weight is at least $p + 2$ and from Eq. (5.16) it follows that

$$p + 2 \leq 1 + 2q + \frac{q(q+1)}{p}. \tag{5.17}$$

From (5.17) it follows that $p^2 - (2q - 1)p - q(q + 1) \leq 0$, hence

$$1 \leq p \leq \frac{2q - 1 + \sqrt{8q^2 + 1}}{2}. \tag{5.18}$$

Moreover, for the simple (p, q) graph it holds that $q \leq p(p - 1)/2$. This implies that $0 \leq p^2 - p - 2q$ and it is true for

$$\frac{\sqrt{1 + 8q} + 1}{2} \leq p. \tag{5.19}$$

Combining Inequalities (5.18) and (5.19) gives

$$\frac{\sqrt{1 + 8q} + 1}{2} \leq p \leq \frac{2q - 1 + \sqrt{8q^2 + 1}}{2}. \tag{5.20}$$

□

Thus we have a bound for the number of vertices for a (p, q) graph to be super $(a, 1)$-VAT.

For example, if $q = 1$, then (5.20) gives $p = 2$. Thus, G is an isolated edge with a trivial super $(4, 1)$-VAT labeling. If $q = 3$, then (5.20) gives $3 \leq p < 7$. There exists a super $(11, 1)$-VAT labeling of C_3, see Fig. 5.3, and a super $(9, 1)$-VAT labeling of the disjoint union of three isolated edges, see Fig. 5.4.

Lemma 5.2.3 ([13]) *For $d \geq 2$, a super (a, d)-VAT graph can contain isolated vertices.*

Fig. 5.3 Super $(11, 1)$-VAT labeling of C_3

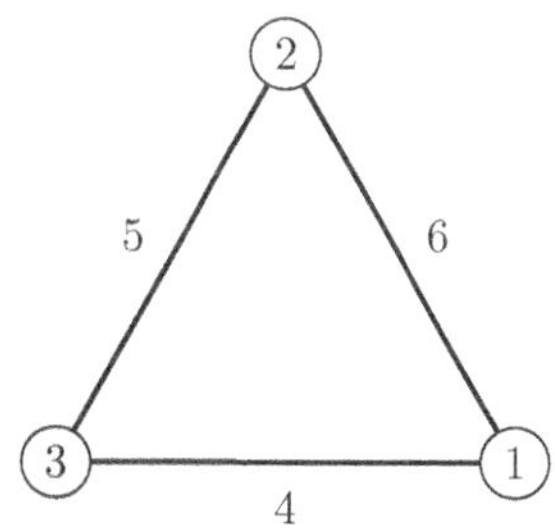

Fig. 5.4 Super $(9, 1)$-VAT labeling of $3P_2$

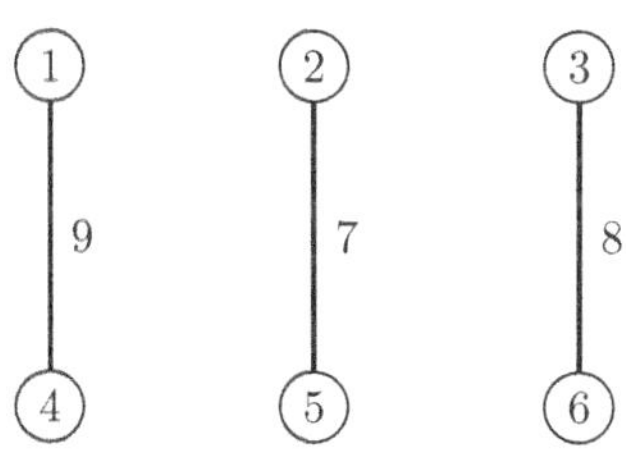

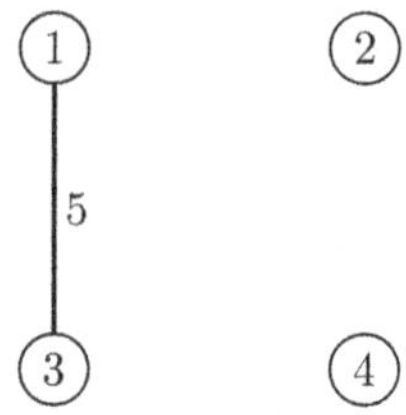

Fig. 5.5 Super (2, 2)-VAT graph with two isolates

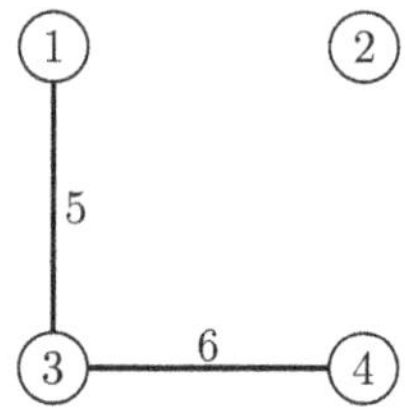

Fig. 5.6 Super (2, 4)-VAT graph with one isolate

Proof Let G be a super (a, d)-VAT (p, q) graph with an isolated vertex v. It is clear that $a \leq wt(v) \leq p$. As $d \geq 2$, from (5.16), for the minimum vertex-weight a of graph G we have

$$a = \frac{1}{2}(p + 1 - (p - 1)d) + 2q + \frac{q(q + 1)}{p} \leq p. \tag{5.21}$$

From (5.21) we get the following quadratic inequation,

$$0 \leq (d + 1)p^2 - (4q + d + 1)p - 2q(q + 1)$$

with the solution

$$p \geq \frac{d + 1 + 4q + \sqrt{(d + 1)(8q^2 + 16q + d + 1) + 16q^2}}{2(d + 1)}. \tag{5.22}$$

Thus we have the lower bound of order of (p, q) graph with a super (a, d)-VAT labeling for given values $q \geq 1$ and $d \geq 2$. □

For example, for $q = 1$, $d = 2$ from (5.22) we obtain $p > 2$ and for $q = 2$, $d = 4$ the (5.22) gives $p > 3$. Figure 5.5 depicts a super (2, 2)-VAT labeling of a graph with two isolates and Fig. 5.6 shows a super (2, 4)-VAT labeling of a graph with one isolate.

5.3 Relationship Between Vertex-Antimagic Edge and (Super) Vertex-Antimagic Total Labelings

A relationship between VAE labeling and VAT labeling is presented in the next theorem.

Theorem 5.3.1 ([28]) *The following statements hold.*

(i) If $d > 1$, then every (a, d)-VAE graph G has an $(a+p+q, d-1)$-VAT labeling.
(ii) Every (a, d)-VAE graph G has an $(a + q + 1, d + 1)$-VAT labeling.

Proof We assume that graph G is (a, d)-VAE with $d > 1$ and let $f : E(G) \to \{1, 2, \ldots, q\}$ be an (a, d)-VAE labeling of G. Then $W = \{wt_f(v) : v \in V(G)\} = \{a, a+d, \ldots, a+(p-1)d\}$ is the set of vertex-weights of G. For $i = 0, 1, \ldots, (p-1)$, let v_i be the vertex with the weight $wt_f(v_i) = a + id$. Define two sets of labels on the vertices

$$f', f'' : V(G) \to \{q + 1, q + 2, \ldots, p + q\}$$

as follows

$$f'(v_i) = q + i + 1$$

$$f''(v_i) = p + 2q + 1 - f'(v_i).$$

Then the labelings f and f' combine to give an $(a + q + 1, d + 1)$-VAT labeling for G, while f and f'' combine to give an $(a + p + q, d - 1)$-VAT labeling for G. □

Next we restate the following lemma that appeared in [273] and which will be useful in the next theorem.

Lemma 5.3.1 ([273]) *Let $\mathfrak{A}$ be a sequence $\mathfrak{A} = \{c, c + 1, \ldots, c + k\}$, k even. Then there exists a permutation $\Pi(\mathfrak{A})$ of the elements of $\mathfrak{A}$ such that $\mathfrak{A} + \Pi(\mathfrak{A}) = \{2c + k/2, 2c + k/2 + 1, \ldots, 2c + 3k/2\}$.*

The following two theorems establish a relationship between VAE labeling and super VAT labeling for regular graphs.

Theorem 5.3.2 ([271]) *Suppose G is a regular (p, q) graph of degree r, where p is odd. If G has an $(a, 1)$-VAE labeling, then G has a super $(a', 1)$-VAT labeling for $a' = r(p + q + 1) + (3 - p)/2 - a$.*

Proof Let G be r-regular and p be odd. If $\alpha : E(G) \to \{1, 2, \ldots, q\}$ is an $(a, 1)$-VAE labeling of G, then the set of vertex-weights contains a sequence $\mathfrak{A} = \{a, a + 1, \ldots, a+p-1\}$. From Lemma 5.3.1 it follows that there exists a permutation $\Pi(\mathfrak{A})$ of the elements of $\mathfrak{A}$ such that $\mathfrak{A} + (\Pi(\mathfrak{A}) - a + q + 1) = \{a + q + (p + 1)/2, a + q + (p + 3)/2, \ldots, a + q + (3p - 1)/2\}$.

If $(\Pi(\mathfrak{A}) - a + q + 1)$ are vertex values under a vertex labeling $\beta : V(G) \to \{q + 1, q + 2, \ldots, q + p\}$, then $\mathfrak{A} + (\Pi(\mathfrak{A}) - a + q + 1)$ gives the set of vertex-weights of G which implies that the total labeling is $(a + q + (p + 1)/2, 1)$-VAT.

We define a map γ on $V(G) \cup E(G)$ as follows.

$$\gamma(u) = p + q + 1 - \beta(u), \qquad \text{for all } u \in V(G)$$

$$\gamma(uv) = p + q + 1 - \alpha(uv), \qquad \text{for all } uv \in E(G).$$

Clearly, γ is a one-to-one map from the set $V(G) \cup E(G)$ into $\{1, 2, \ldots, q + p\}$. For any vertex $v \in V(G)$, we have

$$
\begin{aligned}
wt_\gamma(v) = \gamma(v) + \sum_{vu \in E} \gamma(vu) &= p + q + 1 - \beta(v) + \sum_{vu \in E} (p + q + 1 - \alpha(vu)) \\
&= (r + 1)(p + q + 1) - \beta(v) - \sum_{vu \in E} \alpha(vu) \\
&= (r + 1)(p + q + 1) - wt_{\alpha\beta}(v).
\end{aligned}
$$

We can see that the set of vertex-weights under the labeling γ consists of the consecutive integers $\{wt_\gamma(v) : v \in V(G)\} = \{r(p+q+1)+(3-p)/2-a, r(p+q+1)+(5-p)/2-a, \ldots, r(p+q+1)+(p+1)/2-a\}$ and the vertex labels are the smallest possible labels $1, 2, \ldots, p$.

Hence, γ is a super $(r(p+q+1)+(3-p)/2-a, 1)$-VAT labeling of G. □

Theorem 5.3.3 ([271]) *An (a, d)-VAE labeling of a (p, q) graph G is super $(a', d - 1)$-VAT and super $(a'', d + 1)$-VAT if and only if G is r-regular, where $a' = r(p+q+1) - a + p + (1-p)d$ and $a'' = r(p+q+1) - a + 1 + (1-p)d$.*

Proof Let $\rho : E(G) \to \{1, 2, \ldots, q\}$ be an (a, d)-VAE labeling of G and $W = \{wt_\rho(v) : v \in V(G)\} = \{a, a+d, \ldots, a+(p-1)d\}$ be the set of vertex-weights of G. Let v_i be the vertex of $V(G)$ such that $wt_\rho(v_i) = a+(i-1)d$ for $i = 1, 2, \ldots, p$.

Case A If we label the vertices and edges in G by

$$
\begin{aligned}
\varepsilon(v_i) &= p + q + 1 - i, && \text{for } i = 1, 2, \ldots, p \\
\varepsilon(uv) &= \rho(uv), && \text{for all } uv \in E(G),
\end{aligned}
$$

then the vertex-weights will be $\{p+q+a, p+q+a+d-1, \ldots, p+q+a+(p-1)(d-1)\}$. This means that ε is a $(p+q+a, d-1)$-VAT labeling.

Define a new mapping η by

$$
\begin{aligned}
\eta(u) &= p + q + 1 - \varepsilon(u), && \text{for all } u \in V(G) \\
\eta(uv) &= p + q + 1 - \varepsilon(uv), && \text{for all } uv \in E(G).
\end{aligned}
$$

For any vertex $v \in V(G)$ it holds that

$$
\begin{aligned}
wt_\eta(v) = \eta(v) + \sum_{vu \in E} \eta(vu) &= p + q + 1 - \varepsilon(v) + \sum_{vu \in E} (p + q + 1 - \varepsilon(vu)) \\
&= (r + 1)(p + q + 1) - wt_\varepsilon(v)
\end{aligned}
$$

if and only if G is regular of degree r.

The vertex-weights clearly form the arithmetic progression $\{wt_\eta(v) : v \in V(G)\} = \{r(p+q+1) - a + 1 + (1-p)(d-1), r(p+q+1) - a + 1 + (2-p)(d-1), \ldots, r(p+q+1) - a + 1\}$. Thus η is a super $(r(p+q+1) - a + 1 + (1-p)(d-1), d-1)$-VAT labeling.

Case B If we label the vertices and edges in G by

$$\lambda(v_i) = q + i, \qquad \text{for } i = 1, 2, \ldots, p$$
$$\lambda(uv) = \rho(uv), \qquad \text{for all } uv \in E(G),$$

then the vertex-weights form the arithmetic progression $\{a+q+1, a+q+1+d+1, a+q+1+2(d+1), \ldots, a+q+1+(p-1)(d+1)\}$.

Construct a new mapping μ as follows.

$$\mu(v) = p + q + 1 - \lambda(v), \qquad \text{for all } v \in V(G)$$
$$\mu(uv) = p + q + 1 - \lambda(uv), \qquad \text{for all } uv \in E(G).$$

For any vertex $v \in V(G)$, we have

$$wt_\mu(v) = \mu(v) + \sum_{uv \in E} \mu(uv) = (r+1)(p+q+1) - wt_\lambda(v)$$

if and only if G is r-regular.

Then the set of the vertex-weights is $\{wt_\mu(v) : v \in V(G)\} = \{r(p+q+1)+p-a+(1-p)(d+1), r(p+q+1)+p-a+(2-p)(d+1), \ldots, r(p+q+1)+p-a\}$ and μ is a super $(r(p+q+1)+p-a+(1-p)(d+1), d+1)$-VAT labeling. □

Immediately from Theorems 5.3.1 and 5.3.3, and Lemma 5.3.1, we obtain the following corollary.

Corollary 5.3.1 ([6]) *Let G be a graph of odd order.*

(i) If G is an $(a, 1)$-VAE graph, then G has an $(a', 1)$-VAT labeling for some integer a'.
(ii) If G is a regular $(a, 1)$-VAE graph, then G has a super $(a'', 1)$-VAT labeling for some integer a''.

We examine the connection between VAE and VAT labelings further.

Theorem 5.3.4 ([6]) *Let G be a regular graph on p vertices that can be decomposed into two factors G_1 and G_2. If G_1 is a k-regular graph that admits an (a, d)-VAE labeling and G_2 is a $2r$-regular graph, then G is a super $(a', d+1)$-VAT graph for $a' = a + kp(r+1) + pr(r+2) + r + 1$ and a super $(a'', d-1)$-VAT graph for $d \geq 1$ and $a'' = a + p(k+r+1)(r+1) + r$.*

Proof Let G be a regular graph of order p that can be decomposed into two factors G_1 and G_2, i.e., $V(G) = V(G_1) = V(G_2)$.

Let G_1 be a k-regular factor of G that admits an (a, d)-VAE labeling $\alpha : E(G_1) \to \{1, 2, \dots, kp/2\}$. Thus the vertex-weights of the vertices in G are

$$a, a + d, \dots, a + (p - 1)d.$$

We denote the vertices of the graph G by $v_1, v_2, \dots, v_p$ so that

$$w_\alpha(v_i) = a + (i - 1)d, \quad i = 1, 2, \dots, p. \tag{5.23}$$

Let G_2 be a $2r$-regular factor of G. Since G_2 is even regular, then, according to Theorem 3.1.23, there exists a 2-factorization of G_2. We denote the 2-factors by F_j, $j = 1, 2, \dots, r$. Each factor F_j is a collection of cycles. We order and orient the cycles arbitrarily. Again, by the symbol $e_j^{out}(v_i)$ we denote the unique outgoing arc from the vertex v_i in the factor F_j, and by the symbol $e_j^{in}(v_i)$ we denote the unique incoming arc to the vertex v_i in the factor F_j.

We define the labeling $f : V(G) \cup E(G) \to \{1, 2, \dots, p + (k + 2r)p/2\}$ of the graph G as follows.

$$\begin{aligned}
f(e) &= \alpha(e) + (r + 1)p, && \text{for } e \in E(G_1) \\
f(e_1^{out}(v_i)) &= p + i, && \text{for } i = 1, 2, \dots, p \\
f(e_j^{out}(v_i)) &= 2jp + 1 - f(e_{j-1}^{in}(v_i)), && \text{for } j = 2, 3, \dots, r,\ i = 1, 2, \dots, p \\
f(v_i) &= pr + p + 1 - f(e_r^{in}(v_i)), && \text{for } i = 1, 2, \dots, p.
\end{aligned} \tag{5.24}$$

In contrast to the construction in the proof of Theorem 5.1.4, now the labels in each factor F_j are consecutive integers. It is easy to verify that f is a bijection and the vertices are labeled by the integers $1, 2, \dots, p$. For the vertex-weights under the labeling f we have

$$\begin{aligned}
wt_f(v_i) = &(w_\alpha(v_i) + k(r + 1)p) + \Big(f(e_1^{out}(v_i)) + f(e_1^{in}(v_i))\Big) \\
&+ \Big(f(e_2^{out}(v_i)) + f(e_2^{in}(v_i))\Big) + \dots + \Big(f(e_r^{out}(v_i)) + f(e_r^{in}(v_i))\Big) + f(v_i) \\
= &(w_\alpha(v_i) + k(r + 1)p) + \Big((p + i) + f(e_1^{in}(v_i))\Big) \\
&+ \Big(4p + 1 - f(e_1^{in}(v_i)) + f(e_2^{in}(v_i))\Big) + \cdots \\
&+ \Big(2pr + 1 - f(e_{r-1}^{in}(v_i)) + f(e_r^{in}(v_i))\Big) + pr + p + 1 - f(e_r^{in}(v_i)) \\
= &kp(r + 1) + pr(r + 2) + r + w_\alpha(v_i) + i.
\end{aligned}$$

According to (5.23), the vertex-weights are

$$kp(r+1)+pr(r+2)+r+a+1, kp(r+1)+pr(r+2)+r+a+d+2,$$
$$\dots, kp(r+1)+pr(r+2)+r+a+(p-1)d+p.$$

Thus, the vertex-weights form an arithmetic progression with the difference $(d+1)$ and f is a super $(a', d+1)$-VAT labeling of G with $a' = a + kp(r+1) + pr(r+2) + r + 1$.

Moreover, it is not difficult to check that if in (5.24) we take

$$f(e_1^{out}(v_i)) = 2p + 1 - i, \quad \text{for } i = 1, 2, \dots, p,$$

then, for $d \geq 1$, f is a super $(a'', d-1)$-VAT labeling of G with $a'' = a + p(k + r + 1)(r + 1) + r$. □

Note, that if there are t pairwise non-isomorphic factors among $F_2, F_3, \dots, F_r$, then by rearranging them we can obtain up to $t!$ different super $(a', d+1)$-VAT or super $(a'', d-1)$-VAT labelings of G.

Combining Theorems 5.1.4 and 5.3.4, we get the following theorem.

Theorem 5.3.5 ([6]) *If G is a $2r$-regular Hamiltonian graph of odd order, then G is super (a, d)-VAT for some integer a and for every $d = 0, 1, \dots, r+1$.*

Proof Let G be a $2r$-regular Hamiltonian graph of odd order. By F_1 we denote the Hamiltonian cycle in G. By Theorem 3.1.23, the graph G is decomposable into two even regular factors G_1 and G_2, such that G_1 contains F_1. Since G_1 is $2k$-regular, $k = 1, 2, \dots, r$, then, according to Theorem 5.1.4, the graph G_1 is (a, k)-VAE. Thus, according to Theorem 5.3.4, the graph G admits an $(a', k+1)$-VAT labeling and an $(a'', k-1)$-VAT labeling. This concludes the proof. □

We can generalize Theorem 5.3.4 also for certain non-regular graphs.

Theorem 5.3.6 ([6]) *Let G be a graph on p vertices that is decomposable into two factors G_1 and G_2. If G_1 is an (a, d)-VAE graph and G_2 is a $2r$-regular graph, then G is an $(a', d+1)$-VAT graph for $a' = a+(2r+1)|E(G_1)|+pr(r+2)+r+1$, and an $(a'', d-1)$-VAT graph for $d \geq 1$ and $a'' = a+(2r+1)|E(G_1)|+pr(r+2)+r+p$.*

Proof The proof follows a similar strategy and uses the same notation as the proof of Theorem 5.3.4. Let G be a graph of order p that can be decomposed into two factors G_1 and G_2. Suppose G_1 admits an (a, d)-VAE labeling α, $\alpha : E(G_1) \rightarrow \{1, 2, \dots, |E(G_1)|\}$. The vertex-weights in the labeling α are

$$a, a+d, \dots, a+(p-1)d.$$

We order and denote the vertices of G by $v_1, v_2, \ldots, v_p$ so that

$$w_\alpha(v_i) = a + (i-1)d, \quad i = 1, 2, \ldots, p.$$

Now one can view the vertex-weights as vertex labels. $G_2 = G - G_1$ is the $2r$-regular factor of G. By Theorem 3.1.23 there exists a 2-factorization of G_2. We denote the 2-factors by F_j, $j = 1, 2, \ldots, r$. Each factor F_j is a collection of cycles, we can order and orient them arbitrarily.

It is easy to check that the labeling $f : V(G) \cup E(G) \to \{1, 2, \ldots, p+|E(G_1)|+rp\}$ given by

$$\begin{aligned} f(e) &= \alpha(e), && \text{for } e \in E(G_1) \\ f(e_1^{out}(v_i)) &= |E(G_1)| + p + i, && \\ f(e_j^{out}(v_i)) &= 2|E(G_1)| + 2jp + 1 - f(e_{j-1}^{in}(v_i)), && \text{for } j = 2, 3, \ldots, r \\ f(v_i) &= 2|E(G_1)| + pr + p + 1 - f(e_r^{in}(v_i)), && \end{aligned}$$

for $i = 1, 2, \ldots, p$ is an $(a', d+1)$-VAT labeling of G, where $a' = a + (2r+1)|E(G_1)| + pr(r+2) + r + 1$. Similarly, if

$$f(e_1^{out}(v_i)) = |E(G_1)| + 2p + 1 - i, \qquad \text{for } i = 1, 2, \ldots, p,$$

then f is an $(a'', d-1)$-VAT labeling of G, where $a'' = a + (2r+1)|E(G_1)| + pr(r+2) + r + p$. □

Again, if there are t pairwise non-isomorphic factors among $F_1, F_2, \ldots, F_r$ by rearranging them we can obtain up to $t!$ different super $(a', d+1)$-VAT and super $(a'', d-1)$-VAT labelings of G. As an example of a non-regular (a, d)-VAE subgraph G_1, that satisfies the condition of Theorem 5.3.6, we can consider, e.g., a path on p vertices [69, 72].

In light of Theorem 5.1.4, we can prove the existence of (a, d)-VAT labeling with differences $d = 2r + 2$.

Theorem 5.3.7 ([6]) *Let G be a $2r$-regular Hamiltonian graph of odd order p. Then G is $(a, 2r+2)$-VAT for $a = 3p(r+1)/2 + (15-r)/2 + (r-2)(pr+p+3)$.*

Proof Let F_1 be a Hamiltonian cycle and F_j, $j = 2, 3, \ldots, r$ be such 2-factors that form together with F_1 the 2-factorization of the $2r$-regular graph G of odd order p. We denote the vertices in G by $v_1, v_2, \ldots, v_p$ in such a way that $v_1 v_2 \ldots v_p v_1$ is the Hamiltonian cycle F_1. Let α be the edge labeling of G defined in the following way.

$$\alpha(v_i v_{i+1}) = \begin{cases} \dfrac{i+3}{2} + \dfrac{(i-1)r}{2} & \text{for } i = 1, 3, \ldots, p \\ \dfrac{p+3+i}{2} + \dfrac{(p-1+i)r}{2} & \text{for } i = 2, 4, \ldots, p-1, \end{cases}$$

where the indices are taken modulo p. For the remaining edges we define the labels as follows.

$$\alpha(e_2^{out}(v_i)) = \frac{3p+11}{2} + \frac{(3p-3)r}{2} - \alpha(v_{i-1}v_i) - \alpha(v_iv_{i+1})$$
$$\alpha(e_j^{out}(v_i)) = (p-1)r + p + 2j - \alpha(e_{j-1}^{in}(v_i)), \qquad \text{for } j = 3, 4, \ldots, r,$$

for every $i = 1, 2, \ldots, p$.

Now α is a bijection $\alpha : E(G) \to \{1, 2, \ldots, pr\}$. It is easy to see that the set of the edge labels in the factor F_j, $j = 1, 2, \ldots, r$, is

$$\{\alpha(e) \colon e \in E(F_j)\} = \{1 + j, r + 2 + j, 2r + 3 + j, \ldots, (p-1)r + p + j\}. \tag{5.25}$$

The vertex-weight of vertex v_i, $i = 1, 2, \ldots, p$ in α is

$$\begin{aligned} w_\alpha(v_i) = &(\alpha(v_{i-1}v_i) + \alpha(v_iv_{i+1})) + \left(\alpha(e_2^{out}(v_i)) + \alpha(e_2^{in}(v_i))\right) \\ &+ \left(\alpha(e_3^{out}(v_i)) + \alpha(e_3^{in}(v_i))\right) + \cdots + \left(\alpha(e_r^{out}(v_i)) + \alpha(e_r^{in}(v_i))\right) \\ =&\alpha(v_{i-1}v_i) + \alpha(v_iv_{i+1}) \\ &+ \left(\frac{3p+11}{2} + \frac{(3p-3)r}{2} - \alpha(v_{i-1}v_i) - \alpha(v_iv_{i+1}) + \alpha(e_2^{in}(v_i))\right) \\ &+ \left((p-1)r + p + 6 - \alpha(e_2^{in}(v_i)) + \alpha(e_3^{in}(v_i))\right) \\ &+ \left((p-1)r + p + 8 - \alpha(e_3^{in}(v_i)) + \alpha(e_4^{in}(v_i))\right) \\ &+ \cdots + \left((p-1)r + p + 2r - \alpha(e_{r-1}^{in}(v_i)) + \alpha(e_r^{in}(v_i))\right) \\ =&\frac{3p(r+1)}{2} + \frac{11-3r}{2} + (r-2)(pr+p+3) + \alpha(e_r^{in}(v_i)) \\ =&b + \alpha(e_r^{in}(v_i)), \end{aligned}$$

where $b = 3p(r+1)/2 + (11-3r)/2 + (r-2)(pr+p+3)$. By (5.25) we have

$$\begin{aligned} \left\{\alpha(e_r^{in}(v_i)) \colon i = 1, 2, \ldots, p\right\} &= \{\alpha(e_r^{out}(v_i)) \colon i = 1, 2, \ldots, p\} \\ &= \{r+1, 2r+2, \ldots, pr+p\} \end{aligned}$$

and thus the set of the vertex-weights $w_\alpha(v_i)$ is $\{b+r+1, b+2r+2, \ldots, b+pr+p\}$. Considering the edge labeling α, the vertex-weights form an arithmetic progression with the difference $r+1$. Now we rename the vertices in G by $u_1, u_2, \ldots, u_p$ so that

$$w_\alpha(u_i) = b + i(r+1), \quad \text{for } i = 1, 2, \ldots, p.$$

Define the total labeling $f : V(G) \cup E(G) \to \{1, 2, \ldots, p + pr\}$ as follows.

$$f(u_i) = i + (i-1)r, \quad \text{for } i = 1, 2, \ldots, p$$
$$f(e) = \alpha(e), \quad \text{for all } e \in E(G).$$

One can see that the total labeling f is an $(a, 2r+2)$-VAT labeling of G, where $a = 3p(r+1)/2 + (15-r)/2 + (r-2)(pr+p+3)$. □

Note that if in the proof of the previous theorem we label the vertices in G so that

$$f(u_i) = p + 1 - i + (p-i)r, \quad \text{for } i = 1, 2, \ldots, p,$$

the resulting labeling f is a $(b + p + pr + 1, 0)$-VAT labeling of G. This means that every even regular Hamiltonian graph of odd order is vertex-magic total. This special case was already proved by Kovář in [167].

5.4 Vertex-Antimagic Total Labelings of Cycles and Paths

Among the graphs for which it is easiest to find VAT labelings are the cycles and paths. In this section we provide labelings for both families of graphs. Applying Inequality (5.8) for cycle C_n and path P_n, we get

$$d \le 6 - \frac{3}{n-1}.$$

Thus $d \le 5$ for all $n \ge 4$ and $d \le 4$ for $n = 3$. The same result has been proved by Tezer and Cahit in [276]. Figure 5.7 shows examples of VAT labelings of C_3 for each feasible value of d, $d > 0$.

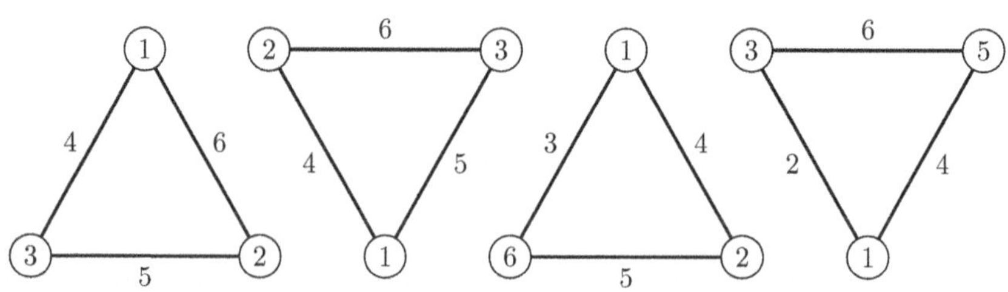

Fig. 5.7 VAT labelings of C_3 for all feasible d, $d > 0$

In Theorem 5.2.2 we proved that every VAT labeling for a graph of the form $G - \{e\}$, where G is regular and where an edge e has the label 1, is obtained from a VAT labeling of G. Since a path P_n is the cycle C_n with an edge removed, then every VAT labeling for the path P_n is obtained from a corresponding VAT labeling for C_n. Note that the converse is not necessarily true.

Theorem 5.4.1 ([28]) *Every odd cycle C_n, $n \geq 3$, has a $((3n+5)/2, 2)$-VAT labeling and a $((5n+5)/2, 2)$-VAT labeling.*

Proof Let C_n be the cycle with $V(C_n) = \{v_i : 1 \leq i \leq n\}$ and $E(C_n) = \{v_i v_{i+1} : 1 \leq i \leq n-1\} \cup \{v_n v_1\}$. Assume that C_n admits an EMT labeling λ with magic constant $\lambda(v_i) + \lambda(v_i v_{i+1}) + \lambda(v_{i+1}) = k$ for all edges $v_i v_{i+1}$ of C_n. If we define a new mapping λ^* by

$$\lambda^*(v_i) = \lambda(v_i v_{i+1})$$

and

$$\lambda^*(v_i v_{i+1}) - \lambda(v_{i+1}),$$

where the subscripts are integers modulo n, then we clearly have k as the vertex-weight at each vertex, and so λ^* is a VMT labeling of G. This means, for cycles (and only for cycles) that an EMT labeling is equivalent to a VMT labeling.

Wallis et al. [289] proved that every odd cycle has an EMT labeling with the magic constant $k = (5n+3)/2$. Moreover, the vertex labels of the considered VMT labeling from [289] constitute an arithmetic progression with difference $d = 1$. Thus, by Theorem 5.2.4, the odd cycle C_n has $((3n+5)/2, 2)$-VAT labeling. To prove that C_n has $((5n+5)/2, 2)$-VAT labeling, we make use of Corollary 5.2.1 and the fact that C_n is a 2-regular graph. It is simple to verify that the minimum vertex-weight is $(5n+5)/2$. □

As an easy consequence of Theorem 5.2.2 we have

Corollary 5.4.1 *For n odd and $n \geq 3$, the path P_n has a $((3n-1)/2, 2)$-VAT labeling.*

Proof The cycle C_n is a 2-regular graph and thus by Theorem 5.4.1 has a $((3n+5)/2, 2)$-VAT labeling in which the label 1 is assigned to an edge e. Theorem 5.2.2 now guarantees that the path P_n has a $((3n-1)/2, 2)$-VAT labeling. □

The following theorems provide examples of VAT labelings with various values of d for cycles C_n and paths P_n.

Theorem 5.4.2 ([28]) *Every cycle C_n, $n \geq 3$, has a $(3n+2, 1)$-VAT labeling and $(2n+2, 1)$-VAT labeling.*

Proof Let the cycle C_n be $(v_1, v_2, \ldots, v_n)$. If we label the vertices and edges in C_n by

$$\begin{aligned} \lambda(v_i) &= i, && \text{for } i = 1, 2, \ldots, n \\ \lambda(v_i v_{i+1}) &= 2n - i, && \text{for } i = 1, 2, \ldots, n-1 \\ \lambda(v_n v_1) &= 2n, \end{aligned}$$

then the vertex-weights will be

$$wt_\lambda(x_i) = \begin{cases} 4n + 1 - i & \text{for } i = 1, 2, \ldots, n-1 \\ 4n + 1 & \text{for } i = n, \end{cases}$$

and these clearly form the arithmetic progression $3n + 2, 3n + 3, \ldots, 4n + 1$. Thus C_n has a $(3n + 2, 1)$-VAT labeling.

Combining this with Corollary 5.2.1, it is easy to see that C_n also has a $(2n + 2, 1)$-VAT labeling. □

Since the cycle C_n has a $(2n+2, 1)$-VAT labeling in which the label 1 is assigned to an edge, by Theorem 5.2.2 we obtain

Corollary 5.4.2 *Every path P_n, $n \geq 3$, has a $(2n - 1, 1)$-VAT labeling.*

Theorem 5.4.3 ([28]) *Every cycle C_n, $n \geq 3$, has a $(2n + 3, 2)$-VAT labeling and $(2n + 2, 2)$-VAT labeling.*

Proof Let the cycle C_n be $(v_1, v_2, \ldots, v_n)$. If we label the vertices and edges in C_n by

$$\begin{aligned} \lambda(v_i) &= 2i - 1, && \text{for } i = 1, 2, \ldots, n \\ \lambda(v_i v_{i+1}) &= 2(n + 1 - i), && \text{for } i = 1, 2, \ldots, n-1 \\ \lambda(v_n v_1) &= 2, \end{aligned}$$

then the vertex-weights are

$$wt_\lambda(v_i) = \begin{cases} 4n + 5 - 2i & \text{for } i = 2, \ldots, n \\ 2n + 3 & \text{for } i = 1, \end{cases}$$

and these form the arithmetic progression $2n + 3, 2n + 5, \ldots, 4n + 1$. Thus C_n has a $(2n + 3, 2)$-VAT labeling.

Combining this with Corollary 5.2.1, it is easy to see that C_n also has a $(2n + 2, 2)$-VAT labeling. □

Since the cycle C_n has a $(2n+2, 2)$-VAT labeling in which the label 1 is assigned to an edge, by Theorem 5.2.2 we have

Corollary 5.4.3 *Every path P_n, $n \geq 3$, has a $(2n-1, 2)$-VAT labeling.*

Theorem 5.4.4 ([28]) *Every cycle C_n, $n \geq 3$, has a $(2n+2, 3)$-VAT labeling and $(n+4, 3)$-VAT labeling.*

Proof As before, the cycle C_n is $(v_1, v_2, \ldots, v_n)$. Label the vertices and edges in C_n as follows:

$$\begin{aligned}
\lambda(v_i) &= i, && \text{for } i = 1, 2, \ldots, n-1\\
\lambda(v_n) &= 2n, &&\\
\lambda(v_i v_{i+1}) &= n + i, && \text{for } i = 1, 2, \ldots, n-1\\
\lambda(v_n v_1) &= n, &&
\end{aligned}$$

then the vertex-weights

$$wt_\lambda(v_i) = 2n - 1 + 3i, \quad \text{for } 1 \leq i \leq n,$$

clearly making a $(2n+2, 3)$-VAT labeling.

Combining this with Corollary 5.2.1, it is easy to see that C_n also has an $(n+4, 3)$-VAT labeling. □

Theorem 5.4.5 ([28]) *Every odd cycle C_n, $n \geq 3$, has an $(n+4, 4)$-VAT labeling and an $(n+3, 4)$-VAT labeling.*

Proof Letting C_n be $(v_1, v_2, \ldots, v_n)$, we label the vertices and edges as follows:

$$\begin{aligned}
\lambda(v_i) &= 2i - 1, && \text{for } i = 1, 2, \ldots, n\\
\lambda(v_i v_{i+1}) &= 1 + i, && \text{for } i \text{ odd}, i \neq n\\
\lambda(v_i v_{i+1}) &= n + 1 + i, && \text{for } i \text{ even}\\
\lambda(v_n v_1) &= n + 1. &&
\end{aligned}$$

Then the vertex-weights are

$$wt_\lambda(v_i) = n + 4i, \quad \text{for } 1 \leq i \leq n,$$

which clearly constitutes an $(n+4, 4)$-VAT labeling for C_n.

Combining this with Corollary 5.2.1, it is easy to see that C_n also has an $(n+3, 4)$-VAT labeling. □

Once again, since the cycle C_n has an $(n+3, 4)$-VAT labeling in which the label 1 is assigned to an edge, by Theorem 5.2.2 we have

Corollary 5.4.4 *Every odd path P_n, $n \geq 3$, has an $(n, 4)$-VAT labeling.*

Theorem 5.4.6 ([28]) *The path P_n has a $(2n-1, 1)$-VAT labeling for any $n \geq 2$.*

Proof Name the vertices in P_n as $v_1, v_2, \ldots, v_n$ and let the set of edges be $E(P_n) = \{v_i v_{i+1} : i = 1, 2, \ldots, n-1\}$. Then attach labels to all the vertices and edges as follows:

$$\lambda(v_i) = \begin{cases} n & \text{for } i = 2 \\ 2n - i & \text{for } i = 3, 4, \ldots, n-1 \\ 2n - 2 & \text{for } i = n \\ 2n - 1 & \text{for } i = 1 \end{cases}$$

$$\lambda(v_i v_{i+1}) = \begin{cases} 1 & \text{for } i = n-1 \\ i & \text{for } i = 2, 3, \ldots, n-2 \\ n - 1 & \text{for } i = 1. \end{cases}$$

Under the labeling λ we have the vertex-weights

$$wt_\lambda(v_i) = \begin{cases} 3n - 1 - i & \text{for } i = n-1, n \\ 2n - 1 + i & \text{for } i = 2, 3, \ldots, n-2 \\ 3n - 2 & \text{for } i = 1. \end{cases}$$

These vertex-weights form the arithmetic progression $2n-1, 2n, \ldots, 3n-2$ and so λ is a $(2n-1, 1)$-VAT labeling. □

We know that a VAT labeling for the path P_n, for $n \geq 3$, provides a corresponding VAT labeling for the cycle C_n. Therefore we have the following corollary.

Corollary 5.4.5 *Every cycle C_n, $n \geq 3$, has a $(2n+2, 1)$-VAT labeling and a $(3n+2, 1)$-VAT labeling.*

Interestingly, this labeling and the labeling produced by Theorem 5.4.2 are both $(2n+2, 1)$-VAT but they are different.

The following problem remains unsolved.

Open Problem 5.4.1 Find an $(a, 4)$-VAT labeling of cycle C_n and path P_n for n even, $n \geq 4$.

Open Problem 5.4.2 For the cycles C_n and the paths P_n, determine if there is an $(a, 5)$-VAT labeling.

5.5 Super Vertex-Antimagic Total Labelings of Certain Families of Connected Graphs

This section is dedicated to the study of super VAT labelings of cycles, paths, generalized Petersen graphs, trees, and unicyclic graphs.

5.5.1 Cycles and Paths

From [27] we have

Theorem 5.5.1 ([27]) *The cycle C_n has a super (a, d)-EAT labeling if and only if one of the following statements hold.*

(i) $d \in \{0, 2\}$ and n is odd, $n \geq 3$.
(ii) $d = 1$ and $n \geq 3$.

The following theorem gives a relation between super (a, d)-EAT and super (a, d)-VAT labeling for cycles.

Theorem 5.5.2 ([272]) *For cycles and only for cycles, a super (a, d)-EAT labeling is equivalent to a super (a', d)-VAT labeling.*

Proof Let the cycle C_n be defined as follows: $V(C_n) = \{v_0, v_1, \ldots, v_{n-1}\}$ and $E(C_n) = \{v_i v_{i+1} : i = 0, 1, \ldots, n-1\}$ with the indices taken modulo n. Suppose that a bijection λ from $V(C_n) \cup E(C_n)$ onto the set $\{1, 2, \ldots, 2n\}$ is super (a, d)-EAT. It means that $\{wt_\lambda(v_i v_{i+1}) : wt_\lambda(v_i v_{i+1}) = \lambda(v_i) + \lambda(v_{i+1}) + \lambda(v_i v_{i+1}), i = 0, 1, \ldots, n-1\} = \{a, a+d, a+2d, \ldots, a+(n-1)d\}$ is the set of edge-weights of C_n.

Define a new mapping α by

$$\alpha(v_i v_{i+1}) = \lambda(v_i), \qquad \text{for } i = 0, 1, \ldots, n-1$$
$$\alpha(v_{i+1}) = \lambda(v_i v_{i+1}), \quad \text{for } i = 0, 1, \ldots, n-1.$$

Thus $\alpha(V(G)) = \{n+1, n+2, \ldots, 2n\}$ and $\alpha(E(G)) = \{1, 2, \ldots, n\}$. Moreover

$$\begin{aligned} wt_\lambda(v_i v_{i+1}) &= \lambda(v_i) + \lambda(v_{i+1}) + \lambda(v_i v_{i+1}) \\ &= \alpha(v_i v_{i+1}) + \alpha(v_{i+1} v_{i+2}) + \alpha(v_{i+1}) = wt_\alpha(v_{i+1}) \end{aligned}$$

for all $i = 0, 1, \ldots, n-1$, i.e., the edge-weight $wt_\lambda(v_i v_{i+1})$ is equivalent to the vertex-weight $wt_\alpha(v_{i+1})$ for all $i = 0, 1, \ldots, n-1$. So, labeling α is (a, d)-VAT.

We construct the dual labeling α' by

$$\alpha'(u) = 2n + 1 - \alpha(u), \qquad \text{for any vertex } u \in V(C_n)$$
$$\alpha'(uv) = 2n + 1 - \alpha(uv), \qquad \text{or any edge } uv \in E(C_n).$$

Since the cycles are regular graphs, then the dual labeling α' is (a', d)-VAT. Again it is readily verified that $\alpha'(V(G)) = \{1, 2, \ldots, n\}$ and $\alpha'(E(G)) = \{n + 1, n + 2, \ldots, 2n\}$. This guarantees that α' is a super (a', d)-VAT labeling. □

In light of Theorems 5.5.1 and 5.5.2, we have

Theorem 5.5.3 ([272]) *The cycle C_n has a super (a, d)-VAT labeling if and only if one of the following statements hold.*

(i) $d \in \{0, 2\}$ and n is odd, $n \geq 3$.
(ii) $d = 1$ and $n \geq 3$.

Next we turn our attention to super (a, d)-VAT labeling of path P_n, $n \geq 3$. Let the path P_n be defined as: $V(P_n) = \{v_1, v_2, \ldots, v_n\}$ and $E(P_n) = \{v_i v_{i+1} : i = 1, 2, \ldots, n - 1\}$. From (5.11) it follows that if P_n, $n \geq 2$, has a super (a, d)-VAT labeling, then $d < 4$.

Theorem 5.5.4 ([272]) *For the path P_n, $n \geq 3$ and $d \in \{0, 1\}$, there is no super (a, d)-VAT labeling.*

Proof The fact that P_n does not have any super VMT labeling was already proved in [180].

Suppose, to the contrary, that γ is a super $(a, 1)$-VAT labeling of P_n. Using Eq. (5.21) we find $a = 3n - 2$. However, the maximum possible weights of the end vertices v_1 and v_n can be obtained as the sum of the largest possible vertex labels and edge labels as follows.

$$wt_\gamma(v_1) = n + (2n - 1) = 3n - 1, \ wt_\gamma(v_n) = (n - 1) + (2n - 2) = 3n - 3 < a$$

or

$$wt_\gamma(v_1) = n + (2n - 2) = a, \ wt_\gamma(v_n) = (n - 1) + (2n - 1) = a.$$

We have a contradiction. Thus P_n does not have any super $(3n - 2, 1)$-VAT labeling. □

Theorem 5.5.5 ([272]) *The path P_n, $n \geq 3$, has a super $(a, 2)$-VAT labeling if and only if n is odd.*

Proof From (5.21) we have that for a super $(a, 2)$-VAT labeling of P_n the smallest vertex-weight is $a = (5n - 3)/2$.

If n is even this contradicts the fact that a is an integer.

For n odd we define the bijection $\beta : V(P_n) \cup E(P_n) \to \{1, 2, \ldots, n\} \cup \{n + 1, n + 2, \ldots, 2n - 1\}$ in the following way:

$$\beta(v_1) = n,$$

$$\beta(v_i) = i - 1 \qquad \text{for } i = 2, 3, \ldots, n$$

$$\beta(v_i v_{i+1}) = \begin{cases} \frac{3n+i}{2} & \text{for } i \text{ odd} \\ n + \frac{i}{2} & \text{for } i \text{ even.} \end{cases}$$

The vertex-weights form the arithmetic progression $(5n - 3)/2, (5n + 1)/2, \ldots, (9n - 7)/2$. Thus P_n has a super $((5n - 3)/2, 2)$-VAT labeling for n odd. □

Theorem 5.5.6 ([272]) *Every path P_n, $n \geq 3$, has a super $(a, 3)$-VAT labeling.*

Proof We discuss two cases.

Case A: n Odd We construct a labeling φ in which the vertices receive the labels

$$\varphi(v_1) = 1,$$

$$\varphi(v_n) = n,$$

$$\varphi(v_i) = n - i + 1, \quad \text{for } i = 2, 3, \ldots, n - 1$$

and the edges receive the labels

$$\varphi(v_i v_{i+1}) = \begin{cases} 2n - 1 - i & \text{for } i \text{ odd} \\ 2n + 1 - i & \text{for } i \text{ even.} \end{cases}$$

We can see that the labeling φ is super labeling and the vertex-weights form the arithmetic progression with difference $d = 3$, namely, $2n - 1, 2n + 2, \ldots, 5n - 4$.

Case B: n Even Define the labeling $\psi : V(P_n) \cup E(P_n) \to \{1, 2, \ldots, 2n - 1\}$, where

$$\psi(v_1) = n - 2,$$

$$\psi(v_n) = n,$$

$$\psi(v_i) = \begin{cases} 2i - 3 & \text{for } i = 2, 3, \ldots, \frac{n}{2} + 1 \\ 2(n - i) & \text{for } i = \frac{n}{2} + 2, \frac{n}{2} + 3, \ldots, n - 1 \end{cases}$$

and

$$\psi(v_i v_{i+1}) = \begin{cases} n + 2i - 1 & \text{for } i = 1, 2, \ldots, \frac{n}{2} \\ 3n - 2i & \text{for } i = \frac{n}{2} + 1, \frac{n}{2} + 2, \ldots, n - 1. \end{cases}$$

We conclude that the total labeling ψ extends to a super $(2n - 1, 3)$-VAT. □

We summarize the results for paths as follows.

Theorem 5.5.7 ([272]) *The path P_n has a super (a, d)-VAT labeling if and only if either one of the following conditions is satisfied.*

(i) $d = 2$ *and* n *is odd,* $n \geq 3$.
(ii) $d = 3$ *and* $n \geq 3$.

5.5.2 Generalized Petersen Graphs

Vertex-antimagic total labelings for generalized Petersen graphs have been studied by Ngurah, Baskoro, and Simamjuntak in [203]. They proved that

Theorem 5.5.8 ([203]) *Every generalized Petersen graph $P(n, m)$, $n \geq 3$, $1 \leq m < n/2$, admits an $(8n + 3, 2)$-VAT labeling.*

Theorem 5.5.9 ([203]) *Every generalized Petersen graph $P(n, m)$, n odd and $m \in \{2, 3, 4\}$, admits an $(a, 1)$-VAT labeling with $a = (15n + 5)/2$ and $a = (21n + 5)/2$.*

They propose the following conjecture.

Conjecture 5.5.1 ([203]) There is an $(a, 1)$-VAT labeling of generalized Petersen graph $P(n, m)$ for n odd and $1 \leq m < n/2$.

Next we will consider super VAT labeling for generalized Petersen graphs.
Since $P(n, m)$ is regular of degree $r = 3$, by Theorems 3.1.8 and 5.3.3, we have

Corollary 5.5.1 *For n even, $n \geq 4$, $1 \leq m \leq n/2 - 1$, every generalized Petersen graph $P(n, m)$ has a super $(a', 2)$-VAT labeling and a super $(a'', 0)$-VAT labeling.*

The next theorem gives a super $(a, 1)$-VAT labeling of $P(n, m)$ for n odd and for all feasible values of m and so proves the Conjecture 5.5.1.

Theorem 5.5.10 ([272]) *For n odd, $n \geq 3$, $1 \leq m < n/2$, every generalized Petersen graph $P(n, m)$ has a super $(a, 1)$-VAT labeling.*

Proof Consider two cycles of $P(n, m)$, an outer-cycle $u_0, u_1, \ldots, u_{n-1}$ and an inner-cycle $v_0, v_m, v_{2m}, \ldots, v_{(n-1)m}$. Rename the inner cycle vertices such that $v_0^* = v_0, v_1^* = v_m, v_2^* = v_{2m}, \ldots, v_{n-1}^* = v_{(n-1)m}$. Then we have the inner-cycle $v_0^*, v_1^*, \ldots, v_{n-1}^*$.

Define a total labeling β for the outer-cycle and the inner-cycle as follows.

$$\beta(v_i^*) = i + 1, \qquad \text{for } i = 0, 1, \ldots, n - 1$$

$$\beta(u_i) = n + 1 + i, \quad \text{for } i = 0, 1, \ldots, n - 1$$

$$\beta(u_i u_{i+1}) = \begin{cases} 3n - \frac{i}{2} & \text{for } i \text{ even} \\ \frac{5n-i}{2} & \text{for } i \text{ odd} \end{cases}$$

$$\beta(v_i^* v_{i+1}^*) = \begin{cases} 4n - \frac{i}{2} & \text{for } i \text{ even} \\ \frac{7n-i}{2} & \text{for } i \text{ odd.} \end{cases}$$

We can see that

$$\beta(u_{i-1}u_i) + \beta(u_i) + \beta(u_i u_{i+1}) = \frac{13n+3}{2}$$

and

$$\beta(v_{i-1}^* v_i^*) + \beta(v_i^*) + \beta(v_i^* v_{i+1}^*) = \frac{15n+3}{2},$$

for $i = 0, 1, \ldots, n-1$, where all the subscripts are taken modulo n.

If we complete the labels for spokes by

$$\beta(y_i x_i) = 4n + 1 + i, \quad \text{for } i = 0, 1, \ldots, n-1,$$

then the vertex-weights of $P(n, m)$ are

$$wt_\beta(u_i) = \frac{21n+5}{2} + i$$
$$wt_\beta(v_i) = \frac{23n+5}{2} + i$$

for $i = 0, 1, \ldots, n-1$.

Thus, the total labeling β is super $((21n+5)/2, 1)$-VAT. □

Theorem 5.5.11 ([271]) *For $n \equiv 0 \pmod 4$, $n \geq 8$, the generalized Petersen graph $P(n, 2)$ has a super $((19n+6)/2, 2)$-VAT labeling and admits a super $((15n+8)/2, 4)$-VAT labeling.*

Proof It was shown in [187] that for $n \equiv 0 \pmod 4$, $n \geq 8$, the generalized Petersen graph $P(n, 2)$ has a $(3n/2+3, 3)$-VAE labeling. By using Theorem 5.3.3 and by direct computation for $p = 2n$ and $q = 3n$, we can see that this theorem is valid. □

Note that for $m = 1$ the generalized Petersen graph $P(n, 1)$ is known also as a prism. Hence from Theorem 5.1.2 and Theorem 5.1.3 it follows.

Theorem 5.5.12 ([32]) *If n is even, $n \geq 4$, then the generalized Petersen graph $P(n, 1)$ has a $((7n+4)/2, 1)$-VAE labeling and a $((3n+6)/2, 3)$-VAE labeling.*

Theorem 5.5.13 ([32]) *If n is odd, $n \geq 3$, then the generalized Petersen graph $P(n, 1)$ has a $((5n+5)/2, 2)$-VAE labeling.*

For $P(n, 1)$, by Theorem 5.3.3, we can formulate the following two corollaries for $d = 3$ and $d = 4$.

Corollary 5.5.2 *For n odd, $n \geq 3$, every generalized Petersen graph $P(n, 1)$ has a super $(a', 3)$-VAT labeling.*

Corollary 5.5.3 *For n even, $n \geq 4$, every generalized Petersen graph $P(n, 1)$ has a super $(a', 4)$-VAT labeling.*

Xirong, Yuansheng, Yue, and Huijun proved the following result.

Theorem 5.5.14 ([295]) *If n is even, $n \geq 6$, then the generalized Petersen graph $P(n, 3)$ has a $((3n+6)/2, 3)$-VAE labeling.*

With respect to Theorem 5.3.3 we have the following corollary.

Corollary 5.5.4 *For n even, $n \geq 6$, every generalized Petersen graph $P(n, 3)$ has a super $((19n+6)/2, 2)$-VAT and a super $((15n+8)/2, 4)$-VAT labeling.*

The following theorem for the generalized Petersen graph $P(n, 2)$ was proved in [296].

Theorem 5.5.15 ([296]) *The generalized Petersen graph $P(n, 2)$ admits a $((3n+6)/2, 3)$-VAE labeling for $n \equiv 2 \pmod 4$, $n \geq 10$.*

According to Theorem 5.3.3 we obtain

Corollary 5.5.5 *For $n \equiv 2 \pmod 4$, $n \geq 10$ the generalized Petersen graph $P(n, 2)$ has a super $((19n+6)/2, 2)$-VAT and a super $((15n+8)/2, 4)$-VAT labeling.*

For further investigation we suggest the following.

Open Problem 5.5.1 For the generalized Petersen graph $P(n, m)$, find (if there exists) a construction of a super (a, d)-VAT labeling.

(i) For n even, $n \geq 4$, $4 \leq m \leq n/2 - 1$, and $d \in \{3, 4\}$.
(ii) For n odd, $n \geq 3$, $2 \leq m < n/2$, and $d \in \{0, 2, 3, 4\}$.

5.5.3 Trees and Unicyclic Graphs

Let G be a graph, where $p = q$. From Eq. (5.16) we have that $a = (7p+3-(p-1)d)/2$. If p is even, then a is an integer only for d odd.

Theorem 5.5.16 ([272]) *For every cycle with at least one tail and even number of vertices there is no super $(a, 1)$-VAT labeling.*

Proof Let G with p vertices be a cycle with at least one tail. Suppose that α is a super $(a, 1)$-VAT labeling of G for $a = 3p + 2$, see (5.16). By assumption, G has at least one vertex of degree 1, say v_t. Then the maximum possible vertex-weight of v_t can be obtained by the biggest value of vertex and the biggest value of edge, i.e., $wt_\alpha(v_t) = p + 2p = 3p$. However, then $wt_\alpha(v_t) < a$ and we have a contradiction. □

Now, we consider a super (a, d)-VAT labeling for tree, where $q = p - 1 \geq 1$. Applying Eq. (5.16), we have $a = (7p - 5 - (p - 1)d)/2$. If p is even, then a is an integer only for d odd.

Theorem 5.5.17 ([272]) *For every tree with even number of vertices there is no super $(a, 1)$-VAT labeling.*

Proof Let G be a tree with $q = p - 1$ and p be even. G has at least two vertices of degree one, say v_t and v_s. Suppose, to the contrary, that β is a super $(a, 1)$-VAT labeling of G for $a = 3p - 2$. Considering the extreme values of the labeling of vertices and edges, the largest vertex-weights for v_t and v_s are

$$wt_\beta(v_t) = p + (2p - 1) = 3p - 1, \; wt_\beta(v_s) = (p - 1) + (2p - 2) = 3p - 3 < a,$$

or

$$wt_\beta(v_t) = p + (2p - 2) = a, \; wt_\beta(v_s) = (p - 1) + (2p - 1) = a.$$

It is obvious that both cases give a contradiction. □

Let v_0 denote the central vertex of a star $K_{1,n}$, $n \geq 1$, and v_i, $1 \leq i \leq n$, be its leaves. In light of Theorem 5.5.17, the star $K_{1,n}$ for n odd has no super $(a, 1)$-VAT labeling. More generally, Sugeng, Miller, Lin, and Bača [272] have proved the following theorem.

Theorem 5.5.18 ([272]) *For star $K_{1,n}$, $n \geq 3$, there is no super (a, d)-VAT labeling for any d.*

Proof Suppose that φ is a super (a, d)-VAT labeling of star $K_{1,n}$. From Inequality (5.11) it follows that $d \leq (3n^2 + 3n - 4)/(2n)$. The smallest vertex-weight of the central vertex v_0 under the labeling φ is

$$\min(wt_\varphi(v_0)) = 1 + (n + 2) + (n + 3) + \cdots + (2n + 1) = \frac{3n^2 + 3n + 2}{2}$$

and the largest vertex-weight of a leave v_i is

$$\max(wt_\varphi(v_i)) = (n + 1) + (2n + 1) = 3n + 2.$$

Clearly, $\min(wt_\varphi(v_0)) - \max(wt_\varphi(v_i)) \leq d$ and thus

$$(n+1)(3n^2 - 9n + 4) \leq 0.$$

The last inequality holds only for two integers $n = 1$ and $n = 2$. This means that $K_{1,n}$ has a super $(4, 1)$-VAT labeling only for $n = 1$, i.e., $\alpha(v_0) = 1$, $\alpha(v_1) = 2$, $\alpha(v_0v_1) = 3$, and a super (a, d)-VAT labeling for $n = 2$, see Theorem 5.5.7. □

5.6 Super Vertex-Antimagic Total Labeling of Disconnected Graphs

Several results for vertex-antimagic total labelings of disconnected graphs are known. Parestu, Silaban, and Sugeng in [206] and [207] proved that the union of suns $S_{n_1} \cup S_{n_2} \cup \cdots \cup S_{n_t}$ admits an (a, d)-VAT labelings for $d \in \{1, 2, 3, 4, 6\}$. Rahim, Ali, and Javaid in [215] studied (a, d)-VAT labelings of disjoint union of cycles and disjoint union of sun graphs. Sugeng and Silaban [274] showed that the disjoint union of any number of odd cycles of orders $n_1, n_2, \ldots, n_t$, each at least 5, has a super $\left(\sum_{i=1}^{t} 3n_i + 2, 1\right)$-VAT labeling.

5.6.1 Disjoint Union of Regular Graphs

Next we deal with the existence of a super (a, d)-VAT labeling for a disjoint union of m copies of regular graph G, denoted by mG. Suppose that an r-regular (p, q) graph admits a super (a, d)-VAT labeling. As G is r-regular, then $r = \Delta = \delta$, $q = rp/2$ and the Inequality (5.11) gives the following upper bound on the value of d

$$d \leq 1 + \frac{r^2(p-2)}{2(p-1)}. \tag{5.26}$$

In [13] Ali et al. presented the following relationship between the minimum vertex-weight of graph G and the minimum vertex-weight of graph mG under a super (a, d)-VAT labeling of G.

Theorem 5.6.1 ([13]) *Let m be a positive integer and a_G be the minimum possible vertex-weight of a super (a, d)-VAT r-regular (p, q) graph G. Then the minimum possible vertex-weight of the graph mG is given by*

$$a_{mG} = ma_G - (m-1)(r + d + 1)/2.$$

Proof From (5.16), it follows that, for a super (a, d)-VAT r-regular (p, q) graph, the minimum possible vertex-weight is

$$a_G = \frac{p+1}{2} - \frac{(p-1)d}{2} + \frac{p(r^2+4r)+2r}{4}.$$

If

$$a_{mG} = \frac{1+mp}{2} - \frac{(mp-1)d}{2} + \frac{mp(r^2+4r)+2r}{4}$$

is the minimum possible vertex-weight of mG, then the equation

$$a_{mG} = ma_G - \frac{(m-1)(r+d+1)}{2} \tag{5.27}$$

gives a relationship between a_G and a_{mG}. □

According to (5.27), if r and d have the same parity and m is even, then the minimum possible vertex-weight of mG is not an integer. Thus, as consequence of Theorem 5.6.1, we obtain the following corollary.

Corollary 5.6.1 ([13]) *Let G be an r-regular (p, q) graph. If r and d have the same parity and $m \geq 2$ is even, then there is no super (a, d)-VAT labeling of mG.*

With respect to Theorem 3.4.1, it follows that

Theorem 5.6.2 ([13]) *Let m be a positive integer. If a graph G is an r-regular graph that admits a super VMT labeling and $(m-1)(r+1)/2$ is an integer, then the graph mG has a super $(a, 2)$-VAT labeling.*

Proof According to Theorem 3.4.1, the graph mG admits a super VMT labeling $\lambda : V(mG) \cup E(mG) \to \{1, 2, \dots, mp + mq\}$ with the constant vertex-weights a^0_{mG}. This means that there exists a vertex labeling, say λ_V, $\lambda_V : V(mG) \to \{1, 2, \dots, mp\}$, and an edge labeling, say λ_E, $\lambda_E : E(mG) \to \{mp + 1, mp + 2, \dots, mp + mq\}$, such that the vertex-weight, associated with the labeling λ_E, for each vertex $v \in V(mG)$ is $wt_{\lambda_E}(v) = a^0_{mG} - \lambda_V(v)$, $1 \leq \lambda_V(v) \leq mp$, thus $a^0_{mG} - mp \leq wt_{\lambda_E}(v) \leq a^0_{mG} - 1$.

Consider now a new labeling $\beta : V(mG) \cup E(mG) \to \{1, 2, \dots, mp + mq\}$ such that

$$\beta(u) = mp + 1 - \lambda_V(u), \qquad \text{for } u \in V(mG)$$

$$\beta(uv) = \lambda_E(uv), \qquad \text{for } uv \in E(mG).$$

The new labeling β induces the vertex-weight $wt_\beta(v) = wt_{\lambda_E}(v) + \beta(x) = a^0_{mG} + mp + 1 - 2\lambda_V(v)$ for each vertex $v \in V(mG)$. Thus the set of vertex-weights induced by β is $\{a^0_{mG} - mp + 1, a^0_{mG} - mp + 3, \ldots, a^0_{mG} + mp - 1\}$ and β is a super $(a^0_{mG} - mp + 1, 2)$-VAT labeling of mG. □

Immediately we obtain

Corollary 5.6.2 ([13]) *Let G be even regular super (a, d)-VAT graph, $d = 0, 2$. Then mG is super (b, d)-VAT if and only if m is odd.*

Now, we concentrate on 2-regular graphs which admit (super) $(a, 1)$-VAT labelings.

Theorem 5.6.3 ([13]) *Let G be a 2-regular (super) $(a, 1)$-VAT graph. Then mG, $m \geq 1$, also admits a (super) $(b, 1)$-VAT labeling.*

Proof Let λ be a (super) $(a, 1)$-VAT labeling of a 2-regular graph G of order p

$$\lambda : V(G) \cup E(G) \to \{1, 2, \ldots, 2p\}.$$

Let u, w be the vertices adjacent to the vertex v. The set of vertex-weights is

$$\{\lambda(v) + \lambda(vu) + \lambda(vw) : v \in V(G)\} = \{a, a + 1, \ldots, a + p - 1\}.$$

By the symbol v_i we denote the vertex corresponding to the vertex v in the ith copy of G in mG. Analogously, let $v_i u_i$ denote the edge corresponding to the edge vu in the ith copy of G in mG.

We define a labeling β of mG in the following way:

$$\beta(v_i) = m\lambda(v) + 1 - i, \qquad \text{for } v \in V(G), i = 1, 2, \ldots, m$$
$$\beta(v_i u_i) = m(\lambda(vu) - 1) + i, \qquad \text{for } vu \in E(G), i = 1, 2, \ldots, m.$$

Obviously β is a total labeling. Let u_i, w_i be the vertices adjacent to the vertex v_i. For the vertex-weight of v_i we get

$$\beta(v_i) + \beta(v_i u_i) + \beta(v_i w_i) = m(\lambda(v) + \lambda(vu) + \lambda(vw)) - 2m + 1 + i.$$

This means that, for $a \leq \lambda(v) + \lambda(vu) + \lambda(vw) \leq a + p - 1$ and $1 \leq i \leq m$, the vertex-weights of mG consist of consecutive integers $m(a-2)+2, m(a-2)+3, \ldots, m(a + p - 2) + 1$. Thus β is a (super) $(m(a - 2) + 2, 1)$-VAT labeling of mG. □

Now we consider super (a, d)-VAT labelings for the disjoint union of m copies of cycle C_n. Swaminathan and Jeyanthi [275] have proved that the mC_n are super VMT if and only if m and n are odd. The same result follows from the work of Figueroa-Centeno et al. [99]. The following theorem is an analogy of Theorem 5.5.3 and extends these results to super (a, d)-VAT labeling of mC_n.

Theorem 5.6.4 ([13]) *The graph mC_n has a super (a, d)-VAT labeling if and only if one of the following conditions is satisfied.*

(i) $d \in \{0, 2\}$ and m, n are odd, $m, n \geq 3$.
(ii) $d = 1$ for every $m \geq 2$ and $n \geq 3$.

Proof If the disjoint union of m copies of C_n is super (a, d)-VAT, then, for $r = 2$ and $p = mn$, from (5.26) it follows that $d \leq 3 - 2/(mn - 1)$. If $m \geq 2$ and $n \geq 3$, then $2/(mn - 1) > 0$ and thus $d < 3$. According to Theorems 5.5.3, 3.4.1, and 5.6.2, we have that mC_n admits a super (a, d)-VAT labeling for $d \in \{0, 2\}$, and m, n odd.

Following Corollary 5.6.1, the family of cycles mC_n admits no super (a, d)-VAT labeling for $d \in \{0, 2\}$ and m even. It remains to consider the case when $d = 1$. It follows from Theorems 5.5.3 and 5.6.3 that mC_n admits a super $(a, 1)$-VAT labeling for every $m \geq 2$ and $n \geq 3$. □

Recall that by a k-factor of a graph we mean its k-regular spanning subgraph. Kovář [167] presented methods of construction of $(a, 1)$-VAT labelings of regular graphs. He proved

Theorem 5.6.5 ([167]) *Let G be a $2r$-regular graph with vertices $v_1, v_2, \ldots, v_n$. Let s be an integer, $s \in \{(rn + 1)(r + 1) + tn : t = 0, 1, \ldots, r\}$. Then there exists an $(s, 1)$-VAT labeling λ of G such that $\lambda(v_i) = s + (i - 1)$.*

The following theorem is a generalization of the results from [167].

Theorem 5.6.6 ([13]) *Let G be a graph decomposable into two edge-disjoint spanning subgraphs G_1 and G_2, where G_1 is a super $(a, 1)$-VAT graph and G_2 is a $2k$-factor of G. Then G is super $(b, 1)$-VAT.*

Combining the previous results gives

Theorem 5.6.7 ([13]) *Let G be an even regular graph that contains a 2-regular (super) $(a, 1)$-VAT factor. Then mG is (super) $(b, 1)$-VAT for every positive integer m.*

According to Theorems 5.6.4 and 5.6.6, it follows that

Theorem 5.6.8 ([13]) *Let G be an even regular Hamilton graph. Then mG is super $(a, 1)$-VAT for every positive integer m.*

To construct a super $(a, 1)$-VAT labeling for odd regular graphs, Ali, Bača, Lin, and Semaničová-Feňovčíková [13] made use of the known results on the relationship between $(a, 1)$-VAT labeling and supermagic labeling, see Theorem 5.2.3. Note that according to Corollary 5.6.1, if G is an odd regular graph and mG is super $(a, 1)$-VAT, then m must be odd.

It is known that if G is a regular supermagic graph, then G is also super $(a, 1)$-VAT. Thus, according to Theorem 2.5.7, the following theorem is true.

Theorem 5.6.9 ([13]) *Let G be a supermagic graph decomposable into k pairwise edge-disjoint r-regular factors.*

(i) If k is even, then mG is super $(a, 1)$-VAT for every positive integer m.
(ii) If k is odd, then mG is super $(a, 1)$-VAT for every odd positive integer m.

Moreover, using Theorem 5.6.6, it is not difficult to see that

Theorem 5.6.10 ([13]) *Let G be an $(r + 2l)$-regular graph. Let G_1 be its supermagic r-regular factor that is decomposable into t pairwise edge-disjoint δ-regular factors.*

(i) If t is even, then mG is super $(a, 1)$-VAT for every positive integer m.
(ii) If t is odd, then mG is super $(a, 1)$-VAT for every odd positive integer m.

Recall that the Möbius ladder M_n, where $6 \leq n \equiv 0 \pmod 2$, is a 3-regular graph consisting of a cycle on n vertices in which all pairs of opposite vertices are joined by an edge. Sedláček [231] proved the following result.

Theorem 5.6.11 ([231]) *Let $n \geq 6$ be an even integer. The Möbius ladder M_n is supermagic if and only if $n \equiv 2 \pmod 4$.*

Thus we get the following result for odd regular graphs.

Corollary 5.6.3 *Let G be an odd regular graph with a spanning subgraph isomorphic to the Möbius ladder. Then mG is super $(a, 1)$-VAT if and only if m is odd.*

Proof Let G be an odd regular graph decomposable into two edge-disjoint factors G_1 and G_2. Let G_1 be isomorphic to the Möbius ladder. Evidently, G_2 is an even regular factor. As the Möbius ladder is decomposable into three edge-disjoint 1-factors, then, according to Theorems 5.6.11, 5.6.10, and Corollary 5.6.1, mG is super $(a, 1)$-VAT if and only if m is odd. □

Similarly, if G is an odd regular graph with a spanning subgraph isomorphic to tM_n, $6 \leq n \equiv 2 \pmod 4$, $t \equiv 1 \pmod 2$, then for m odd, the graph mG admits a super $(a, 1)$-VAT labeling. Analogous results can also be obtained for other families of odd regular graphs containing a spanning subgraph with desired properties.

5.6.2 Disjoint Union of Paths

Let us now consider a disjoint union of m copies of the path P_n and denote it by mP_n. The graph mP_n, $m > 1$, is disconnected, with vertex set $V(mP_n) = \{v_i^j : 1 \leq i \leq n,\ 1 \leq j \leq m\}$ and edge set $E(mP_n) = \{v_i^j v_{i+1}^j : 1 \leq i \leq n-1,\ 1 \leq j \leq m\}$. Ali, Bača and Bashir [12] characterized a super VAT graphs of mP_2 as follows.

Theorem 5.6.12 ([12]) *The graph mP_2, $m \geq 1$, has a super (a, d)-VAT labeling if and only if m is odd and $d = 1$.*

Proof Assume that mP_2 has a super (a, d)-VAT labeling λ. From (5.11), it follows that, for $\delta = \Delta = 1$, $d \leq (3m-2)/(2m-1) < \frac{3}{2}$. For $d = 0$ we suppose, to the contrary, that λ is a super VMT labeling with common vertex-weight k. Clearly, $\lambda(v_1^j) + \lambda(v_1^j v_2^j) = k = \lambda(v_1^j v_2^j) + \lambda(v_2^j)$ and $\lambda(v_1^j) = \lambda(v_2^j)$, for every $1 \leq j \leq m$. This produces a contradiction. Thus, mP_2 does not have any super VMT labeling.

From (5.16) we have that for $d = 1$ the smallest vertex-weight is $a = (5m+3)/2$. If m is even this contradicts the fact that a is an integer.

It remains to investigate whether mP_2, for m odd, admits a super $((5m+3)/2, 1)$-VAT labeling. We construct a total labeling λ_1 as follows.

$$\lambda_1(v_1^j) = \begin{cases} \frac{m-1}{2} + j & \text{for } 1 \leq j \leq \frac{m+1}{2} \\ j - \frac{m+1}{2} & \text{for } \frac{m+3}{2} \leq j \leq m \end{cases}$$

$$\lambda_1(v_2^j) = \begin{cases} \frac{3m-1}{2} + j & \text{for } 1 \leq j \leq \frac{m+1}{2} \\ \frac{m-1}{2} + j & \text{for } \frac{m+3}{2} \leq j \leq m \end{cases}$$

$$\lambda_1(v_1^j v_2^j) = 2m + j \qquad \text{for } 1 \leq j \leq m.$$

Evidently, λ_1 is a super $((5m+3)/2, 1)$-VAT labeling, for m odd, since all verifications are trivial. □

If the disjoint union of m copies of P_n, $n \geq 3$, is super (a, d)-VAT, then, for $p = mn$ and $q = m(n-1)$, it follows from (5.11) that $d < 4$.

Theorem 5.6.13 ([12]) *For the graph mP_3, $m \geq 1$, there is no super VMT labeling.*

Proof Suppose mP_3 has a super VMT labeling with the common vertex-weight k. The maximum possible sum of the vertex-weights on the leaves is the sum of the $2m$ largest vertex labels and all the edge labels

$$\sum_{j=1}^{m} (wt(v_1^j) + wt(v_3^j)) \leq \sum_{i=1}^{2m} (m+i) + \sum_{i=1}^{2m} (3m+i) = 2m(6m+1).$$

Since there are $2m$ leaves, then

$$k \leq 6m + 1. \tag{5.28}$$

The minimum possible sum of vertex-weights on the internal vertices of degree 2 is the sum of the m smallest vertex labels and all the edge labels

$$\sum_{j=1}^{m} wt(v_2^j) \geq \sum_{j=1}^{m} j + \sum_{i=1}^{2m} (3m+i) = \frac{m(17m+3)}{2}.$$

Since there are m internal vertices, then

$$k \geq \frac{17m+3}{2}. \tag{5.29}$$

Inequalities (5.28) and (5.29) imply that

$$\frac{17m+3}{2} \leq k \leq 6m+1,$$

which is a contradiction. □

Theorem 5.6.14 ([12]) *For the graph mP_3, $m > 1$, there is no super $(a, 3)$-VAT labeling.*

Proof Assume that mP_3 admits a super $(a, 3)$-VAT labeling $\lambda : V(mP_3) \cup E(mP_3) \to \{1, 2, \dots, 5m\}$ and $\{a, a+3, \dots, a+(3m-1)3\}$ is the set of the vertex-weights. The smallest possible vertex-weight is achieved by putting the label 1 on a leaf and the label $3m+1$ on its incident edge. Thus $a = 3m+2$.

Suppose that the first $2m$ vertex-weights $3m+2, 3m+5, \dots, 9m-1$ occur on the leaves and the next m vertex-weights $9m+2, 9m+5, \dots, 12m-1$ occur on the internal vertices of mP_3. The largest possible vertex-weight on a leaf can be composed as a sum of the largest vertex label $3m$ and the largest edge label $5m$. Since $8m < 9m-1$, for $m > 1$, the value $9m-1$ is the vertex-weight of an internal vertex. However, there are still m vertex-weights bigger than $9m-1$ but only $m-1$ internal vertices, and so we have a contradiction. □

Theorem 5.6.15 ([12]) *If $m \equiv 1 \pmod 6$, $m \geq 1$, then the graph mP_3 has a super $(a, 2)$-VAT labeling.*

Proof Let h be a positive integer and let $m = 1+6h$. We construct a labeling λ_2 of mP_3 in the following way.

$$\lambda_2(v_1^j) = \begin{cases} h+1-j & \text{for } 1 \leq j \leq h \\ 2m+h+1-j & \text{for } h+1 \leq j \leq m-h \\ 4m+1-h-j & \text{for } m-h+1 \leq j \leq m \end{cases}$$

$$\lambda_2(v_2^j) = \begin{cases} m+2h+2-2j & \text{for } 1 \leq j \leq \frac{m+1}{2} \\ 2m+2h+2-2j & \text{for } \frac{m+3}{2} \leq j \leq m \end{cases}$$

$$\lambda_2(v_3^j) = \begin{cases} \frac{5m+3}{2} - h - j & \text{for } 1 \leq j \leq \frac{m+1}{2} - h \\ \frac{m+3}{2} + h - j & \text{for } \frac{m+3}{2} - h \leq j \leq \frac{m+1}{2} \\ \frac{7m+3}{2} - h - j & \text{for } \frac{m+3}{2} \leq j \leq m \end{cases}$$

$$\lambda_2(v_1^j v_2^j) = 4m + 1 - j \qquad \text{for } 1 \le j \le m$$

$$\lambda_2(v_2^j v_3^j) = \begin{cases} \frac{9m+3}{2} - j & \text{for } 1 \le j \le \frac{m+1}{2} \\ \frac{11m+3}{2} - j & \text{for } \frac{m+3}{2} \le j \le m. \end{cases}$$

We can see that the labeling λ_2 is a bijective function from $V(mP_3) \cup E(mP_3)$ onto the set $\{1, 2, \ldots, 5m\}$. The vertex-weights of mP_3, under the labeling λ_2, constitute the sets

$$\begin{aligned}
W_{\lambda_2}^1 &= \{wt_{\lambda_2}(v_1^j) = 4m + 2 + h - 2j : \text{ if } \ 1 \le j \le h\} \\
&= \{4m - h + 2, 4m - h + 4, \ldots, 4m + h\} \\
W_{\lambda_2}^2 &= \{wt_{\lambda_2}(v_3^j) = 5m + 3 + h - 2j : \text{ if } \ \tfrac{m+3}{2} - h \le j \le \tfrac{m+1}{2}\} \\
&= \{4m + h + 2, 4m + h + 4, \ldots, 4m + 3h\} \\
W_{\lambda_2}^3 &= \{wt_{\lambda_2}(v_1^j) = 6m + h + 2 - 2j : \text{ if } \ h + 1 \le j \le m - h\} \\
&= \{4m + 3h + 2, 4m + 3h + 4, \ldots, 6m - h\} \\
W_{\lambda_2}^4 &= \{wt_{\lambda_2}(v_1^j) = 8m + 2 - h - 2j : \text{ if } \ m - h + 1 \le j \le m\} \\
&= \{6m - h + 2, 6m - h + 4, \ldots, 6m + h\} \\
W_{\lambda_2}^5 &= \{wt_{\lambda_2}(v_3^j) = 7m + 3 - h - 2j : \text{ if } \ 1 \le j \le \tfrac{m+1}{2} - h\} \\
&= \{6m + h + 2, 6m + h + 4, \ldots, 7m - h + 1\} \\
W_{\lambda_2}^6 &= \{wt_{\lambda_2}(v_3^j) = 9m + 3 - h - 2j : \text{ if } \ \tfrac{m+3}{2} \le j \le m\} \\
&= \{7m - h + 3, 7m - h + 5, \ldots, 8m - h\} \\
W_{\lambda_2}^7 &= \{wt_{\lambda_2}(v_2^j) = \tfrac{19m+9}{2} + 2h - 4j : \text{ if } \ 1 \le j \le \tfrac{m+1}{2}\} \\
&= \{\tfrac{15m+5}{2} + 2h, \tfrac{15m+5}{2} + 2h + 4, \ldots, \tfrac{19m+1}{2} + 2h\} \\
W_{\lambda_2}^8 &= \{wt_{\lambda_2}(v_2^j) = \tfrac{23m+9}{2} + 2h - 4j : \text{ if } \ \tfrac{m+3}{2} \le j \le m\} \\
&= \{\tfrac{15m+5}{2} + 2h + 2, \tfrac{15m+5}{2} + 2h + 6, \ldots, \tfrac{19m+1}{2} + 2h - 2\}.
\end{aligned}$$

Hence the set $\bigcup_{i=1}^{8} W_{\lambda_2}^i = \{4m-h+2, 4m-h+4, \ldots, (19m+1)/2+2h\}$ contains an arithmetic progression with the common difference 2. Thus λ_2 is a super $(a, 2)$-VAT labeling. □

A super $(29, 2)$-VAT labeling of $7P_3$ is given in Fig. 5.8 where integers in italic font represent vertex-weights.

Theorem 5.6.16 ([12]) *For the graph mP_4, $m \ge 1$, there is no super VMT labeling.*

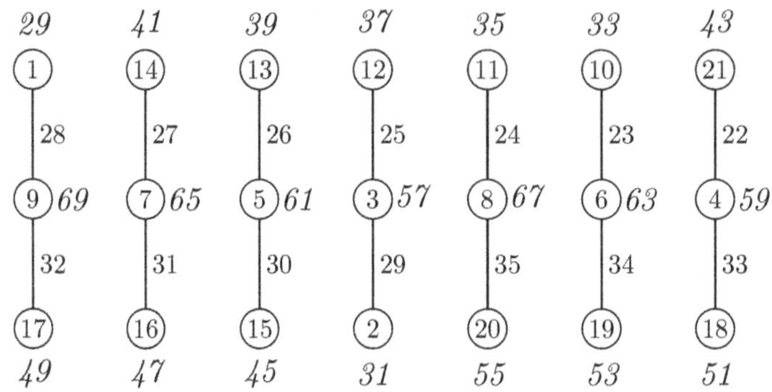

Fig. 5.8 Super (29, 2)-VAT labeling of $7P_3$

Proof Suppose, to the contrary, that mP_4 has a super VMT labeling with the common vertex-weight k. We calculate the minimum possible sum of the vertex-weights on the inner vertices of degree 2; this is achieved by placing the $2m$ smallest vertex labels on the inner vertices and using all edge labels, where the m smallest labels on edges $v_2^j v_3^j$, $1 \leq j \leq m$, will each be added twice. This gives

$$\sum_{j=1}^{m}(wt(v_2^j) + wt(v_3^j)) \geq \sum_{i=1}^{2m} i + 2\sum_{j=1}^{m}(4m + j) + \sum_{i=1}^{2m}(5m + i) = m(23m + 3).$$

Since there are $2m$ inner vertices, we must therefore have

$$k \geq \frac{23m + 3}{2}. \tag{5.30}$$

Calculating the maximum possible sum of the vertex-weights on the outer vertices, we take the $2m$ largest vertex labels and the $2m$ largest edge labels,

$$\sum_{j=1}^{m}(wt(v_1^j) + wt(v_4^j)) \leq \sum_{i=1}^{2m}(2m + i) + \sum_{i=1}^{2m}(5m + i) = 2m(9m + 1).$$

Since there are $2m$ outer vertices, we have

$$k \leq 9m + 1. \tag{5.31}$$

Thus, combining (5.30) and (5.31) gives

$$\frac{23m + 3}{2} \leq k \leq 9m + 1$$

and we have a contradiction. Consequently, a super VMT labeling cannot exist. □

Theorem 5.6.17 ([12]) *If* $m \equiv 3 \pmod 4$, $m \geq 3$, *then the graph* mP_4 *has a super* $(a, 2)$*-VAT labeling.*

Proof Let s be a nonnegative integer and let $m = 3 + 4s$. For $s \geq 0$, define the bijection $\lambda_3 : V(mP_4) \cup E(mP_4) \to \{1, 2, \ldots, 7m\}$ as follows.

$$\lambda_3(v_1^j) = \begin{cases} 3m + 2 + s - j & \text{for } 1 \leq j \leq \frac{m+1}{2} \\ 4m + 2 + s - j & \text{for } \frac{m+3}{2} \leq j \leq m \end{cases}$$

$$\lambda_3(v_2^j) = \begin{cases} \frac{3m+5}{2} + s - 2j & \text{for } 1 \leq j \leq \frac{m+1}{2} \\ \frac{5m+5}{2} + s - 2j & \text{for } \frac{m+3}{2} \leq j \leq m \end{cases}$$

$$\lambda_3(v_3^j) = \begin{cases} \frac{5m+5}{2} + s - 2j & \text{for } 1 \leq j \leq \frac{m+1}{2} \\ \frac{7m+5}{2} + s - 2j & \text{for } \frac{m+3}{2} \leq j \leq m \end{cases}$$

$$\lambda_3(v_4^j) = \begin{cases} 2 + s - j & \text{for } 1 \leq j \leq s + 1 \\ 4m + 2 + s - j & \text{for } s + 2 < j < \frac{m+1}{2} \\ m + 2 + s - j & \text{for } \frac{m+3}{2} \leq j \leq m \end{cases}$$

$$\lambda_3(v_1^j v_2^j) = \begin{cases} \frac{11m+3}{2} - j & \text{for } 1 \leq j \leq \frac{m+1}{2} \\ \frac{13m+3}{2} - j & \text{for } \frac{m+3}{2} \leq j \leq m \end{cases}$$

$$\lambda_3(v_2^j v_3^j) = 5m + 1 - j \qquad \text{for } 1 \leq j \leq m$$

$$\lambda_3(v_3^j v_4^j) = \begin{cases} \frac{13m+3}{2} - j & \text{for } 1 \leq j \leq \frac{m+1}{2} \\ \frac{15m+3}{2} - j & \text{for } \frac{m+3}{2} \leq j \leq m. \end{cases}$$

Then for the vertex-weights of mP_4 we have

$$\begin{aligned}
W_{\lambda_3}^1 &= \{wt_{\lambda_3}(v_4^j) = \tfrac{13m+7}{2} + s - 2j : \text{ if } 1 \leq j \leq s+1\} \\
&= \{\tfrac{13m+3}{2} - s, \tfrac{13m+3}{2} - s + 2, \ldots, \tfrac{13m+3}{2} + s\} \\
W_{\lambda_3}^2 &= \{wt_{\lambda_3}(v_4^j) = \tfrac{17m+7}{2} + s - 2j : \text{ if } \tfrac{m+3}{2} \leq j \leq m\} \\
&= \{\tfrac{13m+3}{2} + s + 2, \tfrac{13m+3}{2} + s + 4, \ldots, \tfrac{15m+1}{2} + s\} \\
W_{\lambda_3}^3 &= \{wt_{\lambda_3}(v_1^j) = \tfrac{17m+7}{2} + s - 2j : \text{ if } 1 \leq j \leq \tfrac{m+1}{2}\} \\
&= \{\tfrac{15m+1}{2} + s + 2, \tfrac{15m+1}{2} + s + 4, \ldots, \tfrac{17m+3}{2} + s\} \\
W_{\lambda_3}^4 &= \{wt_{\lambda_3}(v_1^j) = \tfrac{21m+7}{2} + s - 2j : \text{ if } \tfrac{m+3}{2} \leq j \leq m\} \\
&= \{\tfrac{17m+3}{2} + s + 2, \tfrac{17m+3}{2} + s + 4, \ldots, \tfrac{19m+1}{2} + s\}
\end{aligned}$$

$$W^5_{\lambda_3} = \{wt_{\lambda_3}(v^j_4) = \tfrac{21m+7}{2} + s - 2j : \text{ if } s+2 \le j \le \tfrac{m+1}{2}\}$$
$$= \{\tfrac{19m+1}{2} + s + 2, \tfrac{19m+1}{2} + s + 4, \dots, \tfrac{21m-1}{2} - s\}$$
$$W^6_{\lambda_3} = \{wt_{\lambda_3}(v^j_2) = 12m + 5 + s - 4j : \text{ if } 1 \le j \le \tfrac{m+1}{2}\}$$
$$= \{10m + s + 3, 10m + s + 7, \dots, 12m + s + 1\}$$
$$W^7_{\lambda_3} = \{wt_{\lambda_3}(v^j_2) = 14m + 5 + s - 4j : \text{ if } \tfrac{m+3}{2} \le j \le m\}$$
$$= \{10m + s + 5, 10m + s + 9, \dots, 12m + s - 1\}$$
$$W^8_{\lambda_3} = \{wt_{\lambda_3}(v^j_3) = 14m + 5 + s - 4j : \text{ if } 1 \le j \le \tfrac{m+1}{2}\}$$
$$= \{12m + s + 3, 12m + s + 7, \dots, 14m + s + 1\}$$
$$W^9_{\lambda_3} = \{wt_{\lambda_3}(x^j_3) = 16m + 5 + s - 4j : \text{ if } \tfrac{m+3}{2} \le j \le m\}$$
$$= \{12m + s + 5, 12m + s + 9, \dots, 14m + s - 1\}$$

and

$$\bigcup_{i=1}^{9} W^i_{\lambda_3} = \{\frac{13m+3}{2} - s, \frac{13m+3}{2} - s + 2, \dots, 14m + s + 1\}$$

contains an arithmetic progression with the difference 2. This implies that λ_3 is a super $(a, 2)$-VAT labeling. □

A super (46, 2)-VAT labeling of $7P_4$ is shown in Fig. 5.9. Integers in italic font denote vertex-weights.

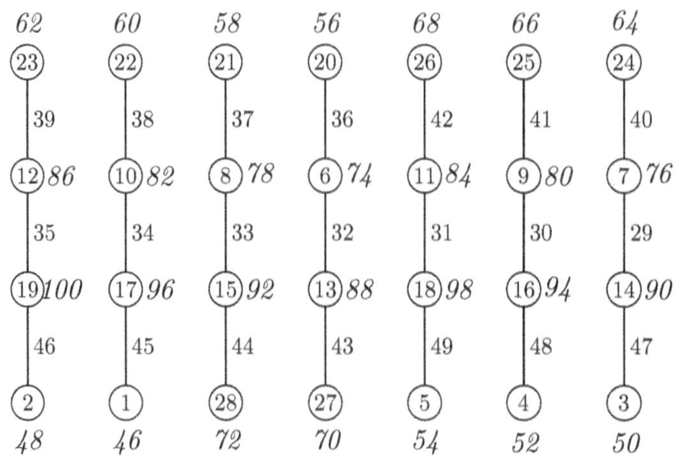

Fig. 5.9 Super (46, 2)-VAT labeling of $7P_4$

For mP_3 and mP_4, Ali et al. [12] tried to find a super $(a, 2)$-VAT labeling also for other values of m and a super $(a, 1)$-VAT labeling for every $m \geq 2$, but so far without success. Thus they propose the following.

Open Problem 5.6.1 ([12]) For the graphs mP_3 and mP_4, determine if there is a super (a, d)-VAT labeling, for every $m \geq 2$ and $d \in \{1, 2\}$.

In the case when $n \geq 5$ and $d < 4$ they do not have any answer. Therefore, for further investigation they propose also the following open problem.

Open Problem 5.6.2 ([12]) For the graph mP_n, $n \geq 5$, and $m > 1$, determine if there is a super (a, d)-VAT labeling for the feasible values of the difference d.

Chapter 6
Edge-Antimagic Total Labelings

6.1 Edge-Antimagic Vertex Labeling

Simanjuntak et al. [256] defined an (a, d)*-edge-antimagic vertex* ((a, d)-EAV) labeling for a (p, q) graph G as an injective mapping f from $V(G)$ onto the set $\{1, 2, \ldots, p\}$ with the property that the edge-weights $\{w(uv) = f(u) + f(v), uv \in E(G)\}$, form an arithmetic sequence with the first term a and difference d, where $a > 0$ and $d \geq 0$ are two fixed integers.

Acharya and Hegde [5], see also [126], introduced the concept of a *strongly* (a, d)*-indexable labeling* which is equivalent to (a, d)-EAV labeling. The relationship between the sequential graphs and the graphs having an (a, d)-EAV labeling is shown in [48]. Note that sequential graphs were defined in Sect. 4.6.

As an illustration, Fig. 6.1 provides an example of a (5, 1)-EAV labeling of the friendship graph f_4, where $\{5, 6, \ldots, 16\}$ is the set of edge-weights.

Assume that a (p, q) graph G has an (a, d)-EAV labeling $f : V(G) \to \{1, 2, \ldots, p\}$ and $\{w(uv) = f(u) + f(v) : uv \in E(G)\} = \{a, a + d, \ldots, a + (q - 1)d\}$ is the set of the edge-weights. The minimum possible edge-weight, under the labeling f, is the sum of two distinct positive integers, and so is at least 3. On the other hand, the maximum edge-weight is no more than $2p - 1$. Thus

$$a + (q - 1)d \leq 2p - 1$$

and

$$d \leq \frac{2p - 4}{q - 1}. \tag{6.1}$$

M. Bača et al., *Magic and Antimagic Graphs*, Developments in Mathematics 60,
https://doi.org/10.1007/978-3-030-24582-5_6

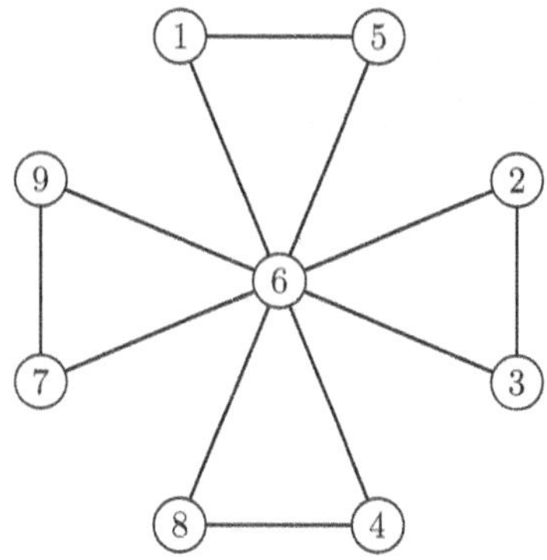

Fig. 6.1 (5, 1)-EAV labeling of f_4

If G is a (p, q) graph that is not a tree, i.e., $p \leq q$, then (6.1) gives $d < 2$. It is not difficult to see that for every connected (p, q) graph, $q \geq 2$, there is no $(a, 0)$-EAV labeling, and so we get

Lemma 6.1.1 ([42]) *Let G be a connected (p, q) graph that is not a tree. If G has an (a, d)-EAV labeling, then $d = 1$.*

Applying Inequality (6.1) to several families of graphs, we obtain

Lemma 6.1.2 ([42]) *The following statements hold.*

(i) For every cycle there is no (a, d)-EAV labeling with $d > 1$.
(ii) For every path there is no (a, d)-EAV labeling with $d > 2$.
(iii) For every complete graph K_n, $n > 3$, there is no (a, d)-EAV labeling.
(iv) For every symmetric complete bipartite graph $K_{n,n}$, $n > 3$, there is no (a, d)-EAV labeling.

For cycle C_n, $n \geq 3$, we have the following lemma.

Lemma 6.1.3 ([256]) *There is no (a, d)-EAV labeling for even cycles.*

In light of Lemmas 6.1.2 part (i) and 6.1.3 we can see that only odd cycles may have an $(a, 1)$-EAV labeling. The existence of this labeling was proved by Simanjuntak, Bertault, and Miller.

Lemma 6.1.4 ([256]) *Every cycle C_{2k+1}, $k \geq 1$, has a $(k + 2, 1)$-EAV labeling.*

For any graph G, if G has an (a, d)-EAV labeling, then we can obtain two new graphs with (a, d)-EAV labelings by removing the edge with the largest, respectively, the smallest weight.

Lemma 6.1.5 ([256]) *Suppose G has an (a, d)-EAV labeling. Let e be the edge with the largest edge-weight and f be the edge with the smallest edge-weight. Then $G - \{e\}$ has an (a, d)-EAV labeling and $G - \{f\}$ has an $(a + d, d)$-EAV labeling.*

Direct consequence of Lemmas 6.1.3 and 6.1.5 gives that every odd path has a $(k + 2, 1)$-EAV labeling. The remaining cases have been established as follows.

Lemma 6.1.6 ([256]) *The following statements hold.*

(i) Every even path P_{2k}, $k \geq 1$, has a $(k+2, 1)$-EAV labeling.
(ii) Every path P_n has a $(3, 2)$-EAV labeling.

For complete graphs K_n, by Lemma 6.1.2 part (iii), there is no (a, d)-EAV labeling for $n > 3$. The cases $n = 1, 2$ are trivial: if $n = 3$, then the graph K_3 is equal to C_3, and C_3 has a $(3, 1)$-EAV labeling, by Lemma 6.1.4. From Lemma 6.1.2 part (iv), we know that every symmetric complete bipartite graph $K_{n,n}$, with $n > 3$, does not have an (a, d)-EAV labeling. However, for complete bipartite graphs $K_{2,2}$ and $K_{3,3}$ in [36] the following result is proved.

Lemma 6.1.7 ([36]) *For symmetric complete bipartite graphs $K_{2,2}$ and $K_{3,3}$, there is no (a, d)-EAV labeling.*

6.2 Building of New Larger (a, d)-Edge-Antimagic Vertex Graphs by Using Adjacency Matrices

Let G be a graph of order n with an (a, d)-EAV labeling f. Label the vertices in G such that $f(v_i) = i$, for $i = 1, 2, \ldots, n$. An $n \times n$ matrix $A_G = [a_{ij}]$, $i, j = 1, 2, \ldots, n$, is called an *adjacency matrix* of G if

$$a_{ij} = \begin{cases} 1 & \text{if } v_i v_j \in E(G) \\ 0 & \text{otherwise.} \end{cases}$$

Let G be an (a, d)-EAV graph with adjacency matrix A_G. Since G is an undirected graph, A_G is a symmetric matrix. Beside that, A_G has another characteristic that shows that A_G is a matrix of an (a, d)-EAV graph. A skew diagonal S_r, $r = 3, 4, \ldots, 2n-1$, of A_G is $\{a_{ij} : i + j = r;\ i, j = 1, 2, \ldots, n\}$, see Fig. 6.2.

A skew diagonal S_r contains all entries of A_G that are related to edges with the weight r. With respect to the symmetry of A_G, every skew diagonal of A_G has either zero or exactly two "1" elements. A skew diagonal that only contains the zero elements is called zero skew diagonal, while a skew diagonal that contains exactly two "1" elements is called nonzero skew diagonal.

Sugeng and Miller [269] explained that the set of edge-weights $\{f(u) + f(v) : u, v \in V(G)\}$ in skew diagonal lines generate a sequence of integers of difference d. If $d = 1$, then the nonzero skew diagonal lines form a band of consecutive integers. If $d = 2$, then the nonzero skew diagonal lines form a band of difference 2 with a zero skew diagonal line in between. We have similar skew diagonal line bands for $d = 3, 4, \ldots$ and denote such a skew diagonal band as d-band.

A *maximal (a, d)-EAV graph* of order n is a graph that has an (a, d)-EAV labeling and has the maximum possible number of edges. If G is a maximal $(a, 1)$-

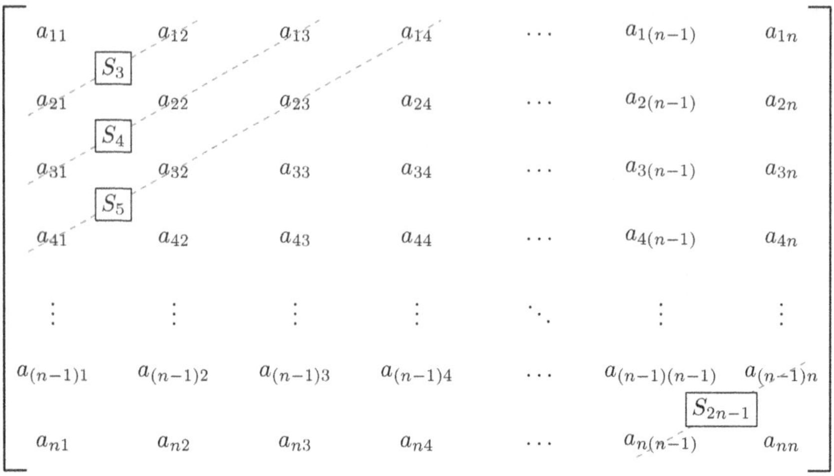

Fig. 6.2 Skew diagonal S_r in a matrix A_G

EAV graph, then $a = 3$. From the adjacency matrix of a maximal (3, 1)-EAV graph, we can see that the first "1" elements will be in the position of (1,2) and (2,1).

Observation 6.2.1 ([269]) *The number of edges of a maximal* (a, d)*-EAV graph of order* n *is* $\lceil (n-1)/d \rceil + \lceil (n-2)/d \rceil$.

Consequently, a maximal (a, d)-EAV graph of order n cannot be connected for $d > 2$ since the maximum number of edges is less than the maximum number of edges for $d = 2$, i.e., $\lceil (n-1)/2 \rceil + \lceil (n-2)/2 \rceil = n - 1$.

In [220] the authors construct adjacency matrices of maximal $(3, d)$-EAV graphs for $d = 1, 2$ by putting "1" elements at the ends of each nonzero skew diagonal. *A triangular book* $B_{n-2}(C_3)$ is the complete tripartite graph $K_{1,1,n-2}$. It is a graph consisting of $n-2$ triangles all sharing a common edge. A double star obtained from two vertex disjoint copies of the star $K_{1,n/2}$ by connecting their centers we call the *twin star graph*, Twin(n). $B_{n-2}(C_3)$ and Twin(n) are maximal (3, 1)-EAV graphs and maximal (3, 2)-EAV graphs of order n, respectively. Figure 6.3 depicts the triangular book graph $B_6(C_3)$ of order 8 with (3, 1)-EAV labeling and its adjacency matrix. Figure 6.4 shows the twin star graph Twin(8) with (3, 2)-EAV labeling and its adjacency matrix.

We can construct new (a^*, d)-EAV graphs from an existing (a, d)-EAV graph by using adjacency matrices manipulation. Here we only consider how adjacency matrix manipulation can be used to construct a new larger maximal $(3, d)$-EAV graph. Given an (a, d)-EAV graph G, there are several ways to obtain a larger (a, d)-EAV graph, such as adding some vertices and edges, combining two (or more) given (a, d)-EAV graphs, and combining two (or more) given (a, d)-EAV graphs and adding some vertices and edges.

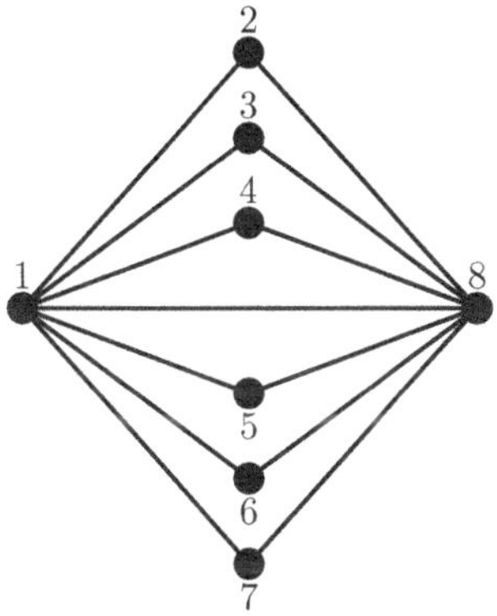

$$\begin{bmatrix} 0 & 1 & 1 & 1 & 1 & 1 & 1 & 1 \\ 1 & 0 & 0 & 0 & 0 & 0 & 0 & 1 \\ 1 & 0 & 0 & 0 & 0 & 0 & 0 & 1 \\ 1 & 0 & 0 & 0 & 0 & 0 & 0 & 1 \\ 1 & 0 & 0 & 0 & 0 & 0 & 0 & 1 \\ 1 & 0 & 0 & 0 & 0 & 0 & 0 & 1 \\ 1 & 0 & 0 & 0 & 0 & 0 & 0 & 1 \\ 1 & 1 & 1 & 1 & 1 & 1 & 1 & 0 \end{bmatrix}$$

Fig. 6.3 Graph $B_6(C_3)$ with (3, 1)-EAV labeling and corresponding adjacency matrix

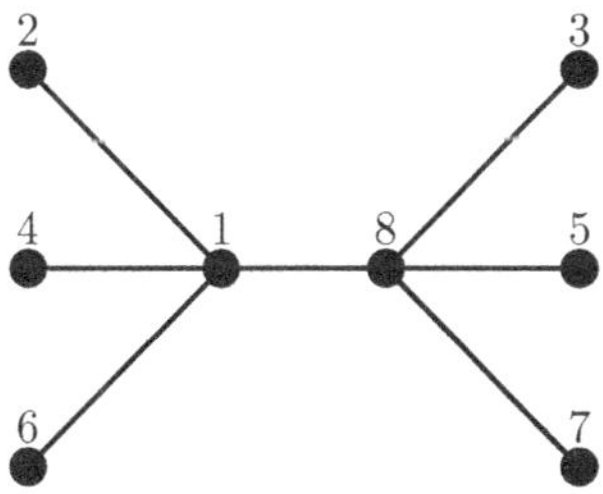

$$\begin{bmatrix} 0 & 1 & 0 & 1 & 0 & 1 & 0 & 1 \\ 1 & 0 & 0 & 0 & 0 & 0 & 0 & 0 \\ 0 & 0 & 0 & 0 & 0 & 0 & 0 & 1 \\ 1 & 0 & 0 & 0 & 0 & 0 & 0 & 0 \\ 0 & 0 & 0 & 0 & 0 & 0 & 0 & 1 \\ 1 & 0 & 0 & 0 & 0 & 0 & 0 & 0 \\ 0 & 0 & 0 & 0 & 0 & 0 & 0 & 1 \\ 1 & 0 & 1 & 0 & 1 & 0 & 1 & 0 \end{bmatrix}$$

Fig. 6.4 Graph Twin(8) with (3, 2)-EAV labeling and corresponding adjacency matrix

6.2.1 Constructing Maximal (3, 1)*-Edge-Antimagic Vertex Graph*

Next we will construct a new, larger maximal (3, 1)-EAV graph from an existing maximal (3, 1)-EAV graph. This can be attained by adding an appropriate number of rows and columns to the adjacency matrix in such a way that the properties of a (3, 1)-EAV graph are preserved. Let us note that the *transpose* A' of a matrix A is the matrix obtained from A by writing its rows as columns.

Theorem 6.2.1 ([220]) *Let G be a maximal* (3, 1)*-EAV graph of order n, $n \geq 2$, with adjacency matrix A_G. Let $t = [t_{i1}]$ be $n \times 1$ matrix with*

$$t_{i1} = \begin{cases} 1 & \textit{if } i = 1, 2 \\ 0 & \textit{if } i = 3, 4, \ldots, n. \end{cases}$$

Then the matrix

$$M = \begin{bmatrix} 0 & t' \\ t & A_G \end{bmatrix}$$

is the adjacency matrix of maximal (3, 1)*-EAV graph of order* $n+1$.

Proof Matrix M contains A_G as its diagonal block matrix starting in position (2, 2). Therefore each vertex v_i in G with label i is now labeled with $i+1$ so that skew diagonals S_r, $r = 5, 6, \ldots, 2n+1$ of M are nonzero diagonals of M since they are nonzero skew diagonals of G. The matrices t' and t insert values into skew diagonals S_3 and S_4 resulting in them being nonzero skew diagonals of M. Now M is a $(n+1) \times (n+1)$ symmetric matrix with nonzero skew diagonal lines forming a band of consecutive integers started with S_3 until S_{2n+1}. □

Since the matrix M in Theorem 6.2.1 is an adjacency matrix of a (3, 1)-EAV graph, it can be considered as A_G. Thus repeating the construction from Theorem 6.2.1 leads to the following corollary.

Corollary 6.2.1 ([220]) *Let G be a maximal* (3, 1)*-EAV graph of order n,* $n \geq 2$, *with adjacency matrix* A_G. *Let* $t_k = [t_{i1}]$, $k = 1, 2, \ldots,$ *be a* $(n+k-1) \times 1$ *matrix with*

$$t_{i1} = \begin{cases} 1 & \text{if } i = 1, 2 \\ 0 & \text{if } i = 3, 4, \ldots, n+k-1 \end{cases}$$

and let M_1 *be a* $(n+1) \times (n+1)$ *matrix with*

$$M_1 = \begin{bmatrix} 0 & t_1' \\ t_1 & A_G \end{bmatrix}.$$

Then the matrix

$$M_k = \begin{bmatrix} 0 & t_k' \\ t_k & M_{k-1} \end{bmatrix}, \quad k = 2, 3, \ldots$$

is the adjacency matrix of a maximal (3, 1)*-EAV graph of order* $n+k$.

Theorem 6.2.1 and Corollary 6.2.1 show a construction of new larger maximal (3, 1)-EAV graphs by adding several columns and rows on the left and top side of the adjacency matrix of an existing maximal (3, 1)-EAV graph. Also several columns and rows can be added on the right and bottom sides of an adjacency matrix.

Theorem 6.2.2 ([220]) *Let G be a maximal $(3, 1)$-EAV graph of order n, $n \geq 2$, with adjacency matrix A_G. Let $t = [t_{i1}]$ and $t^* = [t^*_{i1}]$ be $n \times 1$ matrices with*

$$t_{i1} = \begin{cases} 1 & \text{if } i = 1, 2 \\ 0 & \text{if } i = 3, 4, \ldots, n \end{cases}$$

$$t^*_{i1} = \begin{cases} 0 & \text{if } i = 1, 2, \ldots, n-2 \\ 1 & \text{if } i = n-1, n. \end{cases}$$

Then the matrix

$$M = \begin{bmatrix} 0 & t' & 0 \\ t & A_G & t^* \\ 0 & (t^*)' & 0 \end{bmatrix}$$

is the adjacency matrix of a maximal $(3, 1)$-EAV graph of order $n + 2$.

Proof Matrix M contains A_G as its diagonal block matrix starting in position $(2, 2)$. Therefore each vertex v_i in G with label i is now labeled with $i + 1$ so that skew diagonals S_r, $r = 5, 6, \ldots, 2n + 1$ of M are nonzero diagonals of M since they are nonzero skew diagonals of G. The matrices t' and t insert values into skew diagonals S_{2n+2} and S_{2n+3} resulting in them being nonzero skew diagonals of M. Now M is a $(n+2) \times (n+2)$ symmetric matrix with nonzero skew diagonal lines induce a band of consecutive integers starting with S_3 until S_{2n+3}. □

Theorem 6.2.2 can be done repeatedly which leads to the following corollary.

Corollary 6.2.2 ([220]) *Let G be a maximal $(3, 1)$-EAV graph of order n, $n \geq 2$, with adjacency matrix A_G. Let $t_k = [t_{i1}]$ and $t^*_k = [t^*_{i1}]$, $k = 1, 2, \ldots,$ be $(n + 2k - 2) \times 1$ matrices with*

$$t_{i1} = \begin{cases} 1 & \text{if } i = 1, 2 \\ 0 & \text{if } i = 3, 4, \ldots, n + 2k - 2 \end{cases}$$

$$t^*_{i1} = \begin{cases} 0 & \text{if } i = 1, 2, \ldots, n + 2k - 4 \\ 1 & \text{if } i = n + 2k - 3, n + 2k - 2 \end{cases}$$

and let M_1 be a $(n + 2) \times (n + 2)$ matrix with

$$M_1 = \begin{bmatrix} 0 & t_1' & 0 \\ t_1 & A_G & t_1^* \\ 0 & (t_1^*)' & 0 \end{bmatrix}.$$

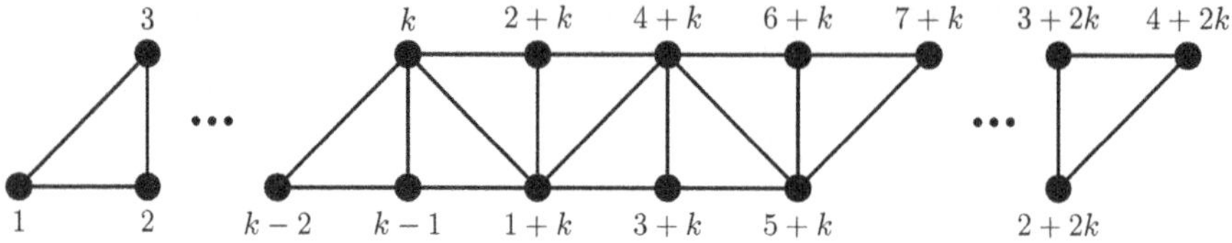

Fig. 6.5 Constructing larger (3, 1)-EAV graphs by using Theorem 6.2.2

Then the matrix

$$M_k = \begin{bmatrix} 0 & t'_k & 0 \\ t_k & M_{k-1} & t^*_k \\ 0 & (t^*_k)' & 0 \end{bmatrix}, \quad k = 2, 3, \ldots$$

is the adjacency matrix of a maximal (3, 1)*-EAV graph of order* $n + 2k$*.*

According to Corollary 6.2.2, the matrix M_k, k even, produces a new maximal (3, 1)-EAV graph of order $4 + 2k$. This graph is a *triangular ladder* $\mathcal{L}_{2+k}$ which can be obtained from the Cartesian product of two paths P_{2+k} and P_2 with $V(P_{2+k}\Box P_2) = \{u_i, v_i : 1 \leq i \leq 2+k\}$ and $E(P_{2+k}\Box P_2) = \{u_i u_{i+1}, v_i v_{i+1} : 1 \leq i \leq 1+k\} \cup \{u_i v_i : 1 \leq i \leq 2+k\}$ by completing the edges $u_{2i-1}v_{2i}$, for $1 \leq i \leq k/2+1$, and $v_{2i}u_{2i+1}$, for $1 \leq i \leq k/2$, see Fig. 6.5.

Corollary 6.2.3 ([220]) *Every triangular ladder* $\mathcal{L}_{2+k}$*,* $k \geq 2$ *even, is a maximal* (3, 1)*-EAV graph.*

Graph $G(H, \mathcal{L}_{2+k})$ is called a *triangular ladder towered graph* if it is obtained from a graph H and the disjoint union of two copies of the triangular ladder $\mathcal{L}_{2+k}$ in such a way that only two different edges in G are mutual with the edges $u_{2+k}v_{2+k}$ in each copy of $\mathcal{L}_{2+k}$.

Let us start with $B_6(C_3)$, see Fig. 6.6. A triangular ladder towered graph $G(B_6(C_3), \mathcal{L}_4)$, see Fig. 6.7, is a maximal (3, 1)-EAV graph. The form of the triangular ladder towered graph $G(B_{n-2}(C_3), \mathcal{L}_{2+k})$, $n \geq 4$ and $k \geq 2$ even, is shown in Fig. 6.8. For any maximal (3, 1)-EAV graph H, the general form of the triangular ladder towered graph $G(H, \mathcal{L}_{2+k})$ is shown in Fig. 6.9.

Corollary 6.2.4 ([220]) *Let* H *be any maximal* (3, 1)*-EAV graph. Then the triangular ladder towered graph* $G(H, \mathcal{L}_{2+k})$*,* $k \geq 2$ *even, is also a maximal* (3, 1)*-EAV graph.*

Fig. 6.6 Graph $B_6(C_3)$

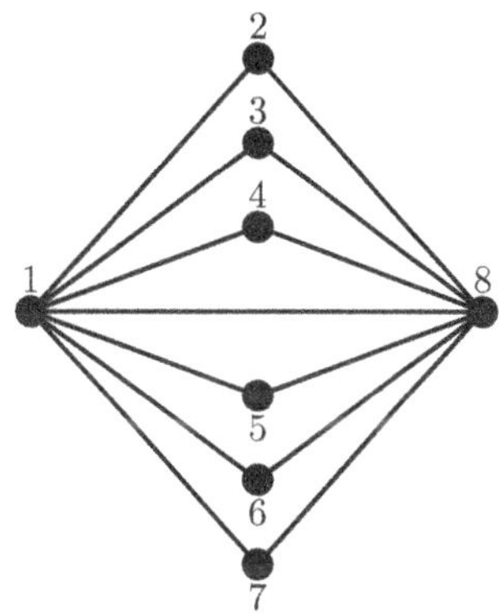

Fig. 6.7 Triangular ladder towered graph $G(B_6(C_3), \mathcal{L}_4)$

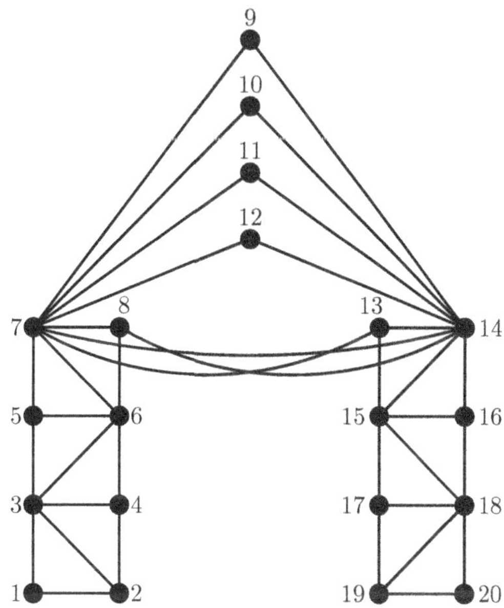

6.2.2 *Constructing Maximal* (3, 2)*-Edge-Antimagic Vertex Graph*

Next we will construct a new, larger maximal (3, 2)-EAV graph from an existing maximal (3, 2)-EAV graph. This can be attained by adding an appropriate number of rows and columns to the adjacency matrix in such a way that the properties of a (3, 2)-EAV graph are preserved. Some of the results presented in this subsection are discussed in detail in [219].

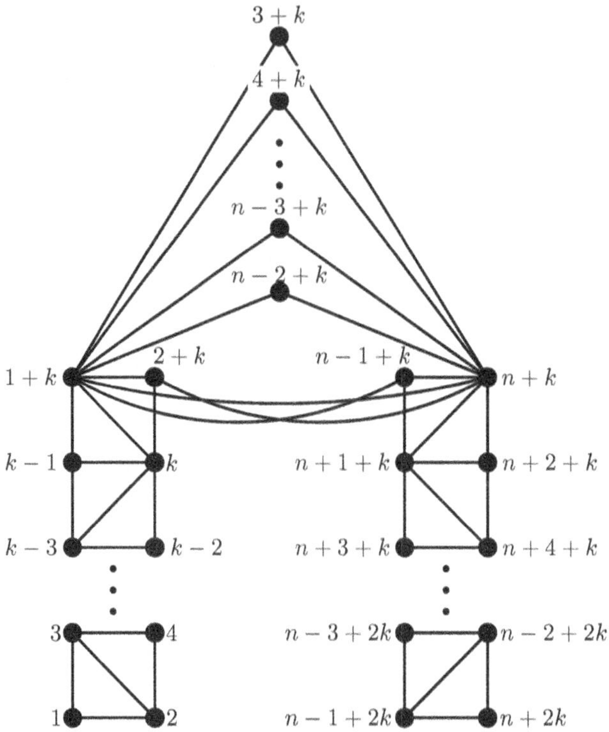

Fig. 6.8 Triangular ladder towered graph $G(B_{n-2}(C_3), \mathcal{L}_{2+k})$, $n \geq 4$ and $k \geq 2$ even

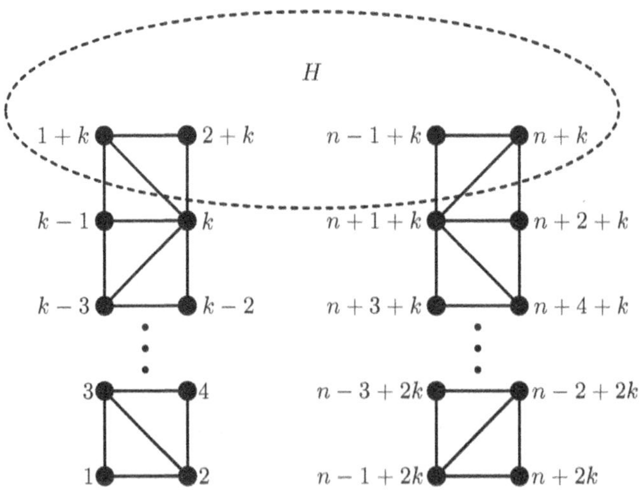

Fig. 6.9 General form of triangular ladder towered graph $G(H, \mathcal{L}_{2+k})$

Theorem 6.2.3 ([220]) *Let G be a maximal $(3, 2)$-EAV graph of order n, $n \geq 1$, with adjacency matrix A_G. Let $s = [s_{i1}]$ and $s^* = [s^*_{i1}]$ be $n \times 1$ matrices with*

$$s_{i1} = \begin{cases} 1 & \text{if } i = 1 \\ 0 & \text{if } i = 2, 3, \ldots, n \end{cases}$$

$$s^*_{i1} = \begin{cases} 0 & \text{if } i = 1, 2, \ldots, n-1 \\ 1 & \text{if } i = n. \end{cases}$$

Then the matrix

$$M = \begin{bmatrix} 0 & s' & 0 \\ s & A_G & s^* \\ 0 & (s^*)' & 0 \end{bmatrix}$$

is the adjacency matrix of a maximal $(3, 2)$-EAV graph of order $n + 2$.

Proof Matrix M contains A_G as its diagonal block matrix starting in position (2, 2). Therefore each vertex v_i in G with label i is now labeled with $i + 1$ so that skew diagonals S_r, $r = 5, 6, \ldots, 2n + 1$ of M are nonzero diagonals of M since they are nonzero skew diagonals of G. The matrices $(s^*)'$ and s^* insert values into skew diagonals S_{2n+2} and S_{2n+3} resulting in them being nonzero skew diagonals of M. Now M is a $(n + 2) \times (n + 2)$ symmetric matrix with nonzero skew diagonal lines forming a band of the arithmetic sequence of difference 2 starting with S_3 until S_{2n+3}. □

Since the matrix M in Theorem 6.2.3 is an adjacency matrix of a $(3, 2)$-EAV graph, it can be considered as A_G. Thus repeating the construction from Theorem 6.2.3 leads to the following corollary.

Corollary 6.2.5 ([220]) *Let G be a maximal $(3, 2)$-EAV graph of order n, $n \geq 1$, with adjacency matrix A_G. Let $s_k = [s_{i1}]$ and $s^*_k = [s^*_{i1}]$, $k = 1, 2, \ldots,$ be $(n + 2k - 2) \times 1$ matrices with*

$$s_{i1} = \begin{cases} 1 & \text{if } i = 1 \\ 0 & \text{if } i = 2, 3, \ldots, n + 2k - 2 \end{cases}$$

$$s^*_{i1} = \begin{cases} 0 & \text{if } i = 1, 2, \ldots, n + 2k - 3 \\ 1 & \text{if } i = n + 2k - 2 \end{cases}$$

and let M_1 be a $(n + 2) \times (n + 2)$ matrix with

$$M_1 = \begin{bmatrix} 0 & s_1' & 0 \\ s_1 & A_G & s_1^* \\ 0 & (s_1^*)' & 0 \end{bmatrix}.$$

Then the matrix

$$M_k = \begin{bmatrix} 0 & s_k' & 0 \\ s_k & M_{k-1} & s_k^* \\ 0 & (s_k^*)' & 0 \end{bmatrix}, \quad k = 2, 3, \dots$$

is the adjacency matrix of a maximal $(3, 2)$*-EAV graph of order* $n + 2k$.

Graph $G(H, P_k)$ is called *path towered graph* if it is obtained from a graph H of order n and the disjoint union of two copies of the path P_k in such a way that an end vertex of each path P_k is adjoined to a vertex of the graph H. Thus $G(H, P_k)$ is graph of order $n + 2k - 2$.

Let us consider the twin star graph Twin(8), see Fig. 6.10a. Then, forming the matrix M_1 by using Corollary 6.2.5 produces the new $(3, 2)$-EAV graph $G(\text{Twin}(8), P_2)$ of order 10, see Fig. 6.10b. Forming the matrix M_k produces the new maximal $(3, 2)$-EAV graph $G(\text{Twin}(8), P_{k+1})$ of order $8 + 2k$, see Fig. 6.10c.

As an immediate consequence of Theorem 6.2.3 and Corollary 6.2.5 we have.

Corollary 6.2.6 ([220]) *Let* H *be any maximal* $(3, 2)$*-EAV graph. Then the path towered graph* $G(H, P_k)$, $k \geq 2$, *is also a maximal* $(3, 2)$*-EAV graph.*

6.2.3 Other Constructions

Sugeng and Miller [269] have proved the following theorem.

Theorem 6.2.4 ([269]) *Let* G_i, $i = 1, 2, \dots, p$, *be an* $(a, 1)$*-EAV graph of order* n_i. *Then there are* $(a, 1)$*-EAV graphs of order* w, *where* $\sum_{i=1}^{p} n_i - 2(p - 1) \leq w \leq \sum_{i=1}^{p} n_i$, *and each contains* G_i *as induced subgraph.*

The proof of Theorem 6.2.4 uses a construction of a new adjacency matrix where its main diagonal contains adjacency matrices of graphs G_i, $i = 1, 2, \dots, p$, to obtain a new adjacency matrix of a maximal $(a, 1)$-EAV graph.

Let $B_{n_i-2}(C_3)$ be the triangular book of order n_i with adjacency matrix A_i, $i = 1, 2, \dots, p$. Then combining the graphs using manipulation of adjacency matrix as the main diagonal block matrices produces new class of maximal $(3, 1)$-EAV graphs with order $\sum_{i=1}^{p} n_i - 2(p - 1)$, see Fig. 6.11, or with order $\sum_{i=1}^{p} n_i - (p - 1)$, see Fig. 6.12. In the first case we obtain a *ladder of triangular books* $LB(n_1 - 2, n_2 - 2, \dots, n_p - 2)$ and in the second case we obtain a *chain of triangular books* $CB(n_1 - 2, n_2 - 2, \dots, n_p - 2)$.

Using the same construction as in Theorem 6.2.4 when the main diagonal of an adjacency matrix contains adjacency matrices of $(b, 2)$-EAV graphs gives the following theorem.

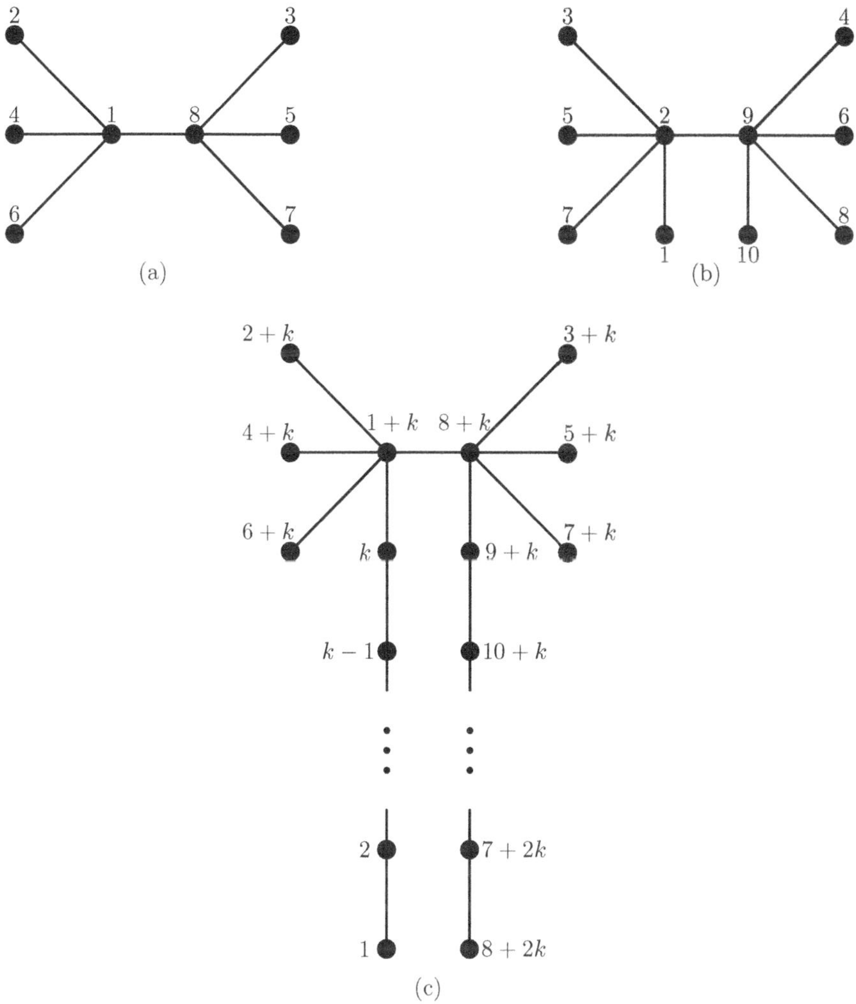

Fig. 6.10 Constructing larger (3, 2)-EAV graph by using Theorem 6.2.3

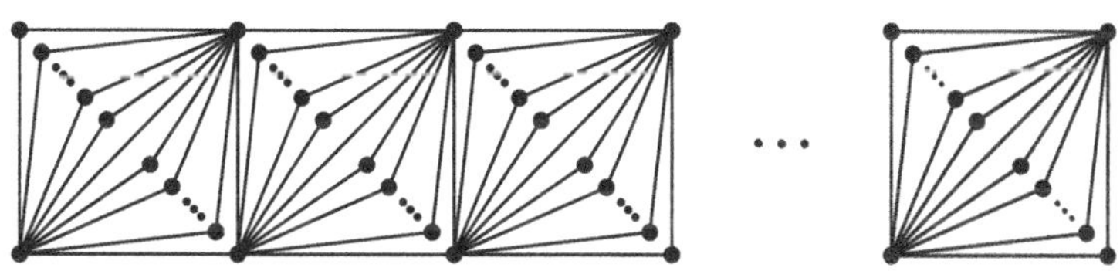

Fig. 6.11 Ladder of triangular books

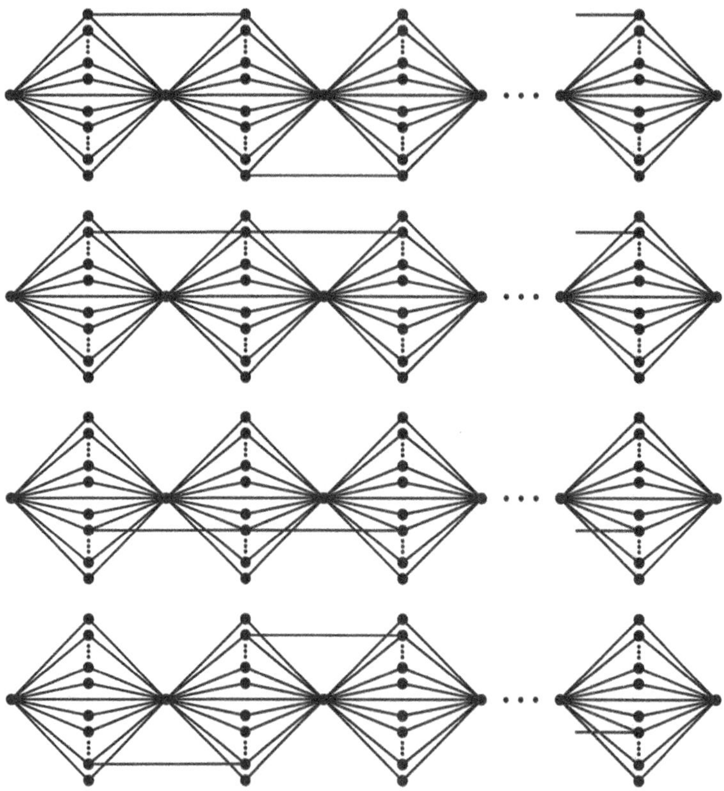

Fig. 6.12 Chain of triangular books

Theorem 6.2.5 ([220]) *Let G_i, $i = 1, 2, \ldots, p$, be $(b, 2)$-EAV graphs of order n_i, respectively. Then there are $(b, 2)$-EAV graphs of order w, where $\sum_{i=1}^{p} n_i - 2(p-1) \le w \le \sum_{i=1}^{p} n_i$, and each contains G_i as induced subgraph.*

Another way to construct a new larger graph that has the same labeling as a given graph was introduced by Cavalier [80]. Using a similar idea gives the following theorem.

Theorem 6.2.6 ([220]) *Let G be a maximal $(3, 2)$-EAV graph of order n with adjacency matrix A_G. Let $s = [s_{i1}]$ and $s^* = [s^*_{i1}]$ be $n \times 1$ matrices with*

$$s_{i1} = \begin{cases} 1 & \textit{if } i = 1 \\ 0 & \textit{if } i = 2, 3, \ldots, n \end{cases}$$

$$s^*_{i1} = \begin{cases} 0 & \textit{if } i = 1, 2, \ldots, n-1 \\ 1 & \textit{if } i = n, \end{cases}$$

and let $\mathbf{0}$ *be the* $n \times 1$ *matrix of all zeros and* $\mathbb{O}$ *be the* $n \times n$ *matrix of all zeros. Then a* $(2pn + 2) \times (2pn + 2)$ *matrix* M *constructed from* $2p$ *copies of* A_G*'s*

$$M = \begin{bmatrix} 0 & s' & \mathbf{0}' & s' & \cdots & \mathbf{0}' & 1 \\ s & A_G & \mathbb{O} & \mathbb{O} & \cdots & \mathbb{O} & \mathbf{0} \\ \mathbf{0} & \mathbb{O} & A_G & \mathbb{O} & \cdots & \mathbb{O} & s^* \\ s & \mathbb{O} & \mathbb{O} & A_G & \cdots & \mathbb{O} & \mathbf{0} \\ \vdots & \vdots & \vdots & \vdots & \ddots & \vdots & \mathbf{0} \\ \mathbf{0} & \mathbb{O} & \mathbb{O} & \mathbb{O} & \cdots & A_G & s^* \\ 1 & \mathbf{0}' & (s^*)' & \mathbf{0}' & \cdots & (s^*)' & 0 \end{bmatrix}$$

is the adjacency matrix of a maximal $(3, 2)$*-EAV graph of order* $2pn + 2$.

Proof According to Theorem 6.2.5 if we add the element 1 as the last element in the first row and as the first element in the last row, and also the matrices s and s^*, then the resulting matrix M with A_G as its main diagonal block matrices forms a new adjacency matrix for $(3, 2)$-EAV graph. □

Theorem 6.2.6 can be done repeatedly and it leads to the following corollary.

Corollary 6.2.7 ([220]) *Let* G *be a maximal* $(3, 2)$*-EAV graph of order* n *with adjacency matrix* $A_G = M_0$ *and let* q_k, $k = 0, 1, 2, \dots$ *be the order of the matrix* M_k. *Let* $s_k = [s_{i1}]$ *and* $s_k^* = [s_{i1}^*]$ *be* $q_k \times 1$ *matrices with*

$$s_{i1} = \begin{cases} 1 & \text{if } i = 1 \\ 0 & \text{if } i = 2, 3, \dots, q_k \end{cases}$$

$$s_{i1}^* = \begin{cases} 0 & \text{if } i = 1, 2, \dots, q_k - 1 \\ 1 & \text{if } i = q_k, \end{cases}$$

and $\mathbf{0}$ *be the* $q_k \times 1$ *matrix of all zeros and* $\mathbb{O}$ *be the* $q_k \times q_k$ *matrix of all zeros and let* M_1 *be a matrix of order* $q_1 = 2pn + 2$ *constructed from* $2p$ *copies of* M_0,

$$M_1 = \begin{bmatrix} 0 & s' & \mathbf{0}' & s' & \cdots & \mathbf{0}' & 1 \\ s & M_0 & \mathbb{O} & \mathbb{O} & \cdots & \mathbb{O} & \mathbf{0} \\ \mathbf{0} & \mathbb{O} & M_0 & \mathbb{O} & \cdots & \mathbb{O} & s^* \\ s & \mathbb{O} & \mathbb{O} & M_0 & \cdots & \mathbb{O} & \mathbf{0} \\ \vdots & \vdots & \vdots & \vdots & \ddots & \vdots & \mathbf{0} \\ \mathbf{0} & \mathbb{O} & \mathbb{O} & \mathbb{O} & \cdots & M_0 & s^* \\ 1 & \mathbf{0}' & (s^*)' & \mathbf{0}' & \cdots & (s^*)' & 0 \end{bmatrix}.$$

Then the matrix M_k constructed from $2p$ copies of M_{k-1}

$$M_k = \begin{bmatrix} 0 & s' & \mathbf{0}' & s' & \cdots & \mathbf{0}' & 1 \\ s & M_{k-1} & \mathbb{O} & \mathbb{O} & \cdots & \mathbb{O} & \mathbf{0} \\ \mathbf{0} & \mathbb{O} & M_{k-1} & \mathbb{O} & \cdots & \mathbb{O} & s^* \\ s & \mathbb{O} & \mathbb{O} & M_{k-1} & \cdots & \mathbb{O} & \mathbf{0} \\ \vdots & \vdots & \vdots & \vdots & \ddots & \vdots & \mathbf{0} \\ \mathbf{0} & \mathbb{O} & \mathbb{O} & \mathbb{O} & \cdots & M_{k-1} & s^* \\ 1 & \mathbf{0}' & (s^*)' & \mathbf{0}' & \cdots & (s^*)' & 0 \end{bmatrix}$$

is the adjacency matrix of a maximal $(3, 2)$*-EAV graph of order* $q_k = 2pq_{k-1} + 2$.

A graph containing only one vertex is a trivial $(3, 2)$-EAV graph. We can combine a finite even number of copies of that trivial graph and construct a new $(3, 2)$-EAV graph with adjacency matrix M_1 by using Corollary 6.2.7. For example, we combine 6 graphs of one vertex and produce a new $(3, 2)$-EAV graph of order 8, see Fig. 6.13a. Then we construct the adjacency matrix M_2 by combining 6 M_1 and produce the new larger $(3, 2)$-EAV graph of order 50, see Fig. 6.13b. Constructing M_3 by combining 6 M_2 will produce a new larger $(3, 2)$-EAV graph, see Fig. 6.14.

6.3 Edge-Antimagic Total Labeling

An (a, d)-*edge-antimagic total* ((a, d)-EAT) labeling of a (p, q) graph G is defined as a one-to-one mapping f from $V(G) \cup E(G)$ onto the set $\{1, 2, \dots, p + q\}$, so that the set of edge-weights $\{f(u) + f(uv) + f(v) : uv \in E(G)\}$ is equal to $\{a, a + d, \dots, a + (q - 1)d\}$, for two integers $a > 0$ and $d \geq 0$.

An (a, d)-EAT labeling f is called *super* if it has the property that the vertex labels are the integers $1, 2, \dots, p$, that is, the smallest possible labels, and $f(E(G)) = \{p + 1, p + 2, \dots, p + q\}$.

Note that when $d = 0$, then (super) $(a, 0)$-EAT labeling is in fact an EMT or super EMT, respectively (see Chap. 4).

The definition of (a, d)-EAT labeling was introduced by Simanjuntak et al. [256], see also [255]. The (a, d)-EAT labeling and super (a, d)-EAT labeling are natural extensions of the notion of an EMT labeling, defined by Kotzig and Rosa [163], and the notion of a super EMT labeling, which was defined by Enomoto et al. [96].

Assume that (p, q) graph G has an (a, d)-EAT labeling f. The sum of all the edge-weights is

$$\sum_{uv \in E(G)} wt(uv) = \sum_{i=0}^{q-1} (a + id) = aq + \frac{q(q - 1)d}{2}. \tag{6.2}$$

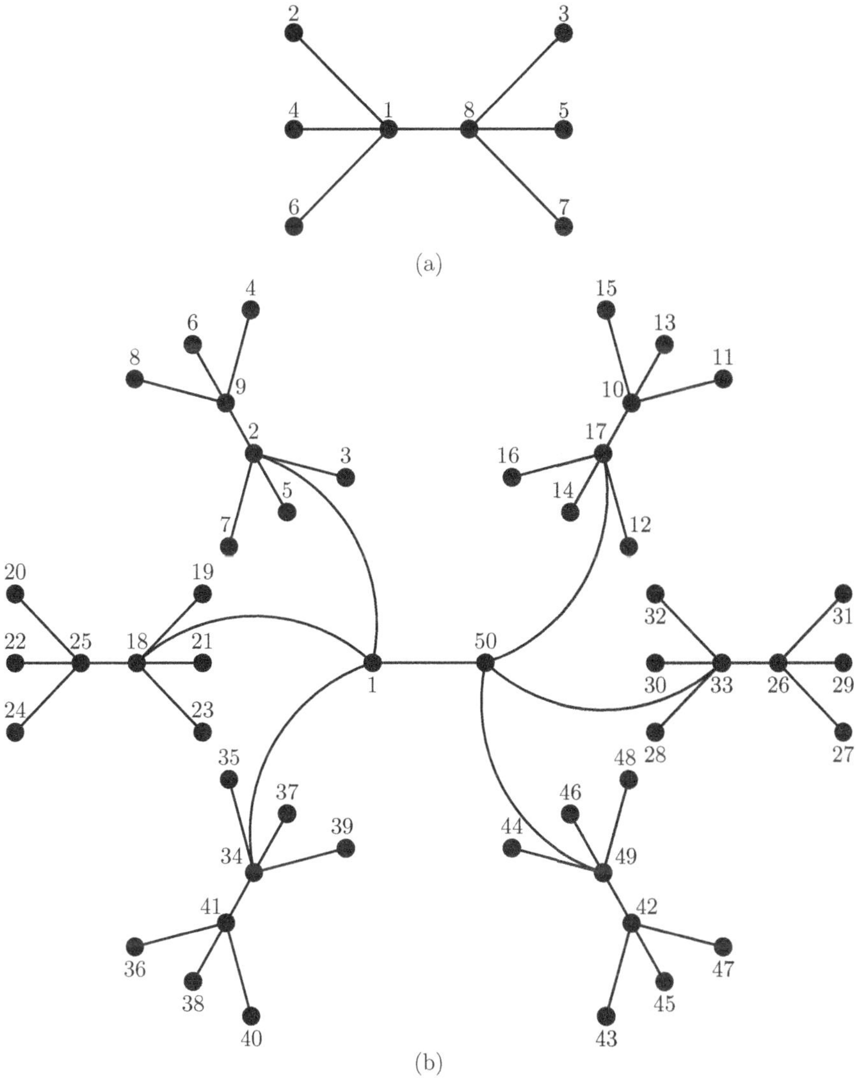

Fig. 6.13 Constructing larger (3, 2)-EAV graph by using Corollary 6.2.7

In the computation of the edge-weights of G, each edge label is used once and the label of vertex u_i is used $\deg(u_i)$ times, $i = 1, 2, \ldots, p$, where $\deg(u_i)$ is the degree of vertex u_i. The sum of all vertex labels and edge labels used to calculate the edge-weights is thus equal to

$$\sum_{j=1}^{p+q} j + \sum_{i=1}^{p}(\deg(u_i) - 1) f(u_i) = \frac{(p+q)(p+q+1)}{2} + \sum_{i=1}^{p}(\deg(u_i) - 1) f(u_i). \tag{6.3}$$

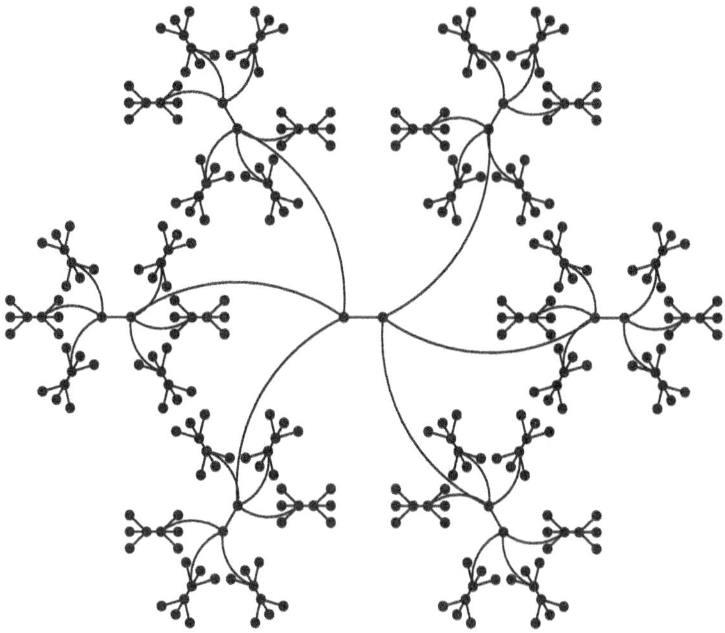

Fig. 6.14 Graph given by adjacency matrix M_3

Combining Eqs. (6.2) and (6.3) gives

$$aq + \frac{q(q-1)d}{2} = \frac{(p+q)(p+q+1)}{2} + \sum_{i=1}^{p} (\deg(u_i) - 1) f(u_i). \tag{6.4}$$

Using parity considerations for the left-hand and the right-hand sides of Eq. (6.4), we obtain

Theorem 6.3.1 ([256]) *A graph with all vertices of odd degree cannot have an (a, d)-EAT labeling with a and d both even.*

Theorem 6.3.2 ([256]) *Let G be a (p, q) graph with all vertices of odd degree. If $q \equiv 0 \pmod 4$ and $p \equiv 2 \pmod 4$, then G has no (a, d)-EAT labeling.*

Theorem 6.3.3 ([256]) *Suppose G is a (p, q) graph whose every vertex has odd degree. Then in the following cases G has no (a, d)-EAT labeling.*

(i) $q \equiv 1 \pmod 4$, $p \equiv 0 \pmod 4$, *and a even.*
(ii) $q \equiv 1 \pmod 4$, $p \equiv 2 \pmod 4$, *and a odd.*
(iii) $q \equiv 2 \pmod 4$, $p \equiv 2 \pmod 4$, *and d odd.*
(iv) $q \equiv 3 \pmod 4$, $p \equiv 0 \pmod 4$, *a even, and d odd.*

For an (a, d)-EAT labeling of a (p, q) graph, the minimum possible edge-weight is at least $1 + 2 + 3$. Consequently, $a \geq 6$. The maximum possible edge-weight is no more than $(p + q - 2) + (p + q - 1) + (p + q) = 3p + 3q - 3$. Thus

$$a + (q - 1)d \leq 3p + 3q - 3,$$

$$d \leq \frac{3p + 3q - 9}{q - 1} \tag{6.5}$$

and we have obtained an upper bound for the parameter d for an (a, d)-EAT labeling of G.

Let (p, q) graph be a super (a, d)-EAT. It is easy to see that the minimum possible edge-weight is at least $p + 4$ and the maximum possible edge-weight is not more than $3p + q - 1$. Thus

$$a + (q - 1)d \leq 3p + q - 1$$

and

$$d \leq \frac{2p + q - 5}{q - 1}. \tag{6.6}$$

For any (p, q) graph, where $p - 1 \leq q$, it follows that $d \leq 3$. In particular if G is connected, then $d \leq 3$.

Next we present some relationships between (a, d)-EAV labeling, (a, d)-EAT labeling, and other kinds of labelings, in particular, edge-magic total labeling.

Theorem 6.3.4 ([36]) *Let G be a (p, q) graph which admits total labeling and whose edge labels constitute an arithmetic progression with difference d. Then the following are equivalent.*

(i) G has an EMT labeling with magic constant k.
(ii) G has a $(k - (q - 1)d, 2d)$-EAT labeling.

In [36] the following theorem is proved.

Theorem 6.3.5 ([36]) *Let a (p, q) graph G have an (a, d)-EAV labeling. Then the following statements hold.*

(i) G has a super $(a + p + 1, d + 1)$-EAT labeling.
(ii) G has a super $(a + p + q, d - 1)$-EAT labeling.

Now we present the following useful lemma.

Lemma 6.3.1 ([273]) *Let $\mathfrak{P}$ be a sequence $\mathfrak{P} = \{c, c + 1, \ldots, c + (k - 3)/2, c + (k - 1)/2, c + (k + 3)/2, c + (k + 5)/2, \ldots, c + k + 1\}$, k odd. Then there exists a sequence $\mathfrak{R}$ of the integers $\{1, 2, \ldots, k + 1\}$, such that the sequence $\mathfrak{P} + \mathfrak{R}$ consists of consecutive integers.*

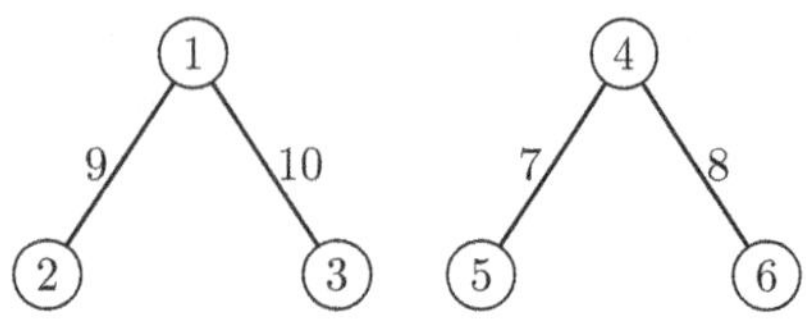

Fig. 6.15 Super (12, 2)-EAT labeling of $2P_3$

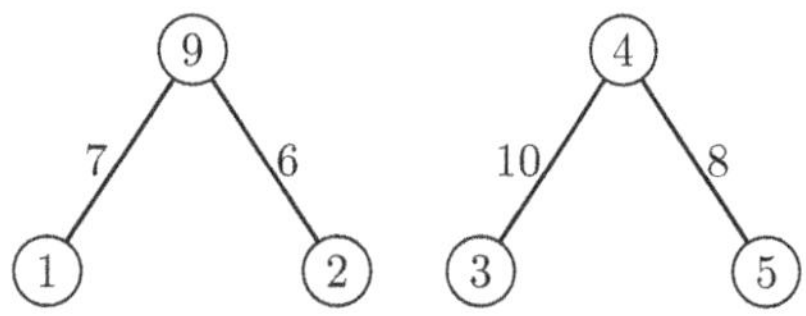

Fig. 6.16 A (17, 0)-EAT labeling of $2P_3$

Figueroa-Centeno, Ichishima, and Muntaner-Batle in [98] proved the following theorem.

Theorem 6.3.6 ([98]) *A (p, q) graph G has an $(a, 1)$-EAV labeling if and only if G has a super EMT (super $(a, 0)$-EAT) labeling, with magic constant $k = a + p + q$.*

Theorems 6.3.6 and 6.3.5 allow us to extend the known results on super EMT labelings onto super $(a - q + 1, 2)$-EAT labeling. However, the condition in Theorem 6.3.5 is only sufficient for the existence of a super $(a, 2)$-EAT labeling from the existence of a super $(a, 0)$-EAT labeling of a graph. For example, let us consider two copies of a path on three vertices. In [98] it is proved that $2P_3$ is not super $(a, 0)$-EAT, but it is super $(a, 2)$-EAT, see Fig. 6.15. Note, that $2P_3$ is $(a, 0)$-EAT, see Fig. 6.16.

According to Theorem 6.3.5 from the results presented in Sect. 6.2 it follows that

Corollary 6.3.1 ([220]) *The triangular book graph $B_{n-2}(C_3)$ and triangular ladder $\mathcal{L}_n$ both of order n and size $2n - 3$ admit a super $(n + 4, 2)$-EAT labeling and a super $(3n, 0)$-EAT labeling.*

Corollary 6.3.2 ([220]) *The triangular ladder towered graph $G(B_{n-2}(C_3), \mathcal{L}_{2+k})$ of order $n+2k$ and size $2n+8k+5$, $k \geq 2$ even, admits a super $(n+2k+4, 2)$-EAT labeling and a super $(3n + 10k + 8, 0)$-EAT labeling.*

Corollary 6.3.3 ([220]) *The ladder of triangular books $LB(n_1 - 2, n_2 - 2, \ldots, n_p - 2)$ of order $\sum_{i=1}^{p} n_i - 2(p - 1)$ admits a super $(\sum_{i=1}^{p} n_i - 2(p - 1) + 4, 2)$-EAT labeling and a super $(3 \sum_{i=1}^{p} n_i - 2(p - 1), 0)$-EMT labeling.*

Corollary 6.3.4 ([220]) *The chain of triangular books $CB(n_1-2, n_2-2, \ldots, n_p-2)$ of order $\sum_{i=1}^{p} n_i - (p - 1)$ admits a super $(\sum_{i=1}^{p} n_i - (p - 1) + 4, 2)$-EAT labeling and a super $(3 \sum_{i=1}^{p} n_i - (p - 1), 0)$-EMT labeling.*

Corollary 6.3.5 ([220]) *The twin star graph* Twin(n) *of order n and size $n - 1$ admits a super $(2n + 2, 1)$-EAT labeling and a super $(n + 4, 3)$-EAT labeling.*

Corollary 6.3.6 ([220]) *The path towered graph $G(H, P_k)$, $k \geq 2$, of order $n + 2k - 2$ and size $n + 2k - 3$ admits a super $(2n + 4k - 2, 1)$-EAT labeling and a super $(n + 2k + 2, 3)$-EAT labeling.*

6.3.1 Super $(a, 1)$-Edge-Antimagic Total Labeling of Regular Graphs

In this section we deal with the existence of super $(a, 1)$-EAT labelings of regular graphs. We also give some constructions of non-regular super $(a, 1)$-EAT graphs.

The construction in the following theorem allows us to find a super $(a, 1)$-EAT labeling of any even regular graph. Notice that the construction does not require the graph to be connected.

Theorem 6.3.7 ([34]) *Let G be a graph on p vertices that can be decomposed into two factors G_1 and G_2. If G_1 is edge-empty or if G_1 is a super $(2p + 2, 1)$-EAT graph and G_2 is a $2r$-regular graph, then G is super $(2p + 2, 1)$-EAT.*

Proof First we start with the case when G_1 is not edge-empty. Since G_1 is a super $(2p + 2, 1)$-EAT graph with p vertices and q edges, there exists a total labeling $f : V(G_1) \cup E(G_1) \to \{1, 2, \ldots, p + q\}$ such that

$$\{f(v) + f(uv) + f(v) \colon uv \in E(G)\} = \{2p + 2, 2p + 3, \ldots, 2p + q + 1\}.$$

By the Petersen Theorem (Theorem 3.1.23) there exists a 2-factorization of G_2. We denote the 2-factors by F_j, $j = 1, 2, \ldots, r$. Let $V(G) = V(G_1) = V(F_j)$ for all j and $E(G) = \bigcup_{j=1}^{r} E(F_j) \cup E(G_1)$. Each factor F_j is a collection of cycles. We order and orient the cycles arbitrarily. Now by the symbol $e_j^{out}(v_i)$ we denote the unique outgoing arc from the vertex v_i in the factor F_j.

We define a total labeling g of G in the following way.

$$g(v) = f(v) \qquad \text{for } v \in V(G)$$

$$g(e) = \begin{cases} f(e) & \text{for } e \in E(G_1) \\ q + (j + 1)p + 1 - f(v_i) & \text{for } e = e_j^{out}(v_i). \end{cases}$$

The vertices are labeled by the first p integers, the edges of G_1 by the next q labels and the edges of G_2 by consecutive integers starting at $p+q+1$. Thus g is a bijection $V(G) \cup E(G) \to \{1, 2, \ldots, p + q + pr\}$ since $|E(G)| = q + pr$.

It is not difficult to verify that g is a super $(2p + 2, 1)$-EAT labeling of G. For the weights of the edges e in $E(G_1)$ is $w_g(e) = w_f(e)$. The weights form the progression $2p + 2, 2p + 3, \ldots, 2p + q + 1$. For convenience we denote by v_k

the unique vertex such that $v_i v_k = e_j^{out}(v_i)$ in F_j. The weights of the edges in F_j, $j = 1, 2, \ldots, r$ are

$$\begin{aligned} w_g(e_j^{out}(v_i)) =& w_g(v_i v_k) = g(v_i) + (q + (j+1)p + 1 - f(v_i)) + g(v_k) \\ =& f(v_i) + q + (j+1)p + 1 - f(v_i) + f(v_k) \\ =& q + (j+1)p + 1 + f(v_k) \end{aligned}$$

for all $i = 1, 2, \ldots, p$ and $j = 1, 2, \ldots, r$. Since F_j is a factor, the set $\{f(v_k): v_k \in F_j\} = \{1, 2, \ldots, p\}$. Hence we have that the set of the edge-weights in the factor F_j is $\{q + (j+1)p + 2, q + (j+1)p + 3, \ldots, q + (j+1)p + p + 1\}$ and thus the set of all edge-weights in G is $\{2p+2, 2p+3, \ldots, q + (r+2)p + 1\}$.

If G_1 is edge-empty it is enough to take $q = 0$ and proceed with the labeling of factors F_j. □

By taking an edge-empty graph G_1 we have the following theorem.

Theorem 6.3.8 ([34]) *All even regular graphs of order p with at least one edge are super* $(2p+2, 1)$*-EAT.*

The construction from Theorem 6.3.7 can be extended also to the case when G_1 is not a factor. One can add isolated vertices to a graph and keep the property of being super $(a, 1)$-EAT. A graph consisting of m isolated vertices is denoted by mK_1. We can obtain the following lemma.

Lemma 6.3.2 ([34]) *If G is a super* $(a, 1)$*-EAT graph, then also* $G \cup mK_1$ *is a super* $(a + m + 2t, 1)$*-EAT graph for all* $t \in \{0, 1, \ldots, m\}$.

Proof Since G is a super $(a, 1)$-EAT graph with p vertices and q edges, there exists such a total labeling $f : V(G) \cup E(G) \to \{1, 2, \ldots, p+q\}$ that

$$\{f(u) + f(uv) + f(v): uv \in E(G)\} = \{a, a+1, \ldots, a+q-1\}.$$

Let t be any fixed integer from $\{0, 1, \ldots, m\}$. Let $(c_1, c_2, \ldots, c_m)$ be any permutation of the integers in $\{1, 2, \ldots, p+m\} \setminus \{t+1, t+2, \ldots, t+p\}$. We denote the vertices of mK_1 by $v_{c_1}, v_{c_2}, \ldots, v_{c_m}$ arbitrarily. Now we define a labeling g of the graph $H = G \cup mK_1$.

$$g(v) = \begin{cases} f(v) + t & \text{for } v \in V(G) \\ i & \text{for } v = v_i, \text{ where } v_i \in mK_1 \end{cases}$$

$$g(e) = f(e) + m \quad \text{for } e \in E(H).$$

Obviously g is a bijection $V(H) \cup E(H) \to \{1, 2, \ldots, p+q+m\}$. The edges are labeled by the q highest labels and the vertices by the first $p+m$ integers. It is easy to verify that g is super $(a + m + 2t, 1)$-EAT labeling of H, since any edge $uv \in E(H)$ is also in $E(G)$.

$$\begin{aligned} w_g(uv) &= g(u) + g(uv) + g(v) = (f(u) + t) + (f(uv) + m) + (f(v) + t) \\ &= w_f(uv) + m + 2t \end{aligned}$$

and the claim follows. □

Notice that we can find $m + 1$ different (up to isomorphism) super $(b, 1)$-EAT labelings of $G \cup mK_1$ but all with the same parity of the smallest edge-weight.

Next we show that also all odd regular graphs with a perfect matching are super $(a, 1)$-EAT.

Lemma 6.3.3 ([34]) *Let k, m be positive integers. Then the graph $kP_2 \cup mK_1$ is super $(2(2k + m) + 2, 1)$-EAT.*

Proof We denote the vertices of the graph $G \cong kP_2 \cup mK_1$ by the symbols $v_1, v_2, \ldots, v_{2k+m}$ in such a way that $E(G) = \{v_i v_{k+m+i} : i = 1, 2, \ldots, k\}$ and the remaining vertices are denoted arbitrarily by the unused symbols.

We define the labeling $f : V(G) \cup E(G) \to \{1, 2, \ldots, 3k + m\}$ in the following way:

$$\begin{aligned} f(v_j) &= j, && \text{for } j = 1, 2, \ldots, 2k + m, \\ f(v_i v_{k+m+i}) &= 3k + m + 1 - i, && \text{for } i = 1, 2, \ldots, k. \end{aligned}$$

It is easy to see that f is a bijection and that the vertices of G are labeled by the smallest possible numbers. For the edge-weights we get

$$\begin{aligned} w_f(v_i v_{k+m+i}) &= f(v_i) + f(v_i v_{k+m+i}) + f(v_{k+m+i}) \\ &= i + (3k + m + 1 - i) + (k + m + i) = 2(2k + m) + 1 + i \end{aligned}$$

for $i = 1, 2, \ldots, k$. Thus f is a super $(2(2k + m) + 2, 1)$-EAT labeling of G. □

Figure 6.17 illustrates a super $(28, 1)$-EAT labeling of the graph $5P_2 \cup 3K_1$.

By taking $m = 0$ and observing that the number of vertices in kP_2 is $2k$, we immediately obtain the following theorem.

Theorem 6.3.9 ([34]) *If G is an odd regular graph on p vertices that has a 1-factor, then G is super $(2p + 2, 1)$-EAT.*

Unfortunately the construction does not solve the existence of $(a, 1)$-EAT labelings for all odd regular graphs, it only works for those that contain a 1-factor. We know that some graphs that arose by Cartesian products also satisfy this property; therefore, we can obtain the following corollary.

Corollary 6.3.7 ([34]) *Let G be a regular graph. Then the Cartesian product $G \square K_2$ is a super $(a, 1)$-EAT graph.*

Fig. 6.17 Super (28, 1)-EAT labeling of $5P_2 \cup 3K_1$

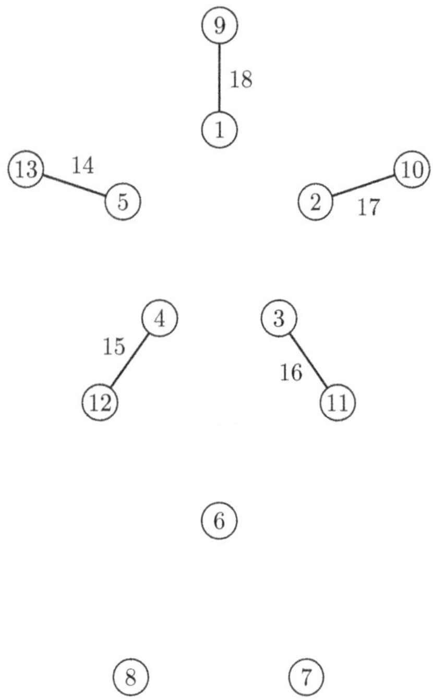

Proof If G is a $(2r+1)$-regular graph, then the Cartesian product $G\Box K_2$ is $(2r+2)$-regular and by Theorem 6.3.8 it is super $(a, 1)$-EAT. If G is $2r$-regular, then $G\Box K_2$ is a $(2r + 1)$-regular graph with a 1-factor and thus according to Theorem 6.3.9 is super $(a, 1)$-EAT. □

Theorem 6.3.7 is not restricted to regular graphs, it can be used also to obtain super $(a, 1)$-EAT labelings of certain non-regular graphs. We illustrate the technique on a couple of examples. First we introduce the following lemmas.

Lemma 6.3.4 ([34]) *Let k, m be positive integers, $k < 2m + 3$. Then the graph $K_{1,k} \cup mK_1$ is super $(2(k + m + 1) + 2, 1)$-EAT.*

Proof We distinguish two subcases according to the parity of k. Let k be an odd positive integer. We denote the vertices of the graph $G \cong K_{1,k} \cup mK_1$ by the symbols $v_1, v_2, \ldots, v_{k+m+1}$ in such a way that $E(G) = \{v_i v_{m+2+(k-1)/2} \colon i = 1, 2, \ldots, k\}$ and the remaining vertices are denoted arbitrarily by the unused symbols. Notice that it is possible to use such notation as $k < 2m + 3$.

We define the labeling $f : V(G) \cup E(G) \to \{1, 2, \ldots, 2k + m + 1\}$ in the following way:

$$f(v_j) = j \qquad \text{for } j = 1, 2, \ldots, k + m + 1$$

$$f(v_i v_{m+2+\frac{k-1}{2}}) = \begin{cases} m + \frac{3k+1}{2} + i & \text{for } i = 1, 2, \ldots, \frac{k+1}{2} \\ m + \frac{k+1}{2} + i & \text{for } i = \frac{k+3}{2}, \frac{k+5}{2}, \ldots, k. \end{cases}$$

For the edge-weights we have

$$w_f(v_i v_{m+2+\frac{k-1}{2}}) = f(v_i) + f(v_i v_{m+2+\frac{k-1}{2}}) + f(v_{m+2+\frac{k-1}{2}})$$

$$= \begin{cases} i + \left(m + \frac{3k+1}{2} + i\right) + \left(m + 2 + \frac{k-1}{2}\right) \\ \qquad = 2m + 2k + 2 + 2i \quad \text{for} \quad i = 1, 2, \dots, \frac{k+1}{2} \\ i + \left(m + \frac{k+1}{2} + i\right) + \left(m + 2 + \frac{k-1}{2}\right) \\ \qquad = 2m + k + 2 + 2i \quad \text{for} \quad i = \frac{k+3}{2}, \frac{k+5}{2}, \dots, k, \end{cases}$$

i.e., the set of the edge-weights is $\{2m + 2k + 4, 2m + 2k + 5, \dots, 2m + 3k + 3\}$. Thus for $2m + 3 > k$, $k \equiv 1 \pmod 2$, f is a super $(2(k + m + 1) + 2, 1)$-EAT labeling of G.

Notice that the edge $v_{(k+1)/2}v_{m+2+(k-1)/2}$ is labeled under the labeling f by the highest label $m + 2k + 1$ and has also the maximal edge-weight $2m + 3k + 3$. Thus it is possible to delete this edge from G and the obtained graph $K_{1,(k-1)} \cup (m + 1)K_1$ will also be super $(2(k + m + 1) + 2, 1)$-EAT. It means that it is possible to construct the required labeling also in the case when the star has even number of pending edges (for k even). □

In Fig. 6.18 we exhibit a super (26, 1)-EAT labeling of the graph $K_{1,7} \cup 4K_1$.

Lemma 6.3.5 ([34]) *Let k, m be positive integers, and let m be even. Let H be an arbitrary 2-regular graph of order k. Then the graph $H \cup mK_1$ is super $(2(k + m) + 2, 1)$-EAT.*

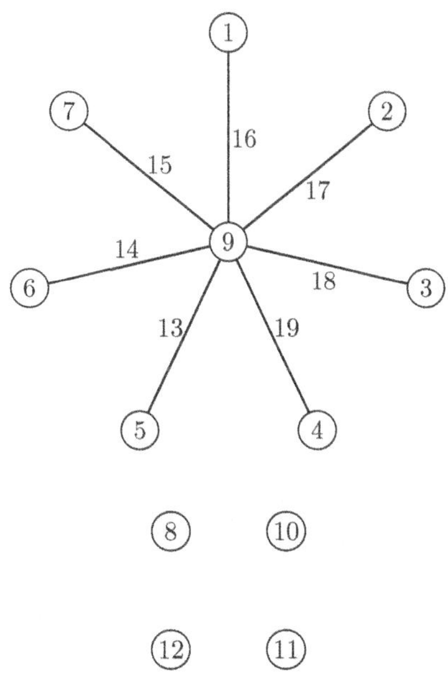

Fig. 6.18 Super (26, 1)-EAT labeling of $K_{1,7} \cup 4K_1$

Proof According to Theorem 6.3.7 the graph H is super $(2k+2, 1)$-EAT. Using Lemma 6.3.2 for $t = m/2$ we get that $H \cup mK_1$ is a super $(2(k+m)+2, 1)$-EAT graph. □

Lemma 6.3.6 ([34]) *Let k, m be positive integers, let m be even. Then the graph $P_k \cup mK_1$ is super $(2(k+m)+2, 1)$-EAT.*

Proof It is known that the path on k vertices is super $(2k+2, 1)$-EAT, see [36]. According to Lemma 6.3.2 for $t = m/2$ we get that the graph $P_k \cup mK_1$ is super $(2(k+m)+2, 1)$-EAT. □

Immediately from the previous lemmas and Theorem 6.3.7 we see that it is possible to "add" certain edges to an even regular graph and obtain a super $(a, 1)$-EAT graph. The edges are added in such a way that the graph induced by these edges is isomorphic to a collection of independent edges, to a star, to a 2-regular graph, or to a path.

Theorem 6.3.10 ([34]) *Let k, m be positive integers. Let G be a graph on p vertices that can be decomposed into two factors G_1 and G_2. The graph G is super $(2p+2, 1)$-EAT if G_2 is a $2r$-regular graph and one of the following statements holds.*

(i) G_1 is the graph $kP_2 \cup mK_1$.
(ii) G_1 is the graph $K_{1,k} \cup mK_1$ for $k < 2m+3$.
(iii) H is an arbitrary 2-regular graph of order k and $G_1 \cong H \cup mK_1$ for even m.
(iv) G_1 is the graph $P_k \cup mK_1$ for even m.

Proof Since the smallest edge-weight in G_1 in Case (i) is $2(2k+m)+2 = 2p+2$, then the claim immediately follows by Lemma 6.3.3 and Theorem 6.3.7. By a similar argument one can prove Cases (ii), (iii), and (iv) using Theorem 6.3.7 and Lemmas 6.3.4, 6.3.5, and 6.3.6, respectively. □

If factor $G_1 \cong 5P_2 \cup 3K_1$ and G_2 is a 2-factor, then the resulting graph G is disjoint union of the prism $C_5 \square P_2$ and the cycle C_3. Figure 6.19 shows a super $(28, 1)$-EAT labeling of the graph $(C_5 \square P_2) \cup C_3$ described in the proof of Theorem 6.3.10.

If $G_1 \cong K_{1,7} \cup 4K_1$ and G_2 is a 2-factor, then the resulting graph G is the set of wheel W_7 and cycle C_5 having the common central vertex. Figure 6.20 illustrates a super $(26, 1)$-EAT labeling of the graph G obtained in the proof of Theorem 6.3.10.

Notice that in Lemmas 6.3.3, 6.3.4, 6.3.5, and 6.3.6 by taking $m = 0$ we obtain an $(2p'+2, 1)$-EAT labeling of the corresponding graph on $p' = p - m$ vertices. Now adding m isolated vertices one can obtain by Lemma 6.3.2 not one, but $m+1$ different super $(a, 1)$-EAT labelings of the graph G_1 in each of the cases of Theorem 6.3.10. This again implies several different super $(2p+2, 1)$-EAT labelings of the graph G in Theorem 6.3.10. There can be significantly more than $m+1$ different labelings, since we may choose various orderings of orientations of the 2-factors F_j of G_2 (as described in the proof of Theorem 6.3.7).

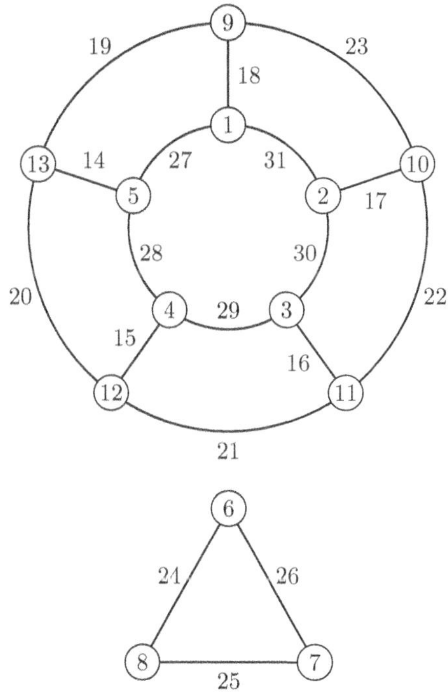

Fig. 6.19 Super (28, 1)-EAT labeling of $(C_5 \square P_2) \cup C_3$

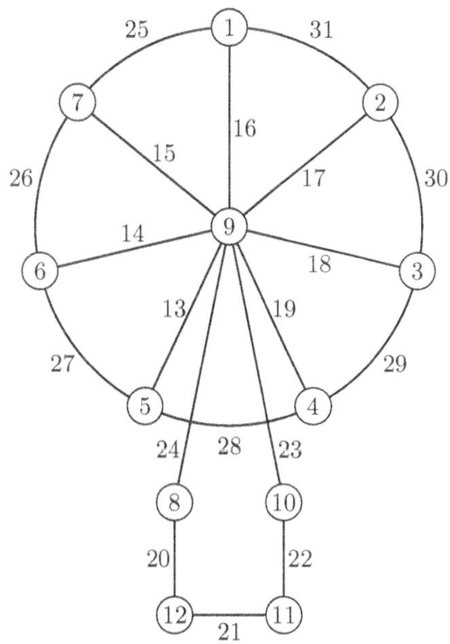

Fig. 6.20 Super (26, 1)-EAT labeling of G

6.3.2 *Super Edge-Antimagic Total Labeling for Certain Families of Connected Graphs*

This section summarizes the known results on super EAT labelings for several families of connected graphs. The next theorem gives a characterization for all cycles.

Theorem 6.3.11 ([27]) *The cycle C_n has a super (a, d)-EAT labeling if and only if one of the following conditions is satisfied.*

(i) $d \in \{0, 2\}$ *and* n *is odd,* $n \geq 3$.
(ii) $d = 1$ *and* $n \geq 3$.

Recall that the *friendship graph* f_n is a set of n triangles having a common central vertex, and otherwise disjoint. If the friendship graph $f_n, n \geq 1$, is super (a, d)-EAT, then, by (6.6) it follows that $d < 3$. The following lemma characterizes $(a, 1)$-edge-antimagicness of friendship graphs.

Lemma 6.3.7 ([37]) *The friendship graph f_n has an $(a, 1)$-EAV labeling if and only if $n \in \{1, 3, 4, 5, 7\}$.*

According to Theorem 6.3.5 from Lemma 6.3.7 it follows

Theorem 6.3.12 ([37]) *For $n \in \{1, 3, 4, 5, 7\}$, the friendship graph f_n has a super $(a, 0)$-EAT labeling and a super $(a, 2)$-EAT labeling.*

In [37] Bača, Lin, Miller, and Youssef proved that

Theorem 6.3.13 ([37]) *Every friendship graph f_n, $n \geq 1$, has a super $(a, 1)$-EAT labeling.*

Figure 6.21 illustrates a super EMT ((26, 0)-EAT) labeling of the friendship graph f_4.

For further investigation they propose the following open problem.

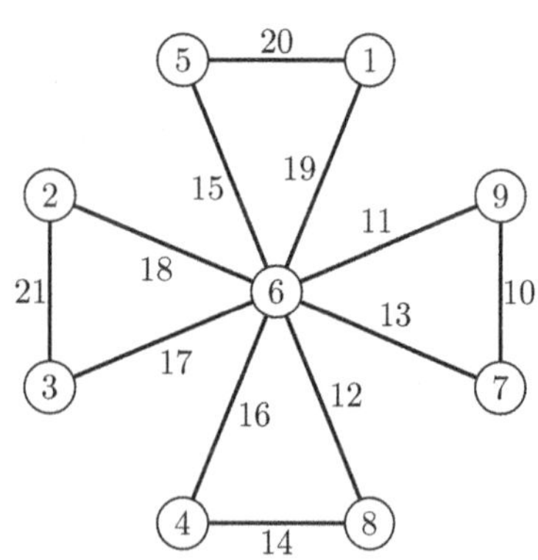

Fig. 6.21 Super (26, 0)-EAT labeling of f_4

Open Problem 6.3.1 ([37]) For the friendship graph f_n, determine if there is a super $(a, 0)$-EAT or a super $(a, 2)$-EAT labeling, for $n > 7$.

Arumugam and Nalliah in [20] investigated the above problem and proved that

Theorem 6.3.14 ([20]) *The friendship graph f_n has no super $(a, 2)$-EAT labeling when n is even and $n \not\equiv 4 \pmod{12}$.*

They also proved that the generalized friendship graph $f_{2,n}$ has a super $(a, 1)$-EAT labeling if and only if n is odd, see also [200].

If the fan F_n, $n \geq 2$, is super (a, d)-EAT, then, by (6.6) it follows that $d < 3$. The next lemma characterizes $(a, 1)$-edge-antimagicness of fans.

Lemma 6.3.8 ([37]) *The fan F_n has a $(3, 1)$-EAV labeling if and only if $2 \leq n \leq 6$.*

In light of Lemmas 6.3.8 and 6.3.1, we obtain the next theorem.

Theorem 6.3.15 ([37]) *The fan F_n is super (a, d)-EAT, if $2 \leq n \leq 6$ and $d \in \{0, 1, 2\}$.*

Let us recall that the proposition that the fan F_n is super EMT if and only if $2 \leq n \leq 6$ was stated by Figueroa-Centeno, Ichishima, and Muntaner-Batle in [98].

Recall that a wheel W_n, $n \geq 3$, is a graph obtained by joining all vertices of cycle C_n to a further vertex c, called the center. If wheel W_n, $n \geq 3$, is super (a, d)-EAT, then, by (6.6) it follows that $d < 2$. Enomoto et al. [96] proved that the wheel W_n is not super EMT. Thus Bača, Lin, Miller, and Youssef have the following result.

Theorem 6.3.16 ([37]) *The wheel W_n has a super (a, d)-EAT labeling if and only if $d = 1$ and $n \not\equiv 1 \pmod{4}$.*

Recall that the generalized prism can be defined as the Cartesian product $C_m \square P_n$ of a cycle on m vertices with a path on n vertices. If the generalized prism is super (a, d)-EAT, then, by (6.6), $d < 3$. In [270] is proved the following.

Lemma 6.3.9 ([270]) *The generalized prism $C_m \square P_n$ has an $(a, 1)$-EAV labeling if m is odd, $m \geq 3$ and $n \geq 2$.*

Now, using Theorem 6.3.5 and Lemma 5.3.1 gives

Theorem 6.3.17 ([270]) *If m is odd, $m \geq 3$, $n \geq 2$, and $d \in \{0, 1, 2\}$, then the generalized prism $C_m \square P_n$ has a super (a, d)-EAT labeling.*

Note that Figueroa-Centeno et al. [98] have shown that the generalized prism $C_m \square P_n$ is super EMT (super $(a, 0)$-EAT), if m is odd and $n \geq 2$. The next theorem shows the super $(a, 1)$-edge-antimagicness of $C_m \square P_n$, for m even.

Theorem 6.3.18 ([270]) *If m is even, $m \geq 4$, and $n \geq 2$, then the generalized prism $C_m \square P_n$ has a super $(a, 1)$-EAT labeling.*

For prism $C_m \square P_2$, Sugeng, Miller, and Bača proved the following.

Theorem 6.3.19 ([270]) *The prism $C_m \square P_2$ has a super (a, d)-EAT labeling if and only if one of the following conditions is satisfied.*

(i) $d \in \{0, 1, 2\}$ *and* m *is odd,* $m \geq 3$.
(ii) $d = 1$ *and* m *is even,* $m \geq 4$.

What can be said about super (a, d)-EAT labeling of $C_m \square P_n$ for the remaining cases if m is even and $d \in \{0, 2\}$? Sugeng, Miller, and Bača propose the following conjecture.

Conjecture 6.3.1 ([270]) If m is even, $m \geq 4$, $n \geq 3$ and $d \in \{0, 2\}$, then the generalized prism $C_m \square P_n$ has a super (a, d)-EAT labeling.

The generalized Petersen graph $P(n, m)$ is defined in Sect. 3.1.3. Figueroa-Centeno et al. [98] have shown that the generalized prism $C_n \square P_k$ is super EMT if n is odd and $k \geq 2$. They proved this by describing a $((n+3)/2, 1)$-EAV labeling of $C_n \square P_k$. We note that the generalized prism $C_n \square P_2$ is the generalized Petersen graph $P(n, 1)$. Fukuchi [107] described a $((n+3)/2, 1)$-EAV labeling for generalized Petersen graph $P(n, 2)$, when n is odd, $n \geq 3$, and proved that $P(n, 2)$ is super EMT. Now, if we use the result of Figueroa-Centeno, Ichishima, and Muntaner-Batle, and result of Fukuchi, and apply Theorem 6.3.5, then we get

Theorem 6.3.20 *Every generalized Petersen graph $P(n, m)$, n odd, $n \geq 3$, $1 \leq m \leq 2$, has a super $((11n+3)/2, 0)$-EAT labeling and a super $((5n+5)/2, 2)$-EAT labeling.*

Furthermore, Ngurah and Baskoro [202] proved that every generalized Petersen graph $P(n, m)$, $n \geq 3$, $1 \leq m < n/2$, has a super $(4n + 2, 1)$-EAT labeling. Bača, Baskoro, Simanjuntak, and Sugeng obtained the following result for $m = (n-1)/2$.

Theorem 6.3.21 ([27]) *Every generalized Petersen graph $P(n, (n-1)/2)$, $n \geq 3$ odd, has a super $((11n+3)/2, 0)$-EAT labeling and a super $((5n+5)/2, 2)$-EAT labeling.*

Figure 6.22 presents a super $(40, 0)$-EAT labeling of the generalized Petersen graph $P(7, 3)$.

Constructions that will produce super $(a, 0)$-EAT labeling and super $(a, 2)$-EAT labeling of $P(n, m)$, for n odd and $3 \leq m \leq (n-3)/2$, have not been found yet. Nevertheless, Bača, Baskoro, Simanjuntak, and Sugeng suggest the following conjecture.

Conjecture 6.3.2 ([27]) There is a super (a, d)-EAT labeling for the generalized Petersen graph $P(n, m)$, for every n odd, $n \geq 9$, $d \in \{0, 2\}$ and $3 \leq m \leq (n-3)/2$.

In [96], Enomoto, Lladó, Nakamigawa, and Ringel proved that a complete bipartite graph $K_{m,n}$ is super EMT if and only if $m = 1$ or $n = 1$. This means that for $n \geq 2$ there is no super $(a, 0)$-EAT labeling of $K_{n,n}$. Using (6.6) for $K_{n,n}$ gives that $d < 2$ if $n \geq 4$, while $d < 3$ if $2 \leq n \leq 3$. It remains to deal with super $(a, 2)$-EAT labelings of $K_{2,2}$ and $K_{3,3}$.

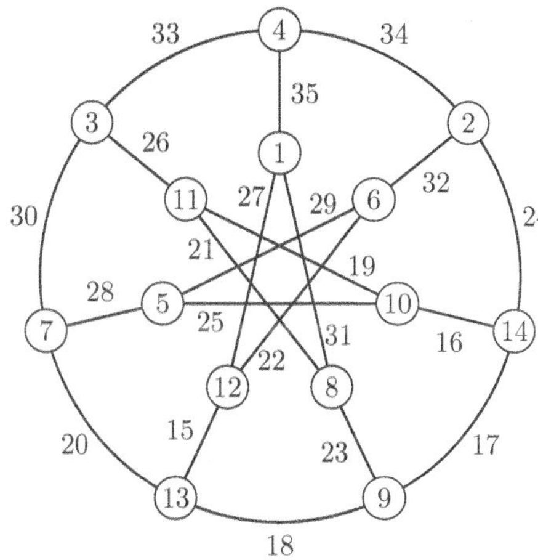

Fig. 6.22 Super (40, 0)-EAT labeling of $P(7, 3)$

Lemma 6.3.10 ([37]) *For complete bipartite graph $K_{n,n}$, $2 \leq n \leq 3$, there is no super $(a, 2)$ EAT labeling.*

The complete bipartite graph $K_{n,n}$ is even regular for n even and odd regular with 1-factor for n odd. Thus according to Theorems 6.3.8, 6.3.9, and Lemma 6.3.10 the next characterization is true.

Theorem 6.3.22 ([37]) *The complete bipartite graph $K_{n,n}$ has a super (a, d)-EAT labeling if and only if $d = 1$ and $n \geq 2$.*

For complete graph K_n, $n \geq 3$, from (6.6) it follows that $d < 2$ if $n \geq 4$, and $d \leq 2$ if $n = 3$. Lemma 6.1.2 part (iii) states that for every complete graph K_n, $n > 3$, there is no (a, d)-EAV labeling. Then, for $d = 1$, from Theorem 6.3.6, it follows that for every complete graph K_n, $n \geq 4$, there is no super $(a, 0)$-EAT labeling. The complete graph K_n is even regular for n odd and odd regular with 1-factor for n even. Thus according to Theorems 6.3.8 and 6.3.9, every complete graph K_n, $n \geq 3$, has a super $(2n + 2, 1)$-EAT labeling. It remains to deal with the case $n = 3$ and $d = 0, 2$.

Complete graph K_3 is the friendship graph f_1 and from Theorem 6.3.12, it follows that f_1 has a super (a, d)-EAT labeling for $d = 0$ and $d = 2$. So, from the previous facts next characterization follows.

Theorem 6.3.23 ([37]) *The complete graph K_n, $n \geq 3$, has a super (a, d)-EAT labeling if and only if one of the following conditions is satisfied.*

(i) $d = 0$ *and* $n = 3$.
(ii) $d = 1$ *and* $n \geq 3$.
(iii) $d = 2$ *and* $n = 3$.

The next theorems completely characterize the super (a, d)-EAT labelings of star $K_{1,n}$ and path P_n.

Theorem 6.3.24 ([273]) *The star $K_{1,n}$ has a super (a, d)-EAT labeling if and only if either one of the following conditions is satisfied.*

(i) $d \in \{0, 1, 2\}$ *and* $n \geq 1$.
(ii) $d = 3$ *and* $1 \leq n \leq 2$.

Theorem 6.3.25 ([38]) *The path P_n, $n \geq 2$, has a super (a, d)-EAT labeling if and only if $d \in \{0, 1, 2, 3\}$.*

Now, we study embedding of paths in the two-dimensional grid and consider a set of elementary transformations which keep the edge-antimagic character of the paths. A tree obtained from some embedding of a path in the two-dimensional grid by a set of elementary transformations is called path-like tree. For definition see Sect. 4.3.5.

Figure 6.23 shows an embedding of path P_{25} with a vertex labeling. We can see that edge-weights, under the vertex labeling, form the set of consecutive integers $\{15, 16, \ldots, 38\}$. This means that the path P_{25} admits a $(15, 1)$-EAV labeling. Using Theorem 6.3.5, the resulting total labeling of the path P_{25} is super $(64, 0)$-EAT or super $(41, 2)$-EAT.

Bača, Lin, and Muntaner-Batle proved the following theorem.

Theorem 6.3.26 ([38]) *All path-like trees are super (a, d)-EAT if and only if $d \in \{0, 1, 2, 3\}$.*

Figure 6.24 exhibits an example of path-like tree, obtained from the embedding of P_{25} in Fig. 6.23. The vertex labeling of the path-like tree is $(15, 1)$-EAV. Again by Theorem 6.3.5, we can obtain a super $(64, 0)$-EAT or a super $(41, 2)$-EAT labeling.

Note that path-like trees have maximum degree at most 4. It is an open problem to decide if a given tree of maximum degree 4 is a path-like tree. Bača, Lin, and Muntaner-Batle in [38] propose the following open problem.

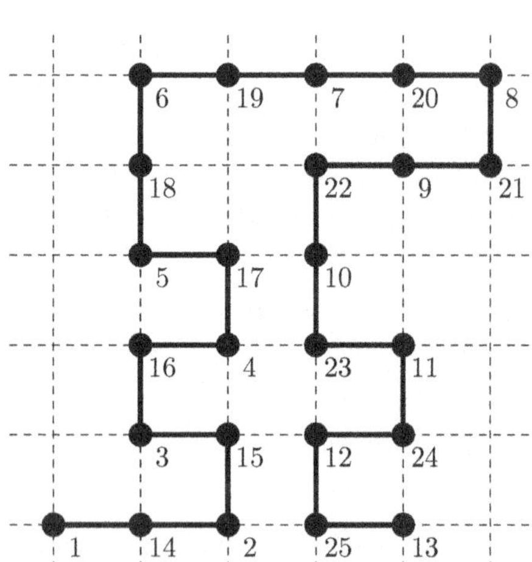

Fig. 6.23 $(15, 1)$-EAV labeling of path P_{25}

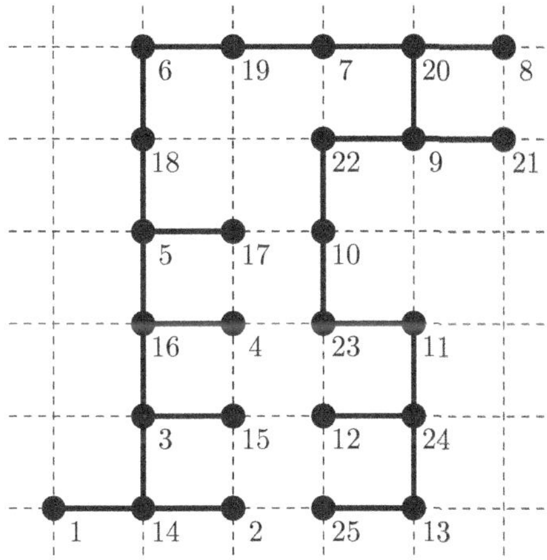

Fig. 6.24 (15, 1)-EAV labeling of path-like tree on 25 vertices

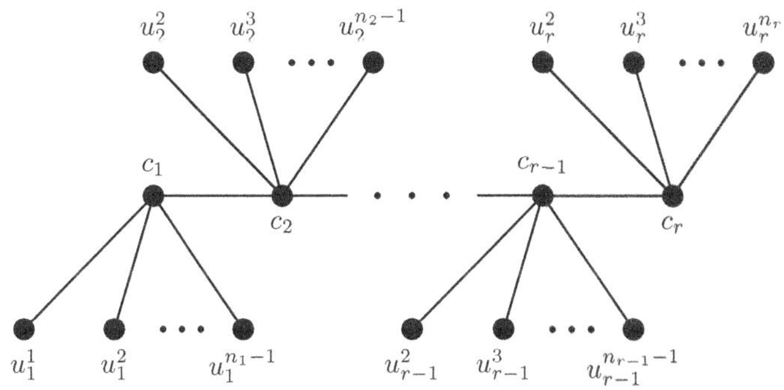

Fig. 6.25 Caterpillar $S_{n_1,n_2,\ldots,n_r}$

Open Problem 6.3.2 ([38]) Determine the complexity of deciding if a given tree of maximum degree 4 is a path-like tree.

A *caterpillar* is a graph derived from a path by hanging any number of leaves from the vertices of the path. The caterpillar can be seen as a sequence of stars $K_{1,n_1} \cup K_{1,n_2} \cup \cdots \cup K_{1,n_r}$, where each K_{1,n_i} is a star with central vertex c_i and n_i leaves for $i = 1, 2, \ldots, r$, and the leaves of K_{1,n_i} include c_{i-1} and c_{i+1}, for $i = 2, 3, \ldots, r-1$. We denote the caterpillar as $S_{n_1,n_2,\ldots,n_r}$, where the vertex set is $V(S_{n_1,n_2,\ldots,n_r}) = \{c_i : 1 \leq i \leq r\} \cup \bigcup_{i=2}^{r-1}\{u_i^j : 2 \leq j \leq n_i - 1\} \cup \{u_1^j : 1 \leq j \leq n_1 - 1\} \cup \{u_r^j : 2 \leq j \leq n_r\}$, and the edge set is $E(S_{n_1,n_2,\ldots n_r}) = \{c_i c_{i+1} : 1 \leq i \leq r-1\} \cup \bigcup_{i=2}^{r-1}\{c_i u_i^j : 2 \leq j \leq n_i - 1\} \cup \{c_1 u_1^j : 1 \leq j \leq n_1 - 1\} \cup \{c_r u_r^j : 2 \leq j \leq n_r\}$, see Fig. 6.25.

According to (6.6), we have that if a caterpillar $S_{n_1,n_2,\ldots,n_r}$ is super (a, d)-EAT, then $d \leq 3$. By using the construction of vertex labeling of caterpillar described by

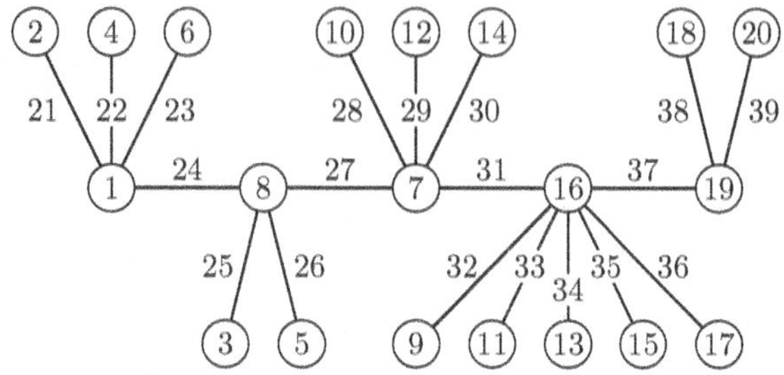

Fig. 6.26 Super (24, 3)-EAT labeling of $S_{4,4,5,7,3}$

Kotzig and Rosa in [163], Sugeng, Miller, Slamin, and Bača proved the following theorem.

Theorem 6.3.27 ([273]) *All caterpillars with p vertices are super (a, d)-EAT for $d \in \{0, 1, 2\}$.*

Let $S_{n_1,n_2,\ldots,n_r}$ be a caterpillar and $N_1 = \sum_{i=1}^{\lceil r/2 \rceil} n_{2i-1}$ and $N_2 = \sum_{i=1}^{\lfloor r/2 \rfloor} n_{2i}$. The remaining theorems in this section give results for super $(a, 3)$-edge-antimagicness of caterpillar $S_{n_1,n_2,\ldots,n_r}$.

Theorem 6.3.28 ([273]) *If r is even and $N_1 = N_2$ or $|N_1 - N_2| = 1$, then the caterpillar $S_{n_1,n_2,\ldots,n_r}$ with p vertices has a super $(a, 3)$-EAT labeling.*

Theorem 6.3.29 ([273]) *If r is odd, and $N_1 = N_2$ or $N_1 = N_2 + 1$, then the caterpillar $S_{n_1,n_2,\ldots,n_r}$ with p vertices has a super $(a, 3)$-EAT labeling.*

Figure 6.26 illustrates a super (24, 3)-EAT labeling of the caterpillar $S_{4,4,5,7,3}$ with $N_1 = 12$ and $N_2 = 11$.

For the caterpillar $S_{n_1,n_2,\ldots,n_r}$, r odd and $N_2 = N_1 + 1$, so far Sugeng, Miller, Slamin, and Bača have not found any super $(a, 3)$-EAT labeling. Consequently, they propose the following open problem.

Open Problem 6.3.3 ([273]) For the caterpillar $S_{n_1,n_2,\ldots,n_r}$, determine if there is a super $(a, 3)$-EAT labeling, for r odd and $N_2 = N_1 + 1$.

We conclude this chapter with the following open problem.

Open Problem 6.3.4 ([273]) For the caterpillar $S_{n_1,n_2,\ldots,n_r}$, determine feasible pairs (N_1, N_2), $N_1 \neq N_2$, and $|N_1 - N_2| \neq 1$, which make a super $(a, 3)$-EAT labeling impossible.

6.3.3 Super Edge-Antimagic Total Labelings of Circulant Graphs

In this section we focus on circulant graphs and we study the existence of the super edge-antimagic total labelings for this family of graphs. The circulant graph was already defined in Sect. 2.8.5.

Let us recall that for $1 \leq a_1 < a_2 < \cdots < a_m \leq n/2$ the *circulant graph* $C_n(a_1, a_2, \ldots, a_m)$ is a regular graph with the vertex-set $V = \{v_0, v_1, \ldots, v_{n-1}\}$ and the edge-set $E = \{v_i v_{i+a_j} : i = 0, 1, \ldots, n-1, \ j = 1, 2, \ldots, m\}$, where indices are taken modulo n.

The circulant graphs are an important class of graphs, which can be used in the design of local area networks, see [59].

It is easy to see that if $a_m < n/2$, then $C_n(a_1, a_2, \ldots, a_m)$ is a $2m$-regular circulant graph with mn edges. On the other hand, if $a_m = n/2$, then the circulant graph is a $(2m-1)$-regular one of size $n(2m-1)/2$.

The circulant graph $C_n(a_1, a_2, \ldots, a_m)$ is connected, see [73], if and only if for the greatest common divisor (gcd) of the numbers $a_1, a_2, \ldots, a_m, n$ is unity, i.e., $\gcd(a_1, a_2, \ldots, a_m, n) = 1$. More precisely, a circulant graph $C_n(a_1, a_2, \ldots, a_m)$ has $h = \gcd(a_1, a_2, \ldots, a_m, n)$ connected components which are isomorphic to $C_{n/h}(a_1/h, a_2/h, \ldots, a_m/h)$. Heuberger [127] proved that a connected circulant graph $C_n(a_1, a_2, \ldots, a_m)$ is bipartite if and only if $a_1, a_2, \ldots, a_m$ are odd and n is even.

Suppose that the circulant graph $C_n(a_1, a_2, \ldots, a_m)$ has a super (a, d)-EAT labeling. It is easy to see that the minimum possible edge-weight is at least $n+4$, i.e., the sum of the two smallest possible vertex labels and the smallest possible edge label. The maximum possible edge-weight is no more than $3|V|+|E|-1$, i.e., the sum of two the largest possible vertex labels and the largest possible edge label. Thus

$$a + (|E|-1)d \leq 3|V| + |E| - 1$$

gives

$$d \leq 1 + \frac{2n-4}{mn-1} \tag{6.7}$$

for $2m$-regular graph and

$$d \leq 1 + \frac{4n-8}{(2m-1)n-2} \tag{6.8}$$

for $(2m-1)$-regular graph.

If $m = 1$ and $n \geq 3$, then from (6.7) we have that $d \leq 2$. If $m \geq 2$ and $n \geq 4$, then $0 < (2n-4)/(mn-1) < 1$ and from (6.7) it follows that $d < 2$. Whenever $n \geq 6$ and $m \geq 3$ thus $0 < (4n-8)/((2m-1)n-2) < 1$. Observe from (6.8) that for $n = 4$ and $m = 2$, respectively for $n \geq 6$ and $m \geq 3$, there is no super $(a, 2)$-EAT labeling for circulant $(2m-1)$-regular graph. However, for $n \geq 6$ even and $m = 2$ circulant 3-regular graph may admit a super $(a, 2)$-EAT labeling. Moreover, from (6.8) it follows that for $n \geq 4$ even and $m = 1$ the circulant graph $C_n(n/2)$ may admit a super (a, d)-EAT labeling for $d \leq 5$.

Hussain et al. [132], see also [131], have constructed the super $(a, 1)$-EAT labeling for Harary graphs C_n^t, $n \geq 4$, and $t \geq 2$. The Harary graph C_n^t is isomorphic to the circulant graph $C_n(1, t)$.

Any circulant graph is either $2m$-regular or $(2m-1)$-regular with 1-factor. Thus, by Theorems 6.3.8 and 6.3.9 we obtain the following result.

Theorem 6.3.30 ([26]) *Let a_j, $j = 1, 2, \dots, m$, be positive integers and $1 \leq a_1 < a_2 < \cdots < a_m \leq n/2$. For every $n \geq 3$ and $m \geq 1$, a circulant graph $C_n(a_1, a_2, \dots, a_m)$ on n vertices admits a super $(2n+2, 1)$-EAT labeling.*

Now we use the properties of circulant graphs and determine the values of m and n for which no super EMT labeling is possible.

Theorem 6.3.31 ([26]) *If n or m are even integers, then for $2m$-regular circulant graph $C_n(a_1, a_2, \dots, a_m)$, $n \geq 3$, $m \geq 1$, there is no super EMT labeling.*

Proof Consider $2m$-regular circulant graph $G \cong C_n(a_1, a_2, \dots, a_m)$ with $a_m < n/2$. Assume that n or m are even integers and G has a super EMT labeling f with common edge-weight k. In the computation of the edge-weights of G, each edge label is used once and each label of vertex $v \in V(G)$ is used $\deg(v)$ times. Thus we have

$$\sum_{v \in V(G)} \deg(v) \cdot f(v) + \sum_{e \in E(G)} f(e) = k|E(G)| \tag{6.9}$$

and for $2m$-regular circulant graph of size mn we get

$$2n + 1 + \frac{mn+1}{2} = k. \tag{6.10}$$

From Eq. (6.10) it follows that, for even product mn the constant k is not an integer, which is a contradiction. □

Theorem 6.3.32 ([26]) *If $n \equiv 0 \pmod 4$, then for $(2m-1)$-regular circulant graph $C_n(a_1, a_2, \dots, a_{m-1}, n/2)$, $n \geq 4$, $m \geq 1$, there is no super EMT labeling.*

Proof Assume to the contrary that for $n \equiv 0 \pmod 4$ the $(2m-1)$-regular circulant graph $C_n(a_1, a_2, \dots, a_{m-1}, n/2)$ has a super EMT labeling f. From Eq. (6.9) we get

$$2n + 1 + \frac{(2m-1)n+2}{4} = k. \tag{6.11}$$

This contradicts the fact that k is an integer. □

Figueroa-Centeno, Ichishima, and Muntaner-Batle in [98] proved that a graph G is super EMT if and only if there exists an $(s, 1)$-EAV labeling. This result will be used for determining additional values of m and n for which no super EMT labeling of circulant graph exists.

The next two theorems indicate when the circulant graphs can never have an EAV labeling with difference $d = 1$ (respectively, super EMT labeling).

Theorem 6.3.33 ([26]) *If $n \geq 3$ and $m > 1$ are both odd integers, then for $2m$-regular circulant graph $C_n(a_1, a_2, \ldots, a_m)$ there is no $(a, 1)$-EAV labeling.*

Proof Suppose to the contrary that for n and m both odd, $n \geq 3$, $m > 1$, the $2m$-regular circulant graph admits an EAV labeling $g : V(G) \to \{1, 2, \ldots, n\}$ with the difference $d = 1$. Thus the sum of all the edge-weights under the vertex labeling g contains each vertex label $\deg(v)$ times and we get the equation

$$\sum_{v \in V(G)} \deg(v) \cdot g(v) = \sum_{i=0}^{|E|-1} (a + i). \tag{6.12}$$

For $2m$-regular circulant graphs, Eq. (6.12) becomes

$$n + 1 = a + \frac{mn - 1}{2}. \tag{6.13}$$

The minimum possible edge-weight is at least 3 and from Eq. (6.13) we have

$$n(2 - m) \geq 3$$

but this is impossible for $m > 1$. □

Theorem 6.3.34 ([26]) *If $n \equiv 2 \pmod 4$, $n \geq 6$ and $m \geq 3$, then for the $(2m-1)$-regular circulant graph $C_n(a_1, a_2, \ldots, a_{m-1}, n/2)$ there is no $(a, 1)$-EAV labeling.*

Proof Assume that $C_n(a_1, a_2, \ldots, a_{m-1}, n/2)$ has an $(a, 1)$-EAV labeling g with the edge-weights $a, a+1, \ldots, a + n(2m-1)/2 - 1$, where $n \equiv 2 \pmod 4$, $n \geq 6$ and $m \geq 3$. For the $(2m-1)$-regular circulant graph, from Eq. (6.12), we have

$$n + 1 = a + \frac{(2m - 1)n - 2}{4}. \tag{6.14}$$

Since $a \geq 3$, then from Eq. (6.14) it follows that $n(5 - 2m) \geq 6$, which is a contradiction for $m \geq 3$. □

It is reasonable to ask whether there exists a super EMT labeling of circulant graph for remaining values of m and n, i.e., for $n \geq 3$ odd and $m = 1$ (if circulant graph is $2m$-regular) and for $n \equiv 2 \pmod 4$ and $m = 1, 2$ (if circulant graph is $(2m-1)$-regular).

Clearly, if $\gcd(a_1, n) = 1$ for n odd and $a_1 \leq (n-1)/2$, then the circulant graph $C_n(a_1)$ is a cycle on n vertices. Otherwise, the circulant graph $C_n(a_1)$ is a disjoint union of $h = \gcd(a_1, n)$ copies of cycle $C_{n/h}$.

Godbold and Slater described in [112] a (super) EMT labeling of cycle C_n, n odd, with magic sum $(5n+3)/2$. Dafik, Miller, Ryan, and Bača, see Theorem 6.3.51 part (i), showed that a disjoint union of h copies of cycle $C_{n/h}$ admits a super EMT labeling if and only if h and n/h are odd. Thus we have

Corollary 6.3.8 ([26]) *For every n odd, $n \geq 3$, the circulant graph $C_n(a_1)$ admits a super EMT labeling.*

Every circulant graph $C_n(n/2)$, for $n \equiv 2 \pmod 4$, is a disjoint union of $n/2$ copies of the path P_2. In [85], see Theorem 6.3.54, it is proved that the disjoint union of an odd number of copies of a path on at least two vertices has a super EMT labeling. According to the previous result we have

Corollary 6.3.9 ([26]) *For every $n \equiv 2 \pmod 4$, $n \geq 6$, the circulant graph $C_n(n/2)$ has a super EMT labeling.*

The next theorem gives a partial result for the existence of a super EMT labeling for connected $(2m-1)$-regular circulant graph $C_n(a_1, n/2)$ for a_1 even and $\gcd(a_1, n/2, n) = \gcd(a_1, n/2) = 1$.

Theorem 6.3.35 ([26]) *If $n \equiv 2 \pmod 4$, $n \geq 6$, a_1 is even and $\gcd(a_1, n/2) = 1$, then the circulant graph $C_n(a_1, n/2)$ has an $((n+6)/4, 1)$-EAV labeling.*

Proof Let $n = 4t+2$, $t \geq 1$, and $a_1 = 2r$, $1 \leq r \leq t$. For $0 \leq i \leq n/2 - 1$ and $1 \leq r \leq t$, we construct a vertex labeling g of $C_n(2r, n/2)$ in the following way:

$$g(v_{2ri}) = \begin{cases} \frac{i}{2} + 1 & \text{if } i \text{ is even} \\ \frac{n+2i}{4} + 1 & \text{if } i \text{ is odd} \end{cases}$$

$$g(v_{n/2+2r(i-1)}) = \begin{cases} \frac{n+i}{2} + 1 & \text{if } i \text{ is even} \\ \frac{3n+2i}{4} + 1 & \text{if } i \text{ is odd} \end{cases}$$

where the indices $2ri$ and $n/2 + 2r(i-1)$ are taken modulo n.

We can see that the vertex labeling g is a bijective function from the vertex-set of $C_n(2r, n/2)$ onto the set $\{1, 2, \ldots, n\}$. If indices are taken modulo n, then for the edge-weights of $C_n(2r, n/2)$ we have

$$w_g(v_{2ri}v_{2r(i+1)}) = \begin{cases} \frac{i}{2} + \frac{n+2(i+1)}{4} + 2 = \frac{n+10}{4} + i & \text{if } 0 \leq i \leq \frac{n}{2} - 2 \\ g(v_{r(n-2)}) + g(v_0) = \frac{n+6}{4} & \text{if } i = \frac{n}{2} - 1 \end{cases}$$

$$w_g(v_{n/2+2r(i-1)}v_{n/2+2ri}) = \begin{cases} \frac{n+i}{2} + \frac{3n+2i+2}{4} + 2 = \frac{5n+10}{4} + i & \text{if } 0 \leq i \leq \frac{n}{2} - 2 \\ g(v_{n/2-4r}) + g(v_{n/2-2r}) = \frac{5n+6}{4} & \text{if } i = \frac{n}{2} - 1 \end{cases}$$

$$w_g(v_{2ri}v_{n/2+2ri}) = \begin{cases} \frac{i}{2} + \frac{3n+2(i+1)}{4} + 2 = \frac{3n+10}{4} + i & \text{if } 0 \le i \le \frac{n}{2} - 3, \\ & i \text{ is even} \\ \frac{3n+6}{4} & \text{if } i = \frac{n}{2} - 1 \\ \frac{n+2i}{4} + \frac{n+i+1}{2} + 2 = \frac{3n+10}{4} + i & \text{if } 1 \le i \le \frac{n}{2} - 2, \\ & i \text{ is odd.} \end{cases}$$

It is a routine procedure to verify that the set of all the edge-weights consists of the consecutive integers $\{(n+6)/4, (n+10)/4, \ldots, (3n+2)/4\} \cup \{(3n+6)/4, (3n+10)/4, \ldots, (5n+2)/4\} \cup \{(5n+6)/4, (5n+10)/4, \ldots, (7n+2)/4\}$, which implies that g is an $((n+6)/4, 1)$-edge-antimagic vertex labeling of $C_n(2r, n/2)$. □

Figure 6.27 shows a (4, 1)-EAV labeling of the circulant graph $C_{10}(4, 5)$. The edge labels are edge-weights under the vertex labeling g described in the proof of Theorem 6.3.35.

For the remaining cases Bača, Bashir, Nadeem, and Shabbir propose the following open problem.

Open Problem 6.3.5 ([26]) For the circulant graph $C_n(a_1, n/2)$, for $n \equiv 2 \pmod 4$, $n \ge 6$, if a_1 is even and $\gcd(a_1, n/2) > 1$ or if a_1 is odd, determine whether there exists an $(a, 1)$-EAV labeling.

In light of Theorem 6.3.5 as an immediate consequence of Theorem 6.3.35 we obtain the following corollary.

Corollary 6.3.10 ([26]) *For every $n \equiv 2 \pmod 4$, $n \ge 6$, the circulant graph $C_n(a_1, n/2)$, with a_1 even and $\gcd(a_1, n/2) = 1$, admits a super EMT labeling.*

Combining Corollaries 6.3.8, 6.3.9, and 6.3.10 we obtain the following theorem.

Theorem 6.3.36 ([26]) *The circulant graph $C_n(a_1, a_2, \ldots, a_m)$ has a super EMT labeling if one of the following conditions is satisfied.*

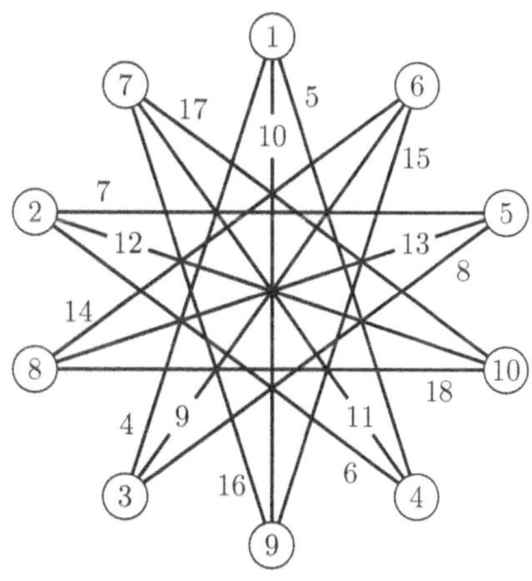

Fig. 6.27 (4, 1)-EAV labeling of $C_{10}(4, 5)$

(i) *n is odd, $n \geq 3$, and $m = 1$ for even regular graph.*
(ii) $n \equiv 2 \pmod 4$, *$n \geq 6$, and $m = 1$ for odd regular graph.*
(iii) $n \equiv 2 \pmod 4$, *$n \geq 6$, $m = 2$ with a_1 even and* $\gcd(a_1, n/2) = 1$ *for odd regular graph.*

If Open Problem 6.3.5 has positive solution, then according to Theorem 6.3.5 and together with Theorem 6.3.36 we have a characterization of all super EMT circulant graphs.

Next we focus on edge-antimagicness of circulant graphs for differences $2 \leq d \leq 5$. It follows from Inequality (6.7) that for $m = 1$ and $n \geq 3$, a 2-regular circulant graph may admit a super $(a, 2)$-EAT labeling. If $\gcd(a_1, n) = 1$, then the circulant graph $C_n(a_1)$ is a cycle on n vertices. If $\gcd(a_1, n) = h \neq 1$, then the graph $C_n(a_1)$ is a disjoint union of h copies of cycle $C_{n/h}$. Bača, Baskoro, Simanjuntak, and Sugeng showed in [27] that for n even there is no super $(a, 2)$-EAT labeling for cycle. Moreover they proved the next result.

Theorem 6.3.37 ([27]) *For n odd, $n \geq 3$, the cycle C_n has a super $((3n+5)/2, 2)$-EAT labeling.*

Dafik, Miller, Ryan, and Bača proved the following theorem.

Theorem 6.3.38 ([85]) *The disjoint union of h copies of cycle $C_{n/h}$, $hC_{n/h}$, has a super $((3n + 5)/2, 2)$-EAT labeling if and only if h and n are odd, $n \geq 3$.*

Thus from Theorems 6.3.37 and 6.3.38 we obtain the following corollary.

Corollary 6.3.11 ([26]) *The circulant graph $C_n(a_1)$ admits a super $((3n+5)/2, 2)$-EAT labeling if and only if n is odd, $n \geq 3$.*

By (6.8) it follows that for $m = 2$ and $n \geq 6$ even the existence of super $(a, 2)$-EAT labeling of $C_n(a_1, n/2)$ is possible. On the other hand, the next theorem shows that for $n \equiv 0 \pmod 4$ such a labeling does not exist.

Theorem 6.3.39 ([26]) *The circulant graph $C_n(a_1, n/2)$ does not admit a super $(a, 2)$-EAT labeling for $n \equiv 0 \pmod 4$.*

Proof Assume to the contrary that $C_n(a_1, n/2)$ admits a super $(a, 2)$-EAT labeling with edge-weights $a, a + 2, \ldots, a + (3n/2 - 1)2$. The sum of all vertex labels and edge labels used to calculate the edge-weights of $C_n(a_1, n/2)$ is equal to the sum of all the edge-weights. Thus

$$\sum_{i=1}^{n} 3i + \sum_{j=1}^{3n/2} (n + j) = \sum_{j=1}^{3n/2} (a + 2j - 2) \tag{6.15}$$

gives

$$5n + 10 = 4a. \tag{6.16}$$

The value of the parameter a must be an integer, thus $n \equiv 0 \pmod 4$ in the Eq. (6.16) which leads to a contradiction. □

Thus, the existence of a super $(a, 2)$-EAT labeling of $C_n(a_1, n/2)$ implies that $n \equiv 2 \pmod 4$.

The following corollary is a consequence of Theorems 6.3.35 and 6.3.5.

Corollary 6.3.12 ([26]) *For every $n \equiv 2 \pmod 4$, $n \geq 6$, the circulant graph $C_n(a_1, n/2)$, with a_1 even and* $\gcd(a_1, n/2) = 1$, *admits a super* $((5n + 10)/4, 2)$-*EAT labeling.*

If Open Problem 6.3.5 has a positive solution, then in light of Theorem 6.3.5 we get the existence of a super EAT labeling of difference $d = 2$ for $C_n(a_1, n/2)$ for $n \equiv 2 \pmod 4$, $n \geq 6$, with a_1 even and $\gcd(a_1, n/2) > 1$, and with a_1 odd.

By (6.8) it follows that for $n \geq 4$ even and $m = 1$ the upper bound for difference d is 5. In this case, the circulant graph $C_n(n/2)$ is a disjoint union of $n/2$ copies of the path P_2.

It was proved in [85], see Theorem 6.3.54, that if k is odd, $k \geq 3$ and $t \geq 2$, then the graph kP_t has a super $((3kt + k + 5)/2, 2)$-EAT labeling.

It is easy to see that for k even and $t = 2$ the minimum edge-weight $a = (3kt + k + 5)/2$ it is not integer. The existence of EAT labeling for $d = 4$ was proved in [85], see Theorem 6.3.56, i.e., the graph kP_2, $k \geq 3$, has a super $((5k+7)/2, 4)$-EAT labeling if and only if k is odd.

From Theorems 6.3.54 and 6.3.56 it follows

Corollary 6.3.13 ([26]) *For every $n \equiv 2 \pmod 4$, $n \geq 6$ and $d \in \{2, 4\}$, the circulant graph $C_n(n/2)$ admits a super (a, d)-EAT labeling.*

The existence of super (a, d)-EAT labelings for the disjoint union of k copies of path P_t and for differences $d = 3$ and $d = 5$ follows from Theorems 6.3.55 and 6.3.57. Thus we get

Corollary 6.3.14 ([26]) *For every $n \geq 4$ even and $d \in \{3, 5\}$, the circulant graph $C_n(n/2)$ has a super (a, d)-edge-antimagic total labeling.*

Other results on super (a, d)-EAT labelings of certain families of connected graphs can be found in [169, 170, 171, 173, 213, 214, 254, 266, 285].

6.3.4 Super Edge-Antimagic Total Labelings of Toroidal Polyhexes

The discovery of fullerene molecules and nanotubes has stimulated much interest in other possibilities for carbons. Classical fullerene is an all-carbon molecule in which the atoms are arranged on a pseudospherical framework made up entirely of pentagons and hexagons. Its molecular graph is a finite trivalent graph embedded on the surface of a sphere with only hexagonal and (exactly 12) pentagonal faces.

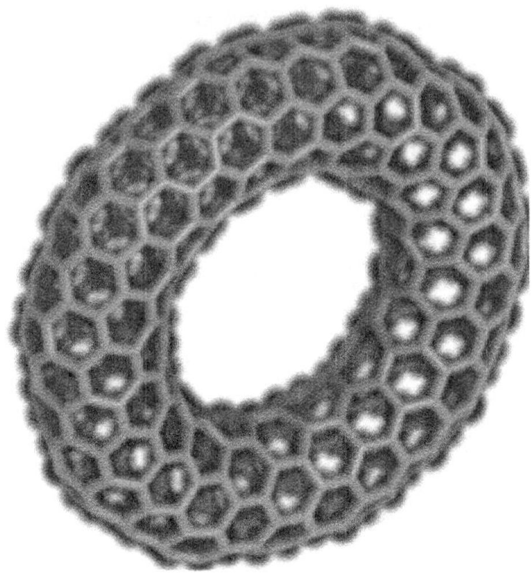

Fig. 6.28 Toroidal polyhex

Deza et al. [88] considered fullerene's extension to other closed surfaces and showed that only four surfaces are possible, namely the sphere, torus, Klein bottle, and projective plane. Unlike spherical fullerenes, toroidal and Klein bottle's fullerenes have been regarded as tessellations of entire hexagons on their surfaces since they necessarily contain no pentagons, see [88] and [157].

A *toroidal polyhex* (toroidal fullerene) is a cubic bipartite graph embedded on the torus such that each face is a hexagon. Note that the torus is a closed surface that can carry graphs of toroidal polyhex such that all vertices have degree 3 and all faces of the embedding are hexagons, see Fig. 6.28.

Some features of toroidal polyhexes with chemical relevance were discussed in [155] and [156]. For example, a systematic coding and classification scheme was given for the enumeration of isomers of toroidal polyhexes, the calculation of the spectrum, and the count for spanning trees. There have been a few work in the enumeration of perfect matchings of toroidal polyhexes by applying various techniques, such as transfer-matrix [158, 229] and permanent of the adjacency matrix [78]. The k-resonance of toroidal polyhexes have been studied in [297, 298, 299].

In this section we give a characterization for the super (a, d)-edge-antimagicness of toroidal polyhexes.

Let L be a regular hexagonal lattice and P_m^n be an $m \times n$ quadrilateral section (with m hexagons on the top and bottom sides and n hexagons on the lateral sides, n is even) cut from the regular hexagonal lattice L. First identify two lateral sides of P_m^n to form a cylinder, and finally identify the top and bottom sides of P_m^n at their corresponding points, see Fig. 6.29. From this we get a toroidal polyhex $\mathcal{H}_m^n$ with mn hexagons.

The graph lying in the interior of the quadrilateral section P_m^n has a proper 2-coloring. The vertices incident with a downward vertical edge and with two upwardly oblique edges can be colored, say black, and the other vertices, say white.

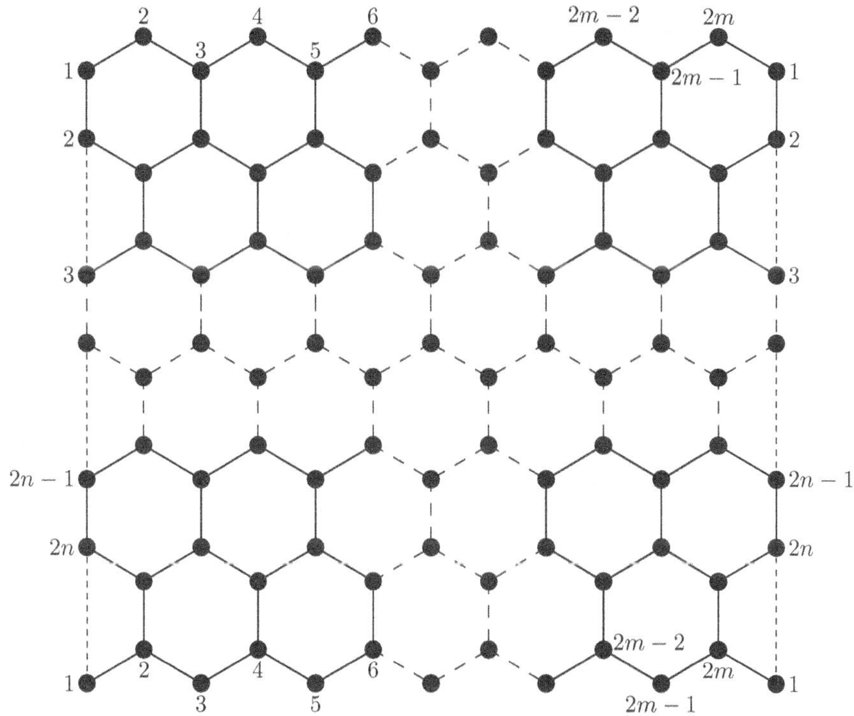

Fig. 6.29 Quadrilateral section P_m^n cuts from the regular hexagonal lattice

Such a 2-coloring is a proper 2-coloring of $\mathcal{H}_m^n$, i.e., the end vertices of each edge receive distinct colors. Hence we have that the toroidal polyhex $\mathcal{H}_m^n$ is a bipartite graph. It is known that $\mathcal{H}_m^n$ has $2mn$ vertices and $3mn$ edges.

We start by a necessary condition for a toroidal polyhex to be super (a, d)-EAT, which will provide an upper bound on the parameter d. From (6.6) it follows that the toroidal polyhex $\mathcal{H}_m^n$ with $2mn$ vertices, $3mn$ edges and mn hexagons is super (a, d)-EAT, then $d < 3$.

The next theorem shows that the toroidal polyhex $\mathcal{H}_m^n$ does not admit any super (a, d)-EAT labeling with $d \in \{0, 2\}$.

Theorem 6.3.40 ([47]) *For the toroidal polyhex $\mathcal{H}_m^n$ with mn hexagons, $m, n \geq 2$, n even, there is no super $(a, 0)$-EAT labeling and no super $(a, 2)$-EAT labeling.*

Proof Let $\mathcal{H}_m^n$ be the toroidal polyhex with mn hexagons. Suppose, to the contrary, that there exists a super (a, d)-EAT labeling $f : V(\mathcal{H}_m^n) \cup E(\mathcal{H}_m^n) \to \{1, 2, \ldots, 5mn\}$, for $d < 3$.

In the computation of the edge-weights of $\mathcal{H}_m^n$, each edge label is used once and each label of vertex is used 3 times. Thus, the sum of all vertex labels and all edge labels, used to calculate the edge-weights, is

$$3 \sum_{v \in V(\mathcal{H}_m^n)} f(v) + \sum_{e \in E(\mathcal{H}_m^n)} f(e) = \frac{3mn(11mn+3)}{2}. \tag{6.17}$$

The sum of all the edge-weights is

$$\sum_{e \in E(\mathcal{H}_m^n)} w_f(e) = 3mna + \frac{3mn(3mn-1)d}{2}. \tag{6.18}$$

From (6.17) and (6.18), we obtain

$$a = \frac{11mn + 3 - (3mn-1)d}{2}. \tag{6.19}$$

For $d = 0$ (respectively, $d = 2$) the Eq. (6.19) gives $a = (11mn + 3)/2$ (respectively, $a = (5mn + 5)/2$).

Since n is even, this contradicts the fact that a is an integer in both cases. This concludes the proof. □

Since $\mathcal{H}_m^n$ is 2-colorable cubic graph, there exist a 1-factor (perfect matching) and a 2-factor (a collection of cycles). Thus from Theorem 6.3.9 it follows that

Corollary 6.3.15 *The toroidal polyhex $\mathcal{H}_m^n$ with mn hexagons, $m, n \geq 2$, n even, is super $(4mn + 2, 1)$-EAT.*

Applying the necessary condition together with Theorem 6.3.40 and Corollary 6.3.15 we have the following characterization.

Theorem 6.3.41 ([47]) *The toroidal polyhex $\mathcal{H}_m^n$ with mn hexagons, $m, n \geq 2$, n even, admits a super (a, d)-EAT labeling if and only if $d = 1$ and $a = 4mn + 2$.*

6.3.5 Super Edge-Antimagic Total Labelings of Disjoint Union of Graphs

Figueroa-Centeno, Ichishima, and Muntaner-Batle proved the following theorem.

Theorem 6.3.42 ([98]) *If G is a (super) EMT bipartite or tripartite graph and m is odd, then mG is (super) EMT.*

The next corollary immediately follows from previous theorem.

Corollary 6.3.16 ([41]) *If T is a (super) EMT tree and m is odd, then mT is (super) EMT.*

Kotzig and Rosa [163] have shown that all cycles are EMT. Thus we have the following corollary.

Corollary 6.3.17 ([41]) *If m is odd and $n > 1$, then the 2-regular graph mC_{2n} is EMT.*

Bača et al. [38] proved that every path on n vertices has a super EMT labeling. From Theorem 6.3.42, it follows.

Corollary 6.3.18 ([41]) *If m is odd, $m \geq 3$ and $n \geq 2$, then the graph mP_n is super EMT.*

With respect to Theorems 6.3.5 and 6.3.6 it means that if G is a super EMT tripartite graph and m is odd, then mG is super $(a, 2)$-EAT. The next theorem extends this result.

Theorem 6.3.43 ([45]) *If G is a (super) $(a, 2)$-EAT tripartite graph and m is odd, then mG is (super) $(a', 2)$-EAT.*

Proof Let G be (super) $(a, 2)$-EAT tripartite (p, q) graph with the partite sets V_1, V_2 and V_3. Then $E(G) = V_1V_2 \cup V_2V_3 \cup V_1V_3$, where the juxtaposition of two partite sets denotes the set of edges between those two sets. Let $f : V(G) \cup E(G) \to \{1, 2, \ldots, p+q\}$ be a (super) $(a, 2)$-EAT labeling of G.

By x_i we denote the element (a vertex or an edge) in the ith copy of mG corresponding to the element $x \in V(G) \cup E(G)$. We define a new labeling g of mG, for m odd, in the following way.

$$g(x_i) = \begin{cases} m\,(f(x) - 1) + i & \text{if } x \in V_1 \cup V_2V_3 \\ m\,(f(x) - 1) + i + \frac{m+1}{2} & \text{if } x \in V_2 \cup V_1V_3 \text{ and } i < \frac{m}{2} \\ m\,(f(x) - 1) + i - \frac{m+1}{2} + 1 & \text{if } x \in V_2 \cup V_1V_3 \text{ and } i > \frac{m}{2} \\ m\,(f(x) - 1) + 2i & \text{if } x \in V_3 \cup V_1V_2 \text{ and } i < \frac{m}{2} \\ m\,(f(x) - 1) + 2i - m & \text{if } x \in V_3 \cup V_1V_2 \text{ and } i > \frac{m}{2}. \end{cases}$$

Let $t \in \{1, 2, \ldots, p+q\}$. We consider the following three cases:

Case A If the number t is assigned by the labeling f to some element of $V_1 \cup V_2V_3$, then the corresponding elements in the copies of G in mG have labels

$$\begin{array}{ll} m(t-1)+1 & \text{in } G_1 \\ m(t-1)+2 & \text{in } G_2 \\ \vdots & \vdots \\ m(t-1)+i & \text{in } G_i \\ \vdots & \vdots \\ mt & \text{in } G_m \end{array}$$

i.e., the numbers $m(t-1)+1, m(t-1)+2, \ldots, mt$.

Case B If the number t is assigned by the labeling f to some element of $V_2 \cup V_1V_3$, then the corresponding elements in the copies of G in mG receive labels

$$\begin{array}{ll} mt + \frac{3-m}{2} & \text{in } G_1 \\ mt + \frac{5-m}{2} & \text{in } G_2 \\ \vdots & \vdots \\ mt & \text{in } G_{\frac{m-1}{2}} \\ m(t-1)+1 & \text{in } G_{\frac{m+1}{2}} \\ m(t-1)+2 & \text{in } G_{\frac{m+3}{2}} \\ \vdots & \vdots \\ mt + \frac{1-m}{2} & \text{in } G_m \end{array}$$

thus the numbers $m(t-1)+1, m(t-1)+2, \dots, mt$.

Case C If the number t is assigned by the labeling f to some element of $V_3 \cup V_1V_2$, then the corresponding elements in the copies of G in mG receive labels

$$\begin{array}{ll} m(t-1)+2 & \text{in } G_1 \\ m(t-1)+4 & \text{in } G_2 \\ \vdots & \vdots \\ mt-1 & \text{in } G_{\frac{m-1}{2}} \\ m(t-1)+1 & \text{in } G_{\frac{m+1}{2}} \\ m(t-1)+3 & \text{in } G_{\frac{m+3}{2}} \\ \vdots & \vdots \\ mt & \text{in } G_m \end{array}$$

hence the corresponding labels are $m(t-1)+1, m(t-1)+2, \dots, mt$.

Thus the set of the labels in mG corresponding to the value t is independent of the labeled element. It means the labeling g is evidently total and assigns the number $1, 2, \dots, m(p+q)$ to the elements of mG. Moreover, if the labeling f is super, then also the smallest possible labels are used to label the vertices in mG and thus g is also super.

In the next part we will calculate the edge-weight of an edge $uv \in E(G_i)$. We again distinguish three cases. If $u \in V_1^i$ and $v \in V_2^i$, if $u \in V_1^i$ and $v \in V_3^i$ and if $u \in V_2^i$ and $v \in V_3^i$. By the symbol V_j^i, $j = 1, 2, 3$ and $i = 1, 2, \dots, m$, we denote the vertex set corresponding to the vertex set V_j in the ith copy of G.

It is easy to verify that in all cases we obtain for the edge-weights

$$g(u_i) + g(v_i) + g(u_i v_i) = \begin{cases} m\,(f(u) + f(v) + f(uv) - 3) + \frac{m+1}{2} + 4i \\ \hfill \text{if } i < \frac{m}{2} \\ m\,(f(u) + f(v) + f(uv) - 3) + \frac{m+1}{2} - 2m + 4i \\ \hfill \text{if } i > \frac{m}{2}. \end{cases}$$

Thus to the edge-weight A of some edge uv in G, $A = f(u) + f(v) + f(uv)$, corresponds to the following edge-weights in mG

$$\begin{array}{ll} m(A-3) + \frac{m+1}{2} + 4 & \text{in } G_1 \\ m(A-3) + \frac{m+1}{2} + 8 & \text{in } G_2 \\ \vdots & \vdots \\ m(A-3) + \frac{m+1}{2} + 2m - 2 & \text{in } G_{\frac{m-1}{2}} \\ m(A-3) + \frac{m+1}{2} + 2 & \text{in } G_{\frac{m+1}{2}} \\ m(A-3) + \frac{m+1}{2} + 6 & \text{in } G_{\frac{m+3}{2}} \\ \vdots & \vdots \\ m(A-3) + \frac{m+1}{2} + 2m & \text{in } G_m. \end{array}$$

It means that the edge-weights are

$$m(A-3) + \tfrac{m+1}{2} + 2, m(A-3) + \tfrac{m+1}{2} + 4, \dots, m(A-1) + \tfrac{m+1}{2}.$$

As f is $(a, 2)$-EAT labeling, then the edge-weights in G are

$$a, a+2, \dots, a + 2(q-1).$$

Thus to the edge-weight $A + 2$ in g the corresponding edge-weights in mG are

$$m(A-1) + \frac{m+1}{2} + 2, m(A-1) + \frac{m+1}{2} + 4, \dots, m(A+1) + \frac{m+1}{2},$$

hence the edge-weights in mG again form an arithmetic sequence with the difference 2 and the initial term $m(a-3) + (m+1)/2 + 2$. This concludes the proof. □

Directly from the previous theorem we get the following result.

Corollary 6.3.19 ([45]) *If G is a (super) $(a, 2)$-EAT bipartite graph and m is odd, then mG is (super) $(a', 2)$-EAT.*

In [141] Ivančo and Lučkaničová proved a more general result than the one in Theorem 6.3.42 for the disjoint union of edge-magic graphs. A mapping $c : V(G) \cup$

$E(G) \to \{1, 2, 3\}$ is called an *e-m-coloring* of a graph G if $\{c(u), c(v), c(uv)\} = \{1, 2, 3\}$ for any edge uv of G. They proved

Theorem 6.3.44 ([141]) *Let m be an odd positive integer. For $i = 1, 2, \dots, m$, let G_i, g_i, and c_i be an edge-magic graph with p_i vertices and q_i edges, an edge-magic total labeling of G_i with its magic number σ_i and an e-m-coloring of G_i, respectively. Suppose that the following conditions are satisfied.*

(i) There is an integer σ such that $\sigma_i = \sigma$ for all $i = 1, 2, \dots, m$.
(ii) If $g_i(x) = g_j(y)$, then $c_i(x) = c_j(y)$ for all $i, j = 1, 2, \dots, m$, $x \in V(G_i) \cup E(G_i)$ and $y \in V(G_j) \cup E(G_j)$.
(iii) There is an integer r such that $r = p_1 + q_1 \geq \cdots \geq p_m + q_m \geq r - 1$.

Then the disjoint union $\bigcup_{i=1}^{m} G_i$ is an edge-magic graph.

Moreover, if all g_i are super edge-magic labelings and $p_1 = p_2 = \cdots = p_m$, then $\bigcup_{i=1}^{m} G_i$ is a super edge-magic graph.

Next theorem shows a similar result for (super) $(a, 2)$-EAT graphs.

Theorem 6.3.45 ([45]) *Let m be an odd positive integer. For $i = 1, 2, \dots, m$, let G_i, f_i, and c_i be an $(a, 2)$-EAT graph with p_i vertices and q_i edges, an $(a, 2)$-EAT labeling of G_i and an e-m-coloring of G_i, respectively. Suppose that the following conditions are satisfied.*

(i) If $f_i(x) = f_j(y)$, then $c_i(x) = c_j(y)$ for all $i, j = 1, 2, \dots, m$, $x \in V(G_i) \cup E(G_i)$ and $y \in V(G_j) \cup E(G_j)$.
(ii) There is an integer r such that $r = p_1 + q_1 \geq \cdots \geq p_m + q_m \geq r - 1$.

Then the disjoint union $\bigcup_{i=1}^{m} G_i$ is an $(a', 2)$-EAT graph.

Moreover, if all f_i are super $(a, 2)$-EAT labelings and $p_1 = p_2 = \cdots = p_m$, then $\bigcup_{i=1}^{m} G_i$ is a super $(a', 2)$-EAT graph.

Proof For $i = 1, 2, \dots, m$, let G_i, f_i, and c_i be an $(a, 2)$-EAT graph with p_i vertices and q_i edges, an $(a, 2)$-EAT labeling of G_i and an e-m-coloring of G_i, respectively.

We define a new labeling g of $\bigcup_{i=1}^{m} G_i$, for m odd, in the following way.

$$g(x_i) = \begin{cases} m\,(f_i(x) - 1) + i & \text{if } c_i(x) = 1 \\ m\,(f_i(x) - 1) + i + \frac{m+1}{2} & \text{if } c_i(x) = 2 \text{ and } i < \frac{m}{2} \\ m\,(f_i(x) - 1) + i - \frac{m+1}{2} + 1 & \text{if } c_i(x) = 2 \text{ and } i > \frac{m}{2} \\ m\,(f_i(x) - 1) + 2i & \text{if } c_i(x) = 3 \text{ and } i < \frac{m}{2} \\ m\,(f_i(x) - 1) + 2i - m & \text{if } c_i(x) = 3 \text{ and } i > \frac{m}{2}. \end{cases}$$

It is easy to check that the labeling g uses each integer $1, 2, \dots, |V(\bigcup_{i=1}^{m} G_i) \cup E(\bigcup_{i=1}^{m} G_i)|$ exactly once. As f_i is an $(a, 2)$-EAT labeling of G_i and c_i is an e-m-coloring of G_i, then analogously as in the proof of Theorem 6.3.43 we show that $\bigcup_{i=1}^{m} G_i$ is an $(a', 2)$-EAT graph. Moreover, if all f_i are super $(a, 2)$-EAT labelings and $p_1 = p_2 = \cdots = p_m$, then $1 \leq g(u) \leq (p_i - 1)m + m = |V(\bigcup_{i=1}^{m} G_i)|$. Thus, g is also a super $(a', 2)$-EAT labeling. □

The next theorem describes a super $(a, 1)$-EAT labeling for the disjoint union of m graphs $G_1, G_2, \ldots, G_m$.

Theorem 6.3.46 ([41]) *Let G_i be a super $(a, 1)$-EAT graph of order p and size q, $i = 1, 2, \ldots, m$. Then the disjoint union $\bigcup_{i=1}^{m} G_i$ is also a super $(b, 1)$-EAT graph.*

Proof Let $G_i, i = 1, 2, \ldots, m$, be a graph with p vertices and q edges. Note that G_i is not necessary isomorphic to G_j for $i \neq j$. Suppose that each $G_i, i = 1, 2, \ldots, m$, admits a super $(a, 1)$-EAT labeling f_i such that

$$\begin{aligned} f_i :\ & V(G_i) \to \{1, 2, \ldots, p\} \\ & E(G_i) \to \{p+1, p+2, \ldots, p+q\} \end{aligned}$$

and

$$\{f_i(u) + f_i(v) + f_i(uv) : uv \in E(G_i)\} = \{a, a+1, \ldots, a+q-1\}.$$

Define the labeling f for the vertices and edges of $\bigcup_{i=1}^{m} G_i$ in the following way:

$$f(x) = \begin{cases} m\,(f_i(x) - 1) + i & \text{if } x \in V(G_i) \\ mf_i(x) + 1 - i & \text{if } x \in E(G_i). \end{cases}$$

It is not difficult to see that the function f assigns the labels

$$\begin{array}{llllll} 1, & m+1, & 2m+1, & \ldots & (p-1)m+1 & \text{to the vertices of } G_1 \\ 2, & m+2, & 2m+2, & \ldots & (p-1)m+2 & \text{to the vertices of } G_2 \\ \vdots & \vdots & \vdots & & \vdots & \vdots \\ i, & m+i, & 2m+i, & \ldots & (p-1)m+i & \text{to the vertices of } G_i \\ \vdots & \vdots & \vdots & & \vdots & \vdots \\ m, & 2m, & 3m, & \ldots & pm & \text{to the vertices of } G_m. \end{array}$$

And for the edge labels of $\bigcup_{i=1}^{m} G_i$ we have

$$\begin{array}{ccccc} (p+1)m, & (p+2)m, & \ldots & (p+q)m & \text{of } G_1 \\ (p+1)m-1, & (p+2)m-1, & \ldots & (p+q)m-1 & \text{of } G_2 \\ \vdots & \vdots & & \vdots & \vdots \\ (p+1)m+1-i, & (p+2)m+1-i, & \ldots & (p+q)m+1-i & \text{of } G_i \\ \vdots & \vdots & & \vdots & \vdots \\ pm+1, & (p+1)m+1, & \ldots & (p+q-1)m+1 & \text{of } G_m. \end{array}$$

It is easy to see that the labeling f is a bijective function which assigns the integer $\{1, 2, \ldots, mp + mq\}$ to the vertices and edges of $\bigcup_{i=1}^{m} G_i$, thus f is a total labeling. Furthermore, f assigns the numbers $1, 2, \ldots, pm$ to the vertices of $\bigcup_{i=1}^{m} G_i$; therefore, the f is a super total labeling.

For the edge-weight of $uv \in E(G_i)$ we have

$$\begin{aligned} f(u) + f(v) + f(uv) &= m\,(f_i(u) - 1) + i + m\,(f_i(v) - 1) + i + mf_i(uv) + 1 - i \\ &= m\,(f_i(u) + f_i(v) + f_i(uv) - 2) + 1 + i. \end{aligned}$$

It means that the edge-weights in the components are

$$\begin{array}{lllll} G_1: & m(a-2)+2, & m(a-1)+2, & \ldots & m(a+q-3)+2 \\ G_2: & m(a-2)+3, & m(a-1)+3, & \ldots & m(a+q-3)+3 \\ \vdots & \vdots & \vdots & & \vdots \\ G_i: & m(a-2)+1+i, & m(a-1)+1+i, & \ldots & m(a+q-3)+1+i \\ \vdots & \vdots & \vdots & & \vdots \\ G_m: & m(a-1)+1, & ma+1, & \ldots & m(a+q-2)+1. \end{array}$$

It is a routine procedure to verify that the edge-weights are distinct and consecutive

$$\{f(u) + f(v) + f(uv) : uv \in E(\bigcup_{i=1}^{m} G_i)\} = \{m(a-2)+2, m(a-2)+3, \ldots, \\ m(a+q-2)+1\}.$$

This implies that $\bigcup_{i=1}^{m} G_i$ has a super $(m(a-2)+2, 1)$-EAT labeling. □

Using Theorem 6.3.46 we can get the following corollary.

Corollary 6.3.20 ([41]) *Let G be a super $(a, 1)$-EAT graph. Then the disjoint union of arbitrary number of copies of G, i.e., mG, $m \geq 1$, also admits a super $(b, 1)$-EAT labeling.*

Moreover for m copies of graph G which is $(a, 1)$-EAT but not super EAT, we can also derive the following result.

Theorem 6.3.47 ([41]) *Let G be an $(a, 1)$-EAT graph. Then mG, $m \geq 1$, is also a $(b, 1)$-EAT graph.*

Proof Let G be a (p, q) graph and let f be an $(a, 1)$-EAT labeling of G

$$f : V(G) \cup E(G) \to \{1, 2, \ldots, p+q\}.$$

For every vertex v in G, we denote by symbol v_i the corresponding vertex of v in the ith copy of G in mG. Analogously, let $u_i v_i$ denote the corresponding edge of uv in the ith copy of G in mG.

We define a labeling g of mG in the following way

$$\begin{aligned} g(v_i) &= m\,(f(v) - 1) + i, && \text{for } v \in V(G), i = 1, 2, \ldots, m \\ g(u_i v_i) &= mf(uv) + 1 - i, && \text{for } uv \in E(G), i = 1, 2, \ldots, m. \end{aligned}$$

Let $t \in \{1, 2, \ldots, p+q\}$, we consider the following two cases.

Case A If the number t is assigned by the labeling f to a vertex of G, then the corresponding vertices in the copies in mG will receive labels

$$\begin{array}{ccccc} m(t-1)+1, & m(t-1)+2, & \ldots\, m(t-1)+i, & \ldots & mt \\ \text{in } G_1 & \text{in } G_2 \quad \ldots & \text{in } G_i & \ldots & \text{in } G_m. \end{array}$$

Case B If the number t is assigned by the labeling f to an edge of G, then the corresponding edges in the copies in mG will have labels

$$\begin{array}{ccccc} mt, & mt-1, & \ldots\, mt+1-i, & \ldots\, m(t-1)+1 \\ \text{in } G_1 & \text{in } G_2 \quad \ldots & \text{in } G_i \quad \ldots & \text{in } G_m. \end{array}$$

It is easy to see that the edge labels and vertex labels in mG are not overlapping, and the maximum used label is $mp + mq$, thus g is a total labeling. Moreover, following the same line of reasoning as in the proof of Theorem 6.3.46, we know that the edge-weights form an arithmetic sequence with the difference 1. This produces the desired result.

□

The next theorem describes a super $(a, 3)$-EAT labeling for the disjoint union of m not necessarily isomorphic graphs.

Theorem 6.3.48 ([41]) *Let G_i be a super $(a, 3)$-EAT graph of order p and size q, $i = 1, 2, \ldots, m$. The disjoint union $\bigcup_{i=1}^{m} G_i$ is a super $(b, 3)$-EAT graph.*

Proof Let G_i, $i = 1, 2, \ldots, m$, be a super $(a, 3)$-EAT (p, q)-graph. Therefore there exists a super $(a, 3)$-EAT labeling f_i for G_i such that

$$\begin{aligned} f_i : \quad & V(G_i) \to \{1, 2, \ldots, p\} \\ & E(G_i) \to \{p+1, p+2, \ldots, p+q\} \end{aligned}$$

and

$$\{f_i(u) + f_i(v) + f_i(uv) : uv \in E(G_i)\} = \{a, a+3, \ldots, a+3(q-1)\}.$$

We define the labeling f for $\bigcup_{i=1}^{m} G_i$ in the following way:

$$f(x) = m\,(f_i(x) - 1) + i, \quad \text{if } x \in V(G_i) \cup E(G_i).$$

It is not difficult to see that the function f assigns the labels

$$\begin{array}{lllll} 1, & m+1, & 2m+1, & \dots & (p-1)m+1 & \text{to the vertices of } G_1 \\ 2, & m+2, & 2m+2, & \dots & (p-1)m+2 & \text{to the vertices of } G_2 \\ \vdots & \vdots & \vdots & & \vdots & \vdots \\ i, & m+i, & 2m+i, & \dots & (p-1)m+i & \text{to the vertices of } G_i \\ \vdots & \vdots & \vdots & & \vdots & \vdots \\ m, & 2m, & 3m, & \dots & pm & \text{to the vertices of } G_m. \end{array}$$

For the edge labels we will have

$$\begin{array}{lllll} pm+1, & (p+1)m+1, & \dots & (p+q-1)m+1 & \text{of } G_1 \\ pm+2, & (p+1)m+2, & \dots & (p+q-1)m+2 & \text{of } G_2 \\ \vdots & \vdots & & \vdots & \vdots \\ pm+i, & (p+1)m+i, & \dots & (p+q-1)m+i & \text{of } G_i \\ \vdots & \vdots & & \vdots & \vdots \\ (p+1)m, & (p+2)m, & \dots & (p+q)m & \text{of } G_m. \end{array}$$

It is easy to see that f is a total labeling and the numbers $1, 2, \dots, pm$ are assigned to the vertices of $\bigcup_{i=1}^{m} G_i$, i.e., f is a super total labeling.

For the edge-weight of $uv \in E(G_i)$ we get

$$\begin{aligned} f(u) + f(v) + f(uv) &= m\,(f_i(u) - 1) + i + m\,(f_i(v) - 1) + i + m\,(f_i(uv) - 1) + i \\ &= m\,(f_i(u) + f_i(v) + f_i(uv) - 3) + 3i. \end{aligned}$$

The edge-weights in the components are

$$\begin{array}{lllll} G_1: & m(a-3)+3, & ma+3, & \dots & m(a+3q-6)+3 \\ G_2: & m(a-3)+6, & ma+6, & \dots & m(a+3q-6)+6 \\ \vdots & \vdots & \vdots & & \vdots \\ G_i: & m(a-3)+3i, & ma+3i, & \dots & m(a+3q-6)+3i \\ \vdots & \vdots & \vdots & & \vdots \\ G_m: & ma, & m(a+3), & \dots & m(a+3q-3). \end{array}$$

So the set of the edge-weights is

$$\{f(u)+f(v)+f(uv): uv \in E(\bigcup_{i=1}^{m} G_i)\} = \{m(a-3)+3, m(a-3)+6, \ldots, m(a+3q-3)\}.$$

This implies that $\bigcup_{i=1}^{m} G_i$ has a super $(m(a-3)+3, 3)$-EAT labeling. □

According to Theorem 6.3.48, immediately we get

Corollary 6.3.21 ([41]) *Let G be a super $(a, 3)$-EAT graph. Then the disjoint union of arbitrary number of copies of G, i.e., mG, $m \geq 1$, also admits a super $(b, 3)$-EAT labeling.*

As the technique of the verification is very similar to the proof of Theorem 6.3.47, we present the following result without the proof.

Theorem 6.3.49 ([41]) *Let G be an $(a, 3)$-EAT graph. Then mG, $m \geq 1$, is also a $(b, 3)$-EAT graph.*

In the literature there are some known conditions for the nonexistence of the (a, d)-EAT labelings for some graphs depending on the order and the size of a graph, see Theorems 4.1.1, 6.3.2, and 6.3.3.

The next lemma is based on the arguments using divisibility.

Lemma 6.3.11 ([45]) *Let G be a (p, q) graph with all vertices of odd degrees and let d be an even integer. Graph G has no (a, d)-EAT labeling if one of the following conditions holds.*

(i) $q \equiv 0 \pmod 4$ *and* $p \equiv 1 \pmod 4$ *or* $p \equiv 2 \pmod 4$.
(ii) $q \equiv 2 \pmod 4$ *and* $p \equiv 0 \pmod 4$ *or* $p \equiv 3 \pmod 4$.

Proof If a (p, q) graph G admits an (a, d)-EAT labeling f, then it is known that the following connection between the order, the size, the degrees of a graph, and the parameters a and d must hold:

$$\sum_{uv \in E(G)} w_f(uv) = \sum_{uv \in E(G)} f(uv) + \sum_{v \in V(G)} \deg(v) f(v).$$

Thus we get

$$aq + \frac{q(q-1)d}{2} = \frac{(p+q)(p+q+1)}{2} + \sum_{v \in V(G)} (\deg(v) - 1) f(v).$$

If G is a graph with all vertices of odd degree, d is even and if one of the conditions in the lemma holds, then using the parity considerations on the left-hand and on the right-hand side of the formula we get a contradiction. □

Thus for an even number of copies of a graph we have

Theorem 6.3.50 ([45]) *Let d, k be positive integers, d be even and k be odd. Let G be a graph with all vertices of odd degrees. If the size and the order of G have a different parity, then the graph $2kG$ has no (a, d)-EAT labeling.*

Proof Consider a graph $2kG$, where G is a graph with all vertices of odd degrees. Let k be odd. If the size of $2kG$ is odd and the order is even, then

$$|E(2kG)| = 2kq \equiv 2 \pmod 4$$
$$|V(2kG)| = 2kp \equiv 0 \pmod 4.$$

If the size is even and the order is odd, then

$$|E(2kG)| = 2kq \equiv 0 \pmod 4$$
$$|V(2kG)| = 2kp \equiv 2 \pmod 4.$$

Thus according to Lemma 6.3.11 the graph $2kG$ is not (a, d)-EAT for d even. □

For example, let us consider a star $K_{1,n}$. In [273] it is proved, see also Theorem 6.3.24, that every star is super $(a, 0)$-EAT and super $(a, 2)$-EAT. As the star is a bipartite graph, then an odd number of copies of a star $K_{1,n}$ is super $(a, 0)$-EAT and super $(a, 2)$-EAT according to Theorems 6.3.42 and 6.3.43. However, for $n \equiv 1 \pmod 2$ using Theorem 6.3.50 we get that $(4k + 2)$ copies of $K_{1,n}$ is neither $(a, 0)$-EAT nor $(a, 2)$-EAT. Thus also the graph G is (a, d)-EAT for $d \equiv 0 \pmod 2$, in many cases there exist no such labeling of even number of copies of G. This indicates that there exists no general construction of (a, d)-EAT labeling for even number of copies of a graph for d even.

6.3.6 Super Edge-Antimagic Total Labeling for Certain Families of Disconnected Graphs

This section summarizes the known results on super edge-antimagic total labelings for the disjoint union of multiple copies of cycles, paths, stars, caterpillars, complete graphs, and complete s-partite graphs.

Some results for edge-antimagicness of disconnected graphs are already known. The super (a, d)-EAT labelings for $P_n \cup P_{n+1}$, $nP_2 \cup P_n$, and $nP_2 \cup P_{n+2}$ are described by Sudarsana, Ismaimuza, Baskoro, and Assiyatun in [268], see also [266]. In [54] Baskoro and Ngurah showed that graph mP_3 is super EMT (super $(a, 0)$-EAT). Ivančo and Lučkaničová [141] have described constructions of super EMT labelings for disconnected graphs $nC_k \cup mP_k$.

In Theorem 6.3.11, it is proved that the cycle C_n has a super (a, d)-EAT labeling if and only if either $d \in \{0, 2\}$ and n is odd, $n \geq 3$, or $d = 1$ and $n \geq 3$. Dafik,

Miller, Ryan, and Bača showed that the graph mC_n has a super (a, d)-EAT labeling for all feasible values of the parameters m, n and d. These results are summarized in the following theorem.

Theorem 6.3.51 ([85]) *The graph mC_n has a super (a, d)-EAT labeling if and only if one of the following conditions is satisfied.*

(i) $d \in \{0, 2\}$ and m, n are odd, $m, n \geq 3$.
(ii) $d = 1$, for every $m \geq 2$ and $n \geq 3$.

Theorem 6.3.25 states that the path P_n, $n \geq 2$, has a super (a, d)-EAT labeling if and only if $d \in \{0, 1, 2, 3\}$.

From (6.6), it follows that if mP_n is super (a, d)-EAT, for $p = mn$ and $q = (n-1)m$, then

$$d \leq \frac{2p + q - 5}{q - 1} = 3 + \frac{2m - 2}{mn - m - 1}.$$

If $n = 2$ and $m \geq 2$, then $(2m - 2)/(mn - m - 1) = 2$ and thus $d \leq 5$. If $n \geq 3$ and $m \geq 2$, then $(2m - 2)/(mn - m - 1) \not\geq 1$ and thus $d \leq 3$.

In [85] is proved that

Theorem 6.3.52 ([85]) *If m is odd, $m \geq 3$, and $n \geq 2$, then the graph mP_n has an $(a, 1)$-EAV labeling.*

Theorem 6.3.53 ([85]) *The graph mP_n has a $(m + 2, 2)$-EAV labeling, for every $m \geq 2$ and $n \geq 2$.*

Then applying Theorems 6.3.52, 6.3.53, and 6.3.5 give the following.

Theorem 6.3.54 ([85]) *If m is odd, $m \geq 3$, and $n \geq 2$, then the graph mP_n has a super $(a, 0)$-EAT labeling and a super $(a', 2)$-EAT labeling.*

Theorem 6.3.55 ([85]) *The graph mP_n has a super $(2mn+2, 1)$-EAT labeling and a super $(mn + m + 3, 3)$-EAT labeling, for every $m \geq 2$ and $n \geq 2$.*

For the disjoint union of m copies of the path P_2 is proved that

Theorem 6.3.56 ([85]) *The graph mP_2, $m \geq 3$, has a super $((5m + 7)/2, 4)$-EAT labeling if and only if m is odd.*

Theorem 6.3.57 ([85]) *The graph mP_2 has a super $(2m + 4, 5)$-EAT labeling, for every $m \geq 2$.*

It remains an open problem to investigate whether mP_n has a super (a, d)-EAT labeling, for $d \in \{0, 2\}$ and m even. Figueroa-Centeno et al. [100] have shown that the forest $2P_n$, $n > 1$, has a super EMT (super $(a, 0)$-EAT) labeling if and only if $n \neq 2$ or 3. This labeling is described in 11 cases according to the possible values of the integer n. Baskoro and Ngurah [54] showed that if m is even, $m \geq 4$, then the graph mP_3 admits a super EMT labeling. For further investigation, Dafik, Miller, Ryan, and Bača suggest the following open problem.

Open Problem 6.3.6 ([85]) For mP_n, $m \geq 2$ even, $n \geq 4$, determine if there is a super (a, d)-EAT labeling, with $d \in \{0, 2\}$.

In Theorem 6.3.24, Sugeng, Miller, Slamin, and Bača proved that the star $K_{1,n}$ has a super (a, d)-EAT labeling if and only if either $d \in \{0, 1, 2\}$ and $n \geq 1$, or $d = 3$ and $1 \leq n \leq 2$.

Now, we will focus on super edge-antimagicness of the disjoint union of two stars, denoted by $K_{1,m} \cup K_{1,n}$. With respect to Inequality (6.6), for $p = m + n + 2$ and $q = m + n$, we have that

$$d \leq 3 + \frac{2}{m + n - 1}.$$

We can see that:

(i) If $m \geq 2$ and $n \geq 2$, then there is no super (a, d)-EAT labeling of $K_{1,m} \cup K_{1,n}$ with $d > 3$.
(ii) If $m + n = 3$, then there is no super (a, d)-EAT labeling of $K_{1,m} \cup K_{1,n}$ with $d > 4$.
(iii) If $m + n = 2$, then there is no super (a, d)-EAT labeling of $K_{1,m} \cup K_{1,n}$ with $d > 5$.

Dafik, Miller, Ryan, and Bača proved that

Theorem 6.3.58 ([83]) *The graph $K_{1,m} \cup K_{1,n}$, $m \geq 2$ and $n \geq 2$, has an $(a, 1)$-EAV labeling if and only if either m is a multiple of $n + 1$ or n is a multiple of $m + 1$.*

Then from Theorems 6.3.58 and 6.3.5 follow next two theorems.

Theorem 6.3.59 ([83]) *If either m is a multiple of $n+1$ or n is a multiple of $m+1$, then the graph $K_{1,m} \cup K_{1,n}$, $m \geq 2$ and $n \geq 2$, has a super $(m+n+t+7, 2)$-EAT labeling.*

Theorem 6.3.60 ([83]) *The graph $K_{1,m} \cup K_{1,n}$, $m \geq 2$ and $n \geq 2$, has a super EMT ($(2m + 2n + t + 6, 0)$-EAT) labeling if and only if either m is a multiple of $n + 1$ or n is a multiple of $m + 1$.*

This result was also proved by Ivančo and Lučkaničová in [141].

Figure 6.30 shows the vertex labeling of a disjoint union of stars $K_{1,8}$ and $K_{1,3}$, where edge labels are edge-weights. Thus we can see that the vertex labeling is $(6, 1)$-EAV. By using Theorem 6.3.5 we can obtain a super $(30, 0)$-EAT or a super $(20, 2)$-EAT labeling.

In [83] Dafik, Miller, Ryan, and Bača are not able to give an answer as to whether or not there exists a super $(a, 2)$-EAT labeling of $K_{1,m} \cup K_{1,n}$ for other values of m and n. Therefore, they propose the following open problem.

Open Problem 6.3.7 ([83]) For the graph $K_{1,m} \cup K_{1,n}$, $m \geq n \geq 2$, if m is not a multiple of $n + 1$ determine whether there is a super $(a, 2)$-EAT labeling.

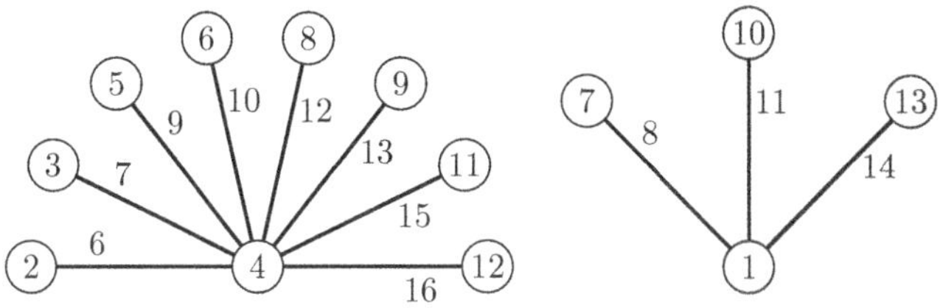

Fig. 6.30 (6, 1)-EAV labeling of $K_{1,8} \cup K_{1,3}$

By using Theorem 6.3.58, with respect to Lemma 5.3.1, the next theorem follows.

Theorem 6.3.61 ([83]) *If $m + n$ is odd, and either m is a multiple of $n + 1$ or n is a multiple of $m + 1$, then the graph $K_{1,m} \cup K_{1,n}$, $m \geq 2$ and $n \geq 2$, has a super $((3(m + n) + 2t + 13)/2, 1)$-EAT labeling.*

In [83] we can find the following result.

Theorem 6.3.62 ([83]) *For the graph $K_{1,m} \cup K_{1,n}$, $m \geq 2$ and $n \geq 2$, there is no $(a, 3)$-EAV labeling.*

The next result is obtained for $K_{1,m} \cup K_{1,n}$ if $m = n$.

Theorem 6.3.63 ([83]) *The graph $K_{1,m} \cup K_{1,m}$, $m \geq 2$, has a $(4, 2)$-EAV labeling.*

In light of Theorem 6.3.5, as an immediate consequence of Theorem 6.3.63 holds the following theorem.

Theorem 6.3.64 ([83]) *The graph $K_{1,m} \cup K_{1,m}$, $m \geq 2$, has a super $(4m + 6, 1)$-EAT and a super $(2m + 7, 3)$-EAT labeling.*

In [83] are proposed the following open problems.

Open Problem 6.3.8 ([83]) For the graph $K_{1,m} \cup K_{1,n}$, $m + n$ even and $m \neq n$, determine if there is a super $(a, 1)$-EAT labeling.

Open Problem 6.3.9 ([83]) For the graph $K_{1,m} \cup K_{1,n}$, if $m \neq n$, determine if there is a super $(a, 3)$-EAT labeling.

Next we focus on investigation of the existence of super (a, d)-EAT labelings for disjoint union of multiple copies of a regular caterpillar. The caterpillar is said to be a *regular*, if every vertex of the path of caterpillar $S_{t_1,t_2,\ldots,t_n}$ has the same number of leaves, i.e., $t_1 = t_2 = \cdots = t_n$. If the disjoint union of m copies of a regular caterpillar $mS_{t_1,t_2,\ldots,t_n}$, $t_1 = t_2 = \cdots = t_n = t$, has a super (a, d)-EAT labeling, then, for $p = mn(t + 1)$ and $q = mn(t + 1) - m$, it follows, from (6.6), that $d \leq 3 + (2m - 2)/(mn(t + 1) - m - 1)$. If $m \geq 2$, $n \geq 2$ and $t \geq 1$, then $(2m - 2)/(mn(t + 1) - m - 1) < 1$, and thus $d < 4$.

The following theorem describes an $(a, 1)$-EAV labeling for the disjoint union of m copies of a regular caterpillar.

Theorem 6.3.65 ([30]) *If mn is odd, $m, n \geq 3$, then the graph $mS_{t_1,t_2,\ldots,t_n}$, for $t_1 = t_2 = \cdots = t_n = t \geq 1$, has a $((mn + 2m + 3)/2, 1)$-EAV labeling.*

According to Theorem 6.3.5 from Theorem 6.3.65 it follows

Theorem 6.3.66 ([30]) *If mn is odd, $m \geq 3$ and $n \geq 3$, then the graph $mS_{t_1,t_2,\ldots,t_n}$, for $t_1 = t_2 = \cdots = t_n = t \geq 1$, has a super $((mn(4t + 5) + 3)/2, 0)$-EAT labeling and a super $((mn(2t + 3) + 5)/2 + m, 2)$-EAT labeling.*

The next theorem is a consequence of Theorem 6.3.65 in light of Lemma 5.3.1.

Theorem 6.3.67 ([30]) *If the product mnt is odd, $m \geq 3$, $n \geq 3$, and $t \geq 1$, then the graph $mS_{t_1,t_2,\ldots,t_n}$, for $t_1 = t_2 = \cdots = t_n = t$, has a super $((mn(3t + 4) +m)/2 + 2, 1)$-EAT labeling.*

The next theorem gives a super $(a, 1)$-EAT labeling, for m even and n odd.

Theorem 6.3.68 ([30]) *If m is even, $m \geq 2$, and n is odd, $n \geq 3$, then the graph $mS_{t_1,t_2,\ldots,t_n}$, for $t_1 = t_2 = \cdots = t_n = t \geq 1$, has a super $(a, 1)$-EAT labeling.*

The next theorem describes an $(a, 2)$-EAV labeling for the disjoint union of m copies of a regular caterpillar when $t = 2$.

Theorem 6.3.69 ([30]) *There is $(m + 2, 2)$-EAV labeling for $mS_{t_1,t_2,\ldots,t_n}$, for $t_1 = t_2 = \cdots = t_n = 2$ and every $m, n \geq 2$.*

With respect to Theorem 6.3.5, the $(m+2, 2)$-EAV labeling from Theorem 6.3.69 can be extended to a super (a, d)-EAT labeling for $d = 1$ and $d = 3$. Thus for $p = 3mn$ and $q = 3mn - m$, the following theorem holds.

Theorem 6.3.70 ([30]) *The graph $mS_{t_1,t_2,\ldots,t_n}$, for $t_1 = t_2 = \cdots = t_n = 2$, has a super $(6mn + 2, 1)$-EAT labeling and a super $(3mn + m + 3, 3)$-EAT labeling, for every $m \geq 2$ and $n \geq 2$.*

We summarize that the graph $mS_{t_1,t_2,\ldots,t_n}$, for $t_1 = t_2 = \cdots = t_n = t$, has a super (a, d)-EAT labeling for

(i) $d \in \{0, 2\}$, if mn is odd and $t \geq 1$,
(ii) $d = 1$, if either mnt is odd, or m is even and n is odd, $t \geq 1$, or $t = 2$ and $m, n \geq 2$,
(iii) $d = 3$, if $m, n \geq 2$ and $t = 2$.

Constructions that will produce a super (a, d)-EAT labelings of $mS_{t_1,t_2,\ldots,t_n}$, for $t_1 = t_2 = \cdots = t_n = t$, $d \in \{0, 1, 2\}$ and every $m, n \geq 2$, have not yet been found. Nevertheless, Bača, Dafik, Miller, and Ryan suggest the following conjecture.

Conjecture 6.3.3 ([30]) There is a super (a, d)-EAT labeling of the graph $mS_{t_1,t_2,\ldots,t_n}$, for $t_1 = t_2 = \cdots = t_n = t \geq 1$, $d \in \{0, 1, 2\}$ and for every $m \geq 2$ and $n \geq 2$.

For the graph $mS_{t_1,t_2,\ldots,t_n}$, for $t_1 = t_2 = \cdots = t_n = t \neq 2$, so far no super $(a,3)$-EAT labeling have found. So, in [30] Bača, Dafik, Miller, and Ryan propose the following open problem.

Open Problem 6.3.10 ([30]) For the graph $mS_{t_1,t_2,\ldots,t_n}$, for $t_1 = t_2 = \cdots = t_n = t$, determine if there is a super $(a,3)$-EAT labeling, for every $m \geq 2$, $n \geq 2$ and $t \neq 2$.

In the case when the graph $mS_{t_1,t_2,\ldots,t_n}$ does not have any restriction on the values of $t_1, t_2, \ldots, t_n$, the problem of finding a super (a,d)-EAT labeling seems to be difficult. For further investigation, Bača, Dafik, Miller, and Ryan suggest the following.

Open Problem 6.3.11 ([30]) Find, if possible, some structural characteristics of a graph $mS_{t_1,t_2,\ldots,t_n}$ which make a super (a,d)-EAT labeling impossible.

The results on super (a,d)-EAT labeling for disjoint union of m copies of complete graph are summarized in the next theorem.

Theorem 6.3.71 ([24]) *The graph mK_n has a super (a,d)-EAT labeling if and only if one of the following conditions is satisfied.*

(i) $d \in \{0,2\}$, $n \in \{2,3\}$, *and m is odd*, $m \geq 3$.
(ii) $d = 1$, $n \geq 2$, *and* $m \geq 2$.
(iii) $d \in \{3,5\}$, $n = 2$, *and* $m \geq 2$.
(iv) $d = 4$, $n = 2$, *and m is odd*, $m \geq 3$.

The next lemma establishes an upper bound on the parameter d for super edge-antimagicness of disjoint union of m copies of complete s-partite graph, denoted by $mK_{s[n]}$.

Lemma 6.3.12 *For the graph $mK_{s[n]}$, $m \geq 2$, $n \geq 2$, and $s \geq 3$, there is no super (a,d)-EAT labeling with $d \geq 2$.*

From Theorems 6.3.8 and 6.3.9 for disjoint union of complete bipartite graph it follows.

Theorem 6.3.72 ([29]) *There is a super $(4mn+2, 1)$-EAT labeling for $mK_{n,n}$, for every $n \geq 1$ and every $m \geq 2$.*

Bača and Brankovic proved the following lemmas.

Lemma 6.3.13 ([29]) *If $mK_{n,n}$ is super (a,d)-EAT, for $d \in \{0,2\}$, then $n = 1$ or $n = 3$ and m is odd, $m \geq 3$.*

Lemma 6.3.14 ([29]) *If m is odd, $m \geq 3$, then $mK_{1,1}$ has a super $((9m+3)/2, 0)$-EAT and a super $((7m+5)/2, 2)$-EAT labeling.*

For $mK_{3,3}$, $m \geq 3$ odd, they have not yet found any super $((33m+3)/2, 0)$-EAT labeling. Therefore, they propose the following open problem.

Open Problem 6.3.12 ([29]) For $mK_{3,3}$, $m \geq 3$ odd, determine if there is a super $((33m + 3)/2, 0)$-EAT labeling.

Bača and Brankovic have not yet found a convenient construction that will produce super $(a, 2)$-EAT labeling for $mK_{3,3}$, $a = (15m + 5)/2$, for all odd m. However, the existence of super $(25, 2)$-EAT labeling for $3K_{3,3}$ led them to suggest the following conjecture.

Conjecture 6.3.4 ([29]) There is a super $((15m + 5)/2, 2)$-EAT labeling for $mK_{3,3}$, for all m odd.

Figure 6.31 illustrates a super $(25, 2)$-EAT labeling of a disjoint union of three copies of the complete bipartite graph $K_{3,3}$.

The next two theorems present super (a, d)-EAT labelings of $mK_{n,n}$, for $d \in \{3, 4, 5\}$.

Theorem 6.3.73 ([29]) *The graph* $mK_{n,n}$, $m \geq 2$, *is super* $(3m + 3, 3)$-*EAT and super* $(2m + 4, 5)$-*EAT if and only if* $n = 1$.

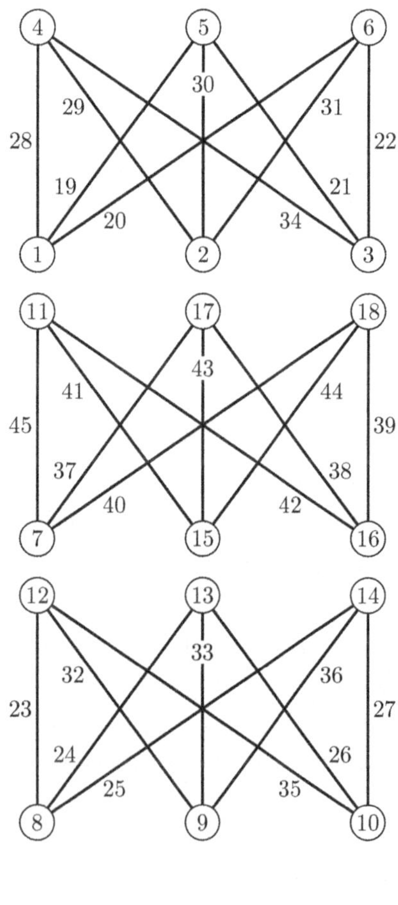

Fig. 6.31 Super (25, 2)-EAT labeling of $3K_{3,3}$

Theorem 6.3.74 ([29]) *The graph $mK_{n,n}$ has a super $(a, 4)$-EAT labeling if and only if $n = 1$, m is odd, $m \geq 3$, and $a = (5m + 7)/2$.*

If $mK_{n,n,n}$, $m \geq 2$ and $n \geq 1$, is super (a, d)-EAT, then, from (6.6), it follows that $d < 3$.

Dafik, Miller, Ryan, and Bača proved the following theorems.

Theorem 6.3.75 ([86]) *The graph $mK_{n,n,n}$ has an $(a, 1)$-EAV labeling if and only if $n = 1$ and m is odd, $m \geq 3$.*

Theorem 6.3.76 ([86]) *For $d \in \{0, 2\}$, the graph $mK_{n,n,n}$ is super (a, d)-EAT if and only if $n = 1$ and m is odd, $m \geq 3$.*

Theorem 6.3.77 ([86]) *The graph $mK_{n,n,n}$ has a super $(6mn+2, 1)$-EAT labeling, for every $m \geq 2$ and $n \geq 1$.*

For the disjoint union of m copies of a complete s-partite graph, Dafik, Miller, Ryan, and Bača proved the following.

Theorem 6.3.78 ([84]) *If either $s \equiv 0, 1 \pmod 4$, $s \geq 4$, $m \geq 2$, $n \geq 1$, or mn is even, $m > 2$, $n \geq 1$, $s \geq 4$, then there is no super $(a, 0)$-EAT labeling for $mK_{s[n]}$.*

For mn odd, $s \equiv 2, 3 \pmod 4$ they propose the following open problem.

Open Problem 6.3.13 ([84]) For the graph $mK_{s[n]}$, mn odd, $m \geq 3$, $n \geq 1$ and $s \equiv 2, 3 \pmod 4$, $s \geq 6$, determine if there is a super $(2mns + (mn^2s(s-1)+6)/4, 0)$-EAT labeling.

The next theorem gives a negative answer for existence a super $(a, 2)$-EAT labeling for the graph $mK_{s[n]}$.

Theorem 6.3.79 ([84]) *If $m \geq 2$, $n = 1$, and $s = 4$, then there is no super $(a, 2)$-EAT labeling for the graph $mK_{s[n]}$.*

The following theorem has been proved by Dafik, Miller, Ryan, and Bača.

Theorem 6.3.80 ([84]) *The graph $mK_{4[n]}$ has a super $(8mn + 2, 1)$-EAT labeling, for every $m \geq 2$ and $n \geq 1$.*

Other results on super (a, d)-EAT labelings of certain families of disconnected graphs can be found in [41] and [267].

6.3.7 Super Edge-Antimagic Total Labeling of Forests

In this section we examine the existence of super (a, d)-EAT labeling of forests, in which every component is a path-like tree. Indeed, we prove that such a labeling exists when the forest has an odd number of components.

Suppose $F \cong \bigcup_{j=1}^{m} T_j$ is a disjoint union of m trees each of order n. If F admits a super (a, d)-EAT labeling, then for $p = mn$, $q = m(n-1)$ and $n \geq 4$, the

Inequality (6.6) gives

$$d \leq 3 + \frac{2m - 2}{m(n-1) - 1} < 4.$$

Consider the path P_n with $V(P_n) = \{w_i : 1 \leq i \leq n\}$ and $E(P_n) = \{w_i w_{i+1} : 1 \leq i \leq n-1\}$. In [38] it is shown that the following labeling

$$f_1(w_i) = \begin{cases} \frac{i+1}{2} & \text{if } i \text{ is odd} \\ \lceil \frac{n}{2} \rceil + \frac{i}{2} & \text{if } i \text{ is even} \end{cases}$$

$$f_1(w_i w_{i+1}) = 2n - i \qquad \text{if } 1 \leq i \leq n-1$$

is a super $\left(2n + \lceil \frac{n}{2} \rceil + 1, 0\right)$-EAT labeling of P_n.

Let $F \cong \bigcup_{j=1}^{m} P_n^j$ be a disjoint union of m paths each on n vertices, $m > 1$, $n \geq 4$, with $V(F) = \{w_i^j : 1 \leq i \leq n,\ 1 \leq j \leq m\}$ and $E(F) = \{w_i^j w_{i+1}^j : 1 \leq i \leq n-1,\ 1 \leq j \leq m\}$. Now, we construct a labeling g_1 of F in the following way.

$$g_1(w_i^j) = \begin{cases} m(f_1(w_i) - 1) + j & \text{if } i \text{ is odd, } 1 \leq j \leq m \\ m(f_1(w_i) - 1) + \frac{m-j}{2} + 1 & \text{if } i \text{ is even, } j \text{ is odd} \\ m f_1(w_i) + \frac{2-j}{2} & \text{if } i \text{ is even, } j \text{ is even} \end{cases}$$

$$g_1(w_i^j w_{i+1}^j) = \begin{cases} m f_1(w_i w_{i+1}) + \frac{1-j}{2} & \text{if } j \text{ is odd, } 1 \leq i \leq n-1 \\ m f_1(w_i w_{i+1}) + \frac{1-j-m}{2} & \text{if } j \text{ is even, } 1 \leq i \leq n-1. \end{cases}$$

Lemma 6.3.15 ([40]) *If m is odd, $m \geq 3$ and $n \geq 4$, then g_1 is a super $(m(2n + \lceil n/2 \rceil) + \frac{3-m}{2}, 0)$-EAT labeling for $F \cong \bigcup_{j=1}^{m} P_n^j$.*

Proof It is not difficult to check that if $f_1(w_i) \in \{1, 2, \ldots, n\}$, then $g_1(V(F)) = \bigcup_{j=1}^{m} \bigcup_{i=1}^{n} \{m f_1(w_i) + 1 - j\} = \{1, 2, \ldots, mn\}$ and if $f_1(w_i w_{i+1}) \in \{n+1, n+2, \ldots, 2n-1\}$, then $g_1(E(F)) = \bigcup_{j=1}^{m} \bigcup_{i=1}^{n} \{m f_1(w_i w_{i+1}) + 1 - j\} = \{mn + 1, mn + 2, \ldots, 2mn - m\}$. Moreover, we can see that for every $1 \leq i \leq n-1$ and $1 \leq j \leq m$ the edge-weight $w_{g_1}(w_i^j w_{i+1}^j) = g_1(w_i^j) + g_1(w_i^j w_{i+1}^j) + g_1(w_{i+1}^j) = m(f_1(w_i) + f_1(w_i w_{i+1}) + f_1(w_{i+1})) + (3 - 3m)/2$. Since $f_1(w_i) + f_1(w_i w_{i+1}) + f_1(w_{i+1}) = 2n + \lceil n/2 \rceil + 1$ for all $1 \leq i \leq n-1$, it follows that the labeling g_1 is super $(m(2n + \lceil n/2 \rceil) + (3 - m)/2, 0)$-EAT. □

Let us remark that Lemma 6.3.15 follows from Theorem 6.3.42. This lemma produces just one convenient total labeling g_1 which will be useful in the second part of this section. Notice that for m even there is only one known result shown by Figueroa-Centeno et al. [100] that the forest $2P_n$, $n > 1$, is super $(a, 0)$-EAT if and only if $n \neq 2$ or 3.

By Lemma 6.3.15 and Theorem 6.3.5 we have the following.

Lemma 6.3.16 ([40]) *If m is odd, $m \geq 3$ and $n \geq 4$, then $F \cong \bigcup_{j=1}^{m} P_n^j$ admits a super $(m(\lceil n/2 \rceil + n) + (5+m)/2, 2)$-EAT labeling.*

Bača et al. [38] produced a super $(2n+2, 1)$-EAT labeling, say f_2, and a super $(n+4, 3)$-EAT labeling, say f_3, for every path P_n, where

$$f_3(w_i) = f_2(w_i) = i$$

for every $1 \leq i \leq n$,

$$f_2(w_i w_{i+1}) = f_1(w_i w_{i+1})$$
$$f_3(w_i w_{i+1}) = n + i$$

for every $1 \leq i \leq n-1$.

Now, we generate the total labelings g_2 and g_3 of $F \cong \bigcup_{j=1}^{m} P_n^j$ in the following way.

$$g_3(w_i^j) = g_2(w_i^j) - m(f_2(w_i) - 1) + j$$

for every $1 \leq i \leq n$, $1 \leq j \leq m$,

$$g_2(w_i^j w_{i+1}^j) = m f_2(w_i w_{i+1}) + 1 - j$$

and

$$g_3(w_i^j w_{i+1}^j) = m(f_3(w_i w_{i+1}) - 1) + j$$

for every $1 \leq i \leq n-1$, $1 \leq j \leq m$.

With the total labelings g_2 and g_3 in hand, we are ready to prove the following lemma.

Lemma 6.3.17 ([40]) *For every $m \geq 2$ and $n \geq 4$ the labeling g_2 is a super $(2mn+2, 1)$-EAT and the labeling g_3 is a super $(mn+m+3, 3)$-EAT for $F \cong \bigcup_{j=1}^{m} P_n^j$.*

Proof By direct computation we obtain that the sets of edge-weights W_1 and W_2 consist of the following arithmetic sequences

$$\begin{aligned} W_1 &= \{w_{g_2}(w_i^j w_{i+1}^j) = m w_{f_2}(w_i w_{i+1}) - 2m + 1 + j : 1 \leq i \leq n-1, 1 \leq j \leq m\} \\ &= \{m(2n - 1 + i) + 1 + j : 1 \leq i \leq n-1, 1 \leq j \leq m\} \\ W_2 &= \{w_{g_3}(w_i^j w_{i+1}^j) = m w_{f_3}(w_i w_{i+1}) - 3m + 3j : 1 \leq i \leq n-1, 1 \leq j \leq m\} \\ &= \{m(n - 2 + 3i) + 3j : 1 \leq i \leq n-1, 1 \leq j \leq m\} \end{aligned}$$

and therefore the desired result follows. □

The purpose of the second part of this section is to study embeddings of paths in the 2-dimensional grid and consider a set of elementary transformations which keep the edge-antimagic character of the paths.

We embed the path P_n in the 2-dimensional grid. We say that a tree T of order n is a *path-like tree* when it can be obtained after a sequence of elementary transformations on an embedding of P_n in the 2-dimensional grid. For definition of path-like tree see Sect. 6.3.2.

Path-like trees were first defined and investigated by Barrientos in [52]. Barrientos proved that all path-like trees are graceful, see [52], Theorem 35, and also noted that path-like trees admit α-labelings. Bača, Lin, and Muntaner-Batle proved in [38] that all path-like trees are super (a, d)-EAT if and only if $d \in \{0, 1, 2, 3\}$. Later Bača and Barrientos [25] proved a stronger result that every α-tree T with $||A| - |B|| \leq 1$, where $\{A, B\}$ is the bipartition of vertex set of T, admits a super (a, d)-EAT labeling for $d \in \{0, 1, 2, 3\}$.

The main goal of this section is to show that a forest, in which every component is a path-like tree, has a super (a, d)-EAT labeling for each feasible value of d. Suppose $F \cong \bigcup_{j=1}^{m} T_j$ is the disjoint union of m path-like trees each of order n, $n \geq 4$. We have mentioned that if the forest F admits a super (a, d)-EAT labeling, then the difference $d < 4$. It remains to investigate the existence of super (a, d)-EAT labeling for $d = 0, 1, 2, 3$.

Consider an embedding of a disjoint union of m paths $P_n^1 \cup P_n^2 \cup \cdots \cup P_n^m$ in the 2-dimensional grid where P_n^j is a path with vertices $V(P_n^j) = \{w_i^j : 1 \leq i \leq n\}$ and edges $E(P_n^j) = \{w_i^j w_{i+1}^j : 1 \leq i \leq n - 1\}$ from which the path-like tree T_j can be obtained, for $j = 1, 2, \ldots, m$. Let $P_n^j = T_j^0, T_j^1, T_j^2, \ldots, T_j^{s_j} = T_j$ be the series of trees obtained by successively applying the appropriate elementary transformations of P_n^j to obtain T_j, for $j = 1, 2, \ldots, m$. Note that we allow a different series of trees $T_j^0, T_j^1, T_j^2, \ldots, T_j^{s_j}$ for different s_j, i.e., the forest F may be a disjoint union of different path-like trees $T_1, T_2, \ldots, T_m$, each of order n. Now, in light of the three previous lemmas we present the following two theorems.

Theorem 6.3.81 ([40]) *Let T_j, $1 \leq j \leq m$, be a path-like tree of order n. If m is odd, $m \geq 3$ and $n \geq 4$, then a forest $F \cong \bigcup_{j=1}^{m} T_j$ admits a super $(m(2n + \lceil n/2 \rceil) + (3 - m)/2, 0)$-EAT labeling and a super $(m(\lceil n/2 \rceil + n) + (5 + m)/2, 2)$-EAT labeling.*

Proof Consider an embedding of a disjoint union of m paths $\bigcup_{j=1}^{m} P_n^j$, and label the vertices of $\bigcup_{j=1}^{m} P_n^j$ using the labeling g_1 as described in Lemma 6.3.15. In order to prove the result, it suffices to show that if $u_0v_0 = w_i^j w_{i+1}^j$, then

$$g_1(w_i^j) + g_1(w_{i+1}^j) = g_1(w_{i-t}^j) + g_1(w_{i+1+t}^j)$$

whenever w_{i-t}^j and $w_{i+1+t}^j \in V(\bigcup_{j=1}^{m} P_n^j)$.

In accordance with the parity of i, j, and t, there are 2^3 cases to consider. However, the cases are similar to each other, and hence, we will only show one case to illustrate how the proof of all cases work. Let i and j be odd, t be even. Then

$$\begin{aligned} g_1(w_i^j) + g_1(w_{i+1}^j) &= m(f_1(w_i) - 1) + j + m(f_1(w_{i+1}) - 1) + \frac{m-j}{2} + 1 \\ &= m(f_1(w_i) + f_1(w_{i+1})) + \frac{2-3m+j}{2} \\ &= m(f_1(w_{i-t}) + f_1(w_{i+1+t})) + \frac{2-3m+j}{2} \\ &= m(f_1(w_{i-t}) - 1) + j + m(f_1(w_{i+1+t}) - 1) + \frac{m-j}{2} + 1 \\ &= g_1(w_{i-t}^j) + g_1(w_{i+1+t}^j). \end{aligned}$$

Thus an elementary transformation keeps the super edge-antimagic character of the forest and the resulting labeling of the $F \cong \bigcup_{j=1}^{m} T_j$ is a super $(m(2n + \lceil n/2 \rceil) + (3 - m)/2, 0)$-EAT.

Now, if we apply Theorem 6.3.5, then the forest $F \cong \bigcup_{j=1}^{m} T_j$ admits a super $(m(\lceil n/2 \rceil + n) + (5 + m)/2, 2)$-EAT labeling. It produces the desired result. □

Figure 6.32 shows an embedding of a disjoint union of three paths $P_{17}^1 \cup P_{17}^2 \cup P_{17}^3$ in the 2-dimensional grid. After a sequence of elementary transformations on every path we obtain the disjoint union of three path-like trees $\bigcup_{j=1}^{3} T_j$. Figure 6.33 depicts a super (82, 2)-EAT labeling of this forest.

Theorem 6.3.82 ([40]) *Let T_j, $1 \le j \le m$, be a path-like tree of order n. For every two integers $m \ge 2$ and $n \ge 4$, the forest $F \cong \bigcup_{j=1}^{m} T_j$ admits a super $(2mn + 2, 1)$-EAT and a super $(mn + m + 3, 3)$-EAT labeling.*

Proof Consider an embedding of $\bigcup_{j=1}^{m} P_n^j$ in the 2-dimensional grid and label the vertices and edges by the total labeling g_2 (g_3) from Lemma 6.3.17. The proof that a sequence of elementary transformations on every path P_n^j produces a path-like tree T_j, $1 \le j \le m$, and keeps the super edge-antimagic character of the forest $F \cong \bigcup_{j=1}^{m} T_j$ is similar to the proof given for total labeling g_1 in the previous theorem, therefore we omit the details. In this case if $u_0 = w_i^j$, $v_0 = w_{i+1}^j$, $u = w_{i-t}^j$, and $v = w_{i+1+t}^j$, then

$$\begin{aligned} g_2(u_0v_0) &= g_2(uv) = mf_2(w_iw_{i+1}) + 1 - j = 2mn - im + 1 - j \\ g_3(u_0v_0) &= g_3(uv) = m(f_3(w_iw_{i+1}) - 1) + j = mn + im - m + j \end{aligned}$$

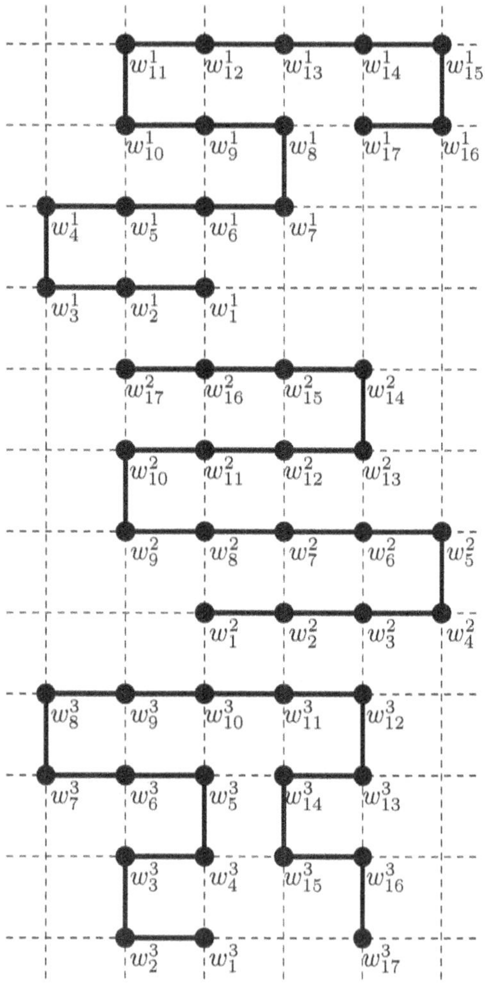

Fig. 6.32 Union of three paths $P^1_{17} \cup P^2_{17} \cup P^3_{17}$

and

$$g_2(u_0) + g_2(u_0v_0) + g_2(v_0) = m(2n + i - 1) + 1 + j$$

$$g_3(u_0) + g_3(u_0v_0) + g_3(v_0) = m(n + 3i - 2) + 3j.$$

The reader will observe that $g_2(u) = g_3(u) = m(i-t-1)+j$ and $g_2(v) = g_3(v) = m(i+t)+j$. It is evident that, $g_\alpha(u_0)+g_\alpha(u_0v_0)+g_\alpha(v_0) = g_\alpha(u)+g_\alpha(uv)+g_\alpha(v)$, for $\alpha = 2$ and 3, and after the elementary transformation the resulting labeling again is super $(2mn+2, 1)$-EAT (super $(mn+m+3, 3)$-EAT). This completes the proof. □

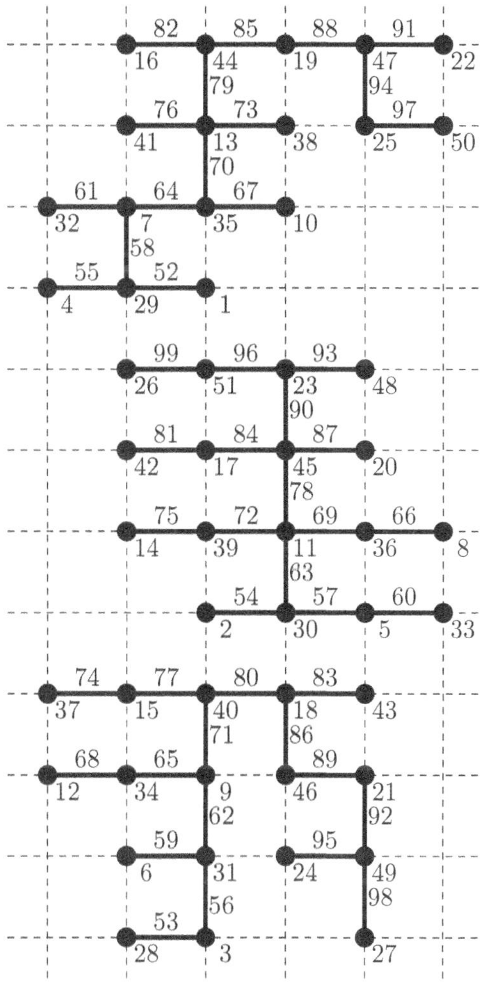

Fig. 6.33 Super (82, 2)-EAT labeling of the forest

In this section we have shown that there exist super (a, d)-EAT labelings of the forest $F \cong \bigcup_{j=1}^{m} T_j$ for $d \in \{0, 2\}$ and m odd, $m \geq 3$. For m even there is as yet no answer for the existence (nonexistence) of super $(a, 0)$-EAT labelings for a non-regular forest $F \cong \bigcup_{j=1}^{m} T_j$. For further investigation, Bača, Lin, and Muntaner-Batle suggest the following open problem.

Open Problem 6.3.14 ([40]) For a forest $F \cong \bigcup_{j=1}^{m} T_j$, $m \geq 2$ even, determine if there is a super (a, d)-EAT labeling with $d \in \{0, 2\}$.

Chapter 7
Graceful and Antimagic Labelings

7.1 Connection Between α-Labeling and Edge-Antimagic Labeling

A (p, q) graph G is said to be labeled by a mapping ϕ if to each vertex $v \in V(G)$ is assigned a nonnegative integer value $\phi(v)$ and to each edge $uv \in E(G)$ is assigned the value $|\phi(u) - \phi(v)|$ called a *weight*. The labeling ϕ is called *graceful* if $\phi : V(G) \to \{1, 2, \ldots, q + 1\}$ is an injection and if all edges of G have assigned distinct labels (weights) from the set $\{1, 2, \ldots, q\}$. A graph is called *graceful* if it admits a graceful labeling. As an example, a graceful labeling of the Petersen graph is depicted in Fig. 7.1, where the values on the edges denote their weights.

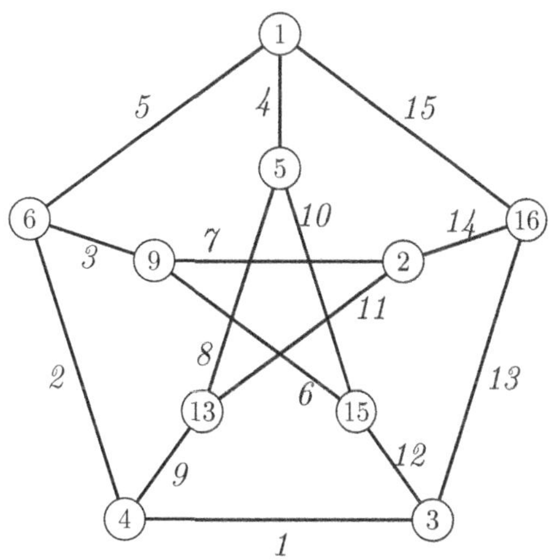

Fig. 7.1 Graceful labeling of the Petersen graph

M. Bača et al., *Magic and Antimagic Graphs*, Developments in Mathematics 60,
https://doi.org/10.1007/978-3-030-24582-5_7

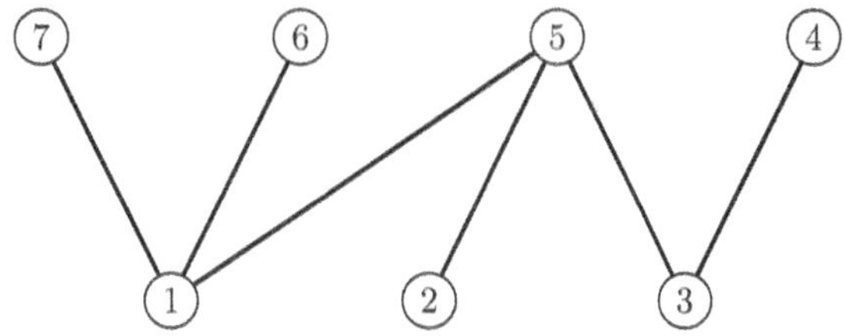

Fig. 7.2 α-labeling of caterpillar

It is known that not every graph is graceful, for instance, we can consider the complete graph K_n, $n \geq 5$, and the cycle C_n, $n \equiv 1$ or $2 \pmod 4$. The smallest graph, in order and size, that is not graceful is $C_3 \cup K_{1,1}$.

Graceful labeling was introduced by Rosa in [224]. The Ringel-Kotzig conjecture that all trees are graceful is a very popular open problem. Among the trees known to be graceful are caterpillars [224], trees with at most 4 end vertices [130], trees with diameter at most 5 [129], and trees with at most 27 vertices [10].

When a graceful labeling ϕ has the property that there exists an integer λ such that for each edge uv either $\phi(u) \leq \lambda < \phi(v)$ or $\phi(v) \leq \lambda < \phi(u)$, ϕ is called an *α- labeling*. The number λ is called the *boundary value* of ϕ.

A graph with an α-labeling is necessarily bipartite and the boundary value must be the smaller of the two vertex labels that yield the edge label 1. A graph that admits an α-labeling is called an *α-graph*. An example of an α-labeling of a caterpillar with the boundary value $\lambda = 3$ is presented in Fig. 7.2.

Various methods for constructing graceful labelings and α-labelings for particular families of trees can be found in [11, 52, 50, 51, 53, 75, 93, 94, 95, 148, 149, 150, 225, 236, 237, 238, 239, 240, 242, 241, 243, 244, 245, 246, 247, 248, 260].

For more information about graceful and α-labelings, the reader is referred to [109].

In this chapter we study a connection between α-labeling and edge-antimagic labeling and we use this connection for generating large classes of edge-antimagic trees from smaller graceful trees.

The first lemma describes a connection between an α-labeling of a tree and an $(a, 1)$-EAV labeling. This result can be found in [25, 42, 198].

Lemma 7.1.1 ([25]) *Let T be a tree of order p. If T admits an α-labeling, then T also admits an $(a, 1)$-EAV labeling.*

In general, the converse of the Lemma 7.1.1 does not hold. Figure 7.3 illustrates a $(5, 1)$-EAV labeling of a tree that is not an α-tree.

As mentioned before, any α-graph is bipartite. Let $\{A, B\}$ be the bipartition of the vertex set of an α-graph. The next theorem establishes a relationship between an α-labeling and an $(a, 2)$-EAV labeling.

Notice that if a tree of size $q = p - 1$ is $(a, 2)$-EAV, then the minimum possible edge-weight is at least 3, $a \geq 3$, and the maximum possible edge-weight is no more than $2p - 1$, $a + 2(p - 2) \leq 2p - 1$. The last inequality holds for $a \leq 3$. Therefore, $a = 3$.

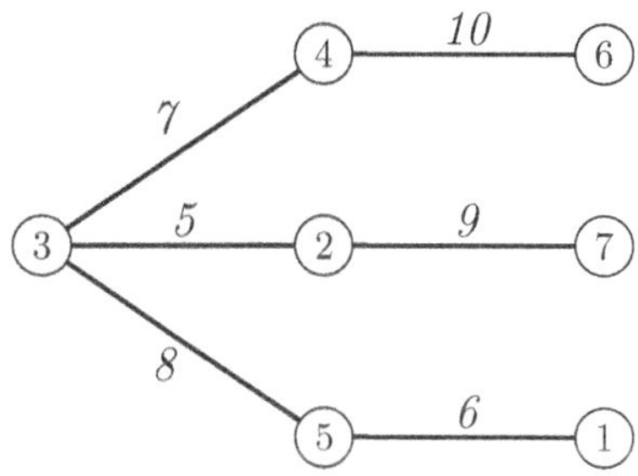

Fig. 7.3 (5, 1)-EAV labeling of a tree

Theorem 7.1.1 ([25]) *A tree T is $(3, 2)$-EAV if and only if T is an α-tree and $||A| - |B|| \leq 1$, where $\{A, B\}$ is the bipartition of the vertex-set of T.*

According to Inequality (6.5), we have that if a tree is super (a, d)-EAT, then $d \leq 3$.

The paper [36] presents relationships between (a, d)-EAV labeling and (a, d)-EAT labeling as follows.

As a consequence of Lemma 7.1.1 and Theorems 7.1.1 and 6.3.5, we have the following theorem.

Theorem 7.1.2 *Every α-graph of order p and size $p - 1$ with $||A| - |B|| \leq 1$ admits a super (a, d)-EAT labeling for all $d \in \{0, 1, 2, 3\}$.*

7.2 Construction of α-Trees

No general method is currently known to allow taking a tree known to be graceful and extending a path from it in an arbitrary position, or to identify an arbitrary vertex with another general tree known to be graceful, in order to produce a larger graceful tree.

There exist various operations that generate large classes of graceful trees from smaller graceful trees. Stanton and Zarnke [261] developed the first nontrivial algorithm for constructing graceful trees. Their method became the basis of many construction methods to follow.

Koh et al. in [160] gave a variation of Stanton and Zarnke's construction. This idea led to many other constructions, see [161] and [162].

In [161], Koh, Rogers, and Tan defined a new graph operation. Let G and H be two graphs and let $\{w_1, w_2, \ldots, w_m\}$ and $\{v_1, v_2, \ldots, v_n\}$ be their corresponding vertex sets. Let v be an arbitrary fixed vertex in H. Based upon the graph G, an isomorphic copy H_i of H is adjoined to each vertex w_i, $i = 1, 2, \ldots, m$ by identifying v^i and w_i, where v^i is the vertex corresponding to v in H_i. All the m copies of H just introduced are pairwise disjoint and no extra edges are added. The obtained graph is denoted by $G \textcircled{v} H$. It is obvious that $|V(G \textcircled{v} H)| = mn$ and $G \textcircled{v} H \not\cong H \textcircled{v} G$ and also $G \textcircled{v} H \not\cong G \textcircled{u} H$ for $v, u \in V(H)$, $v \neq u$, in general.

Note that Koh, Rogers, and Tan used the notation $G\triangle H$ for this operation. They proved the following theorem.

Theorem 7.2.1 ([161]) *If T_1 and T_2 are both graceful trees, then the tree $T_1 ⓥ T_2$ is also graceful.*

The next two theorems use Koh, Rogers, and Tan's graph operation to obtain an α-tree from a smaller graceful tree. The path on k vertices is denoted by P_k.

Theorem 7.2.2 ([46]) *Let T be a graceful tree of order n. If k is an even positive integer, then the tree $P_k ⓥ T$ admits an α-labeling.*

Proof Let T be a tree of order n with the bipartition $\{A, B\}$. Let $f : V(T) \rightarrow \{1, 2, \ldots, n\}$ be a graceful labeling of T with the weights $\{w_f(uv) = |f(u) - f(v)| : uv \in E(T)\} = \{1, 2, \ldots, n-1\}$. Let v be an arbitrary fixed vertex in T. Without loss of generality, we may assume that $v \in A$.

Now, consider k trees $T_1, T_2, \ldots, T_k$, each isomorphic to the tree T, where $\{A^i, B^i\}$ is the bipartition of the vertex set of T_i, for $i = 1, 2, \ldots, k$, such that A^i corresponds to A and B^i corresponds to B, for $i = 1, 2, \ldots, k$. We denote the vertices of the path P_k in such a way that $P_k = w_1 w_2 \ldots w_k$. Thus, according to the definition of a graph $P_k ⓥ T$, the vertex $v^i \in A^i$ is identified with the vertex w_i, for $i = 1, 2, \ldots, k$, see Fig. 7.4.

For $k \equiv 0 \pmod 2$ we define a new labeling g as follows.

$$g(v^i) = \begin{cases} f(v) + \left(k - \frac{i+1}{2}\right) n & \text{for } v \in A^i \cup B^{i+1} \text{ and } i = 1, 3, \ldots, k-1 \\ f(v) + \left(\frac{i}{2} - 1\right) n & \text{for } v \in A^i \cup B^{i-1} \text{ and } i = 2, 4, \ldots, k. \end{cases}$$

It is easy to see that g is a bijection from $V(P_k ⓥ T)$ onto $\{1, 2, \ldots, kn\}$ and $\{g(v) : v \in A^i \cup B^{i-1}, i = 2, 4, \ldots, k\} = \{1, 2, \ldots, kn/2\}$ and $\{g(v) : v \in A^i \cup B^{i+1}, i = 1, 3, \ldots, k-1\} = \{kn/2 + 1, kn/2 + 2, \ldots, kn\}$. Thus the vertex with boundary value $\lambda = kn/2$ belongs to $(A^k \cup B^{k-1})$. To see that g is an α-labeling it is enough to show that the weights of the edges have distinct labels from the set $\{1, 2, \ldots, kn-1\}$. We consider two cases.

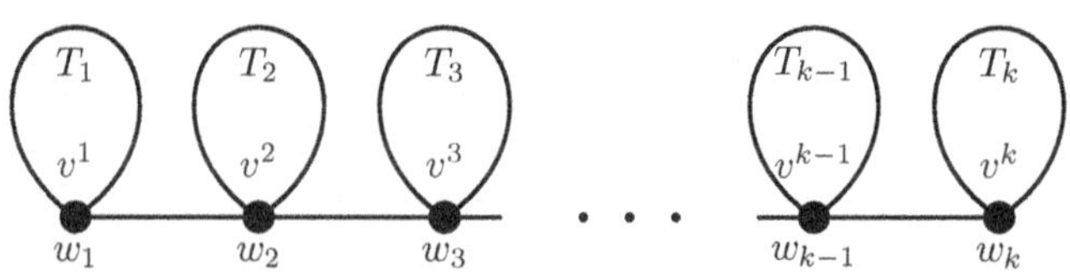

Fig. 7.4 Tree $P_k ⓥ T$

Case A If $e = u^i u^{i+1}, i = 1, 2, \ldots, k-1$, then the following holds

$$w_g(e) = w_g(u^i u^{i+1}) = |g(u^i) - g(u^{i+1})|$$
$$= \begin{cases} |(f(u) + (k - \frac{i+1}{2})n) - (f(u) + (\frac{i+1}{2} - 1)n)| = (k-i)n \\ \qquad\qquad \text{for} \quad i = 1, 3, \ldots, k-1 \\ |(f(u) + (\frac{i}{2} - 1)n) - (f(u) + (k - \frac{i+2}{2})n)| = (k-i)n \\ \qquad\qquad \text{for} \quad i = 2, 4, \ldots, k-2. \end{cases}$$

Case B If $e = u^i v^i \in E(T_i)$, without loss of generality, we can suppose that $u^i \in A^i \cup B^{i+1}$ and $v^i \in B^i \cup A^{i+1}$ where $i = 1, 3, \ldots, k-1$. Then

$$\begin{aligned} w_g(e) = w_g(u^i v^i) &= |g(u^i) - g(v^i)| \\ &= |(f(u) + (k - \tfrac{i+1}{2})n) - (f(v) + (\tfrac{i+1}{2} - 1)n)| \\ &= |(k-i)n + f(u) - f(v)|. \end{aligned}$$

Since f is graceful, i.e., $1 \le |f(u) - f(v)| \le n - 1$, then for $f(u) > f(v)$ we have

$$(k-i)n+1 \le |(k-i)n + f(u) - f(v)| = (k-i)n + (f(u) - f(v)) \le (k-i+1)n - 1,$$

and for $f(u) < f(v)$ we get

$$(k-i-1)n+1 \le |(k-i)n + f(u) - f(v)| = (k-i)n - (f(u) - f(v)) \le (k-i)n - 1.$$

Thus g is an α-labeling of $P_k ⓥ T$.

□

Figure 7.5 depicts the graceful labeling of a caterpillar on 5 vertices with the fixed vertex v. In Fig. 7.6 we exhibit an example of an α-labeling of the tree $P_4 ⓥ T'$ obtained using the construction described in Theorem 7.2.2, where T' is the caterpillar from Fig. 7.5.

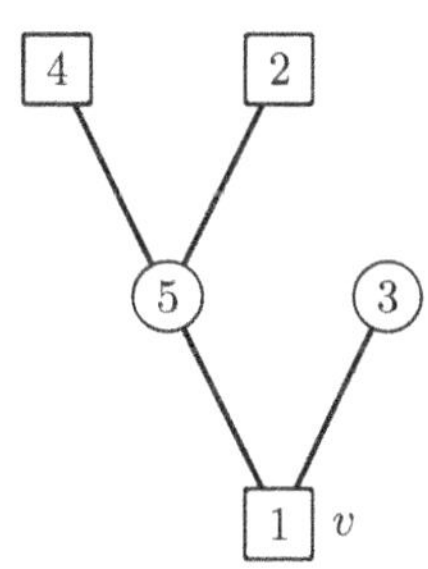

Fig. 7.5 Graceful labeling of a caterpillar T'

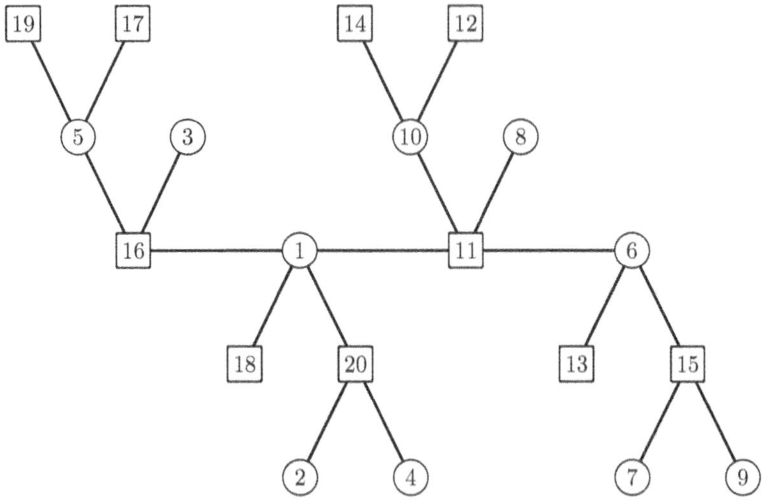

Fig. 7.6 α-labeling of $P_4 \textcircled{v} T'$

Theorem 7.2.3 ([46]) *If T is an α-tree of order n, then $P_k \textcircled{v} T$ admits an α-labeling for every positive integer k.*

Proof Let us assume that f is an α-labeling of T with the bipartition $\{A, B\}$. According to Theorem 7.2.2, for k even, $P_k \textcircled{v} T$ is an α-tree. For k odd we define a new labeling h in the following way.

$$h(v^i) = \begin{cases} f(v) + \frac{k-1}{2}n & \text{for } v \in A^1 \cup B^1 \text{ and } i = 1 \\ f(v) + \frac{k+i-1}{2}n & \text{for } v \in A^i \cup B^{i+1} \text{ and } i = 2, 4, \ldots, k-1 \\ f(v) + \frac{k-i}{2}n & \text{for } v \in A^i \cup B^{i-1} \text{ and } i = 3, 5, \ldots, k. \end{cases}$$

It is easy to see that for the vertex labels we have $\{h(v) : v \in A^i \cup B^{i-1}, i = 3, 5, \ldots, k\} = \{1, 2, \ldots, (k-1)n/2\}$, $\{h(v) : v \in A^1 \cup B^1\} = \{(k-1)n/2 + 1, (k-1)n/2+2, \ldots, (k+1)n/2\}$ and $\{h(v) : v \in A^i \cup B^{i+1}, i = 2, 4, \ldots, k-1\} = \{(k-1)n/2 + 1, (k-1)n/2 + 2, \ldots, kn\}$. Thus h is a bijection from $V(P_k \textcircled{v} T)$ onto $\{1, 2, \ldots, kn\}$. Moreover, as f is an α-labeling with the boundary value λ, then the boundary value of h is $\lambda + (k-1)n/2$.

Analogously, as in Theorem 7.2.2, we can show that h is an α-labeling by proving that the set of the weights of the edges is $\{1, 2, \ldots, kn - 1\}$. □

Figure 7.7 illustrates α-labeling of a caterpillar on seven vertices with the boundary value $\lambda = 4$ and fixed vertex v with the label 3. Figure 7.8 gives an example of an α-labeling of the tree $P_5 \textcircled{v} T''$ obtained using the construction described in Theorem 7.2.3, where T'' is the caterpillar from Fig. 7.7.

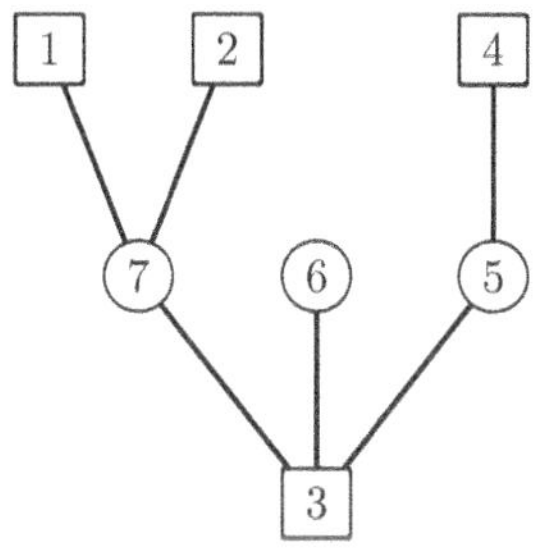

Fig. 7.7 α-labeling of a caterpillar T''

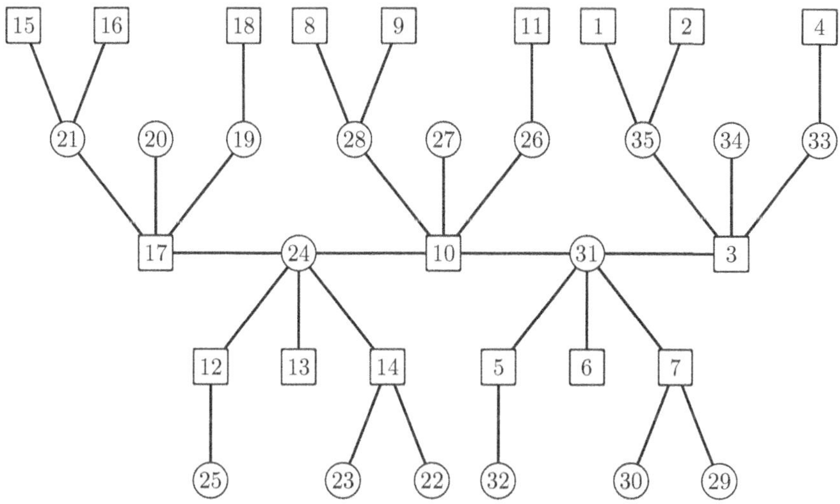

Fig. 7.8 α-labeling of $P_5 ⓥ T''$

In the next theorem we study the case where two isomorphic copies of an α-tree produce a new bigger α-tree by identifying two vertices with the same label.

Theorem 7.2.4 ([25]) *Every α-tree of size q produces an α-tree of size $2q$.*

Proof Let T be an α-tree of size q, with bipartition $\{A, B\}$. Let f be an α-labeling of T that assigns its boundary value λ to a vertex in A. For $i = 1, 2$, X_i is a copy of T. We define a labeling g of the vertices of $X_1 \cup X_2$ as follows.

$$g(v) = \begin{cases} f(v) & \text{if } v \in A_1 \\ q + f(v) & \text{if } v \in B_1 \\ q + \lambda + 1 - f(v) & \text{if } v \in V(X_2). \end{cases}$$

The labeling g assigns the labels $\{1, 2, \dots, \lambda\} \cup \{q+\lambda+1, q+\lambda+2, \dots, 2q+1\}$ to the vertices of X_1. The induced weights of the edges are $\{q+1, q+2, \dots, 2q\}$. The labeling g assigns the labels $\{\lambda, \lambda+1, \dots, \lambda+q\}$ to the vertices of X_2. We can see that the induced weights of edges are $\{1, 2, \dots, q\}$. Both X_1 and X_2 have a vertex labeled λ. In X_1, λ is assigned to a vertex in A_1; in X_2, λ is assigned to a vertex in B_2. Thus, identifying both vertices labeled λ, we have a new tree $\mathcal{T}$ of size $2q$ with an α-labeling of boundary value q. □

7.3 Edge-Antimagic Total Trees

In this section we use the connection between α-labelings and edge-antimagic labelings for generating large classes of edge-antimagic total trees from smaller graceful and α-trees.

The next theorems present methods for generating a tree of type $P_k \textcircled{v} T$ with a super (a, d)-EAT labeling for a positive even integer k.

Theorem 7.3.1 ([46]) *Let T be a graceful tree and let v be an arbitrary fixed vertex in T. If k is an even positive integer, then the tree $P_k \textcircled{v} T$ admits a super (a, d)-EAT labeling for all $d \in \{0, 1, 2, 3\}$.*

Proof Let T be a graceful tree and v be an arbitrary fixed vertex in T. It follows from Theorem 7.2.2 that if k is an even positive integer, then the tree $P_k \textcircled{v} T$ admits an α-labeling. If $\{A, B\}$ is a bipartition of the vertex set of $P_k \textcircled{v} T$, then it is not difficult to see that $|A| = |B|$ and according to Theorem 7.1.2 we have that $P_k \textcircled{v} T$ admits a super (a, d)-EAT labeling for all $d \in \{0, 1, 2, 3\}$. □

Theorem 7.3.2 ([46]) *Let T be an α-tree and let $||A| - |B|| \le 1$, $\{A, B\}$ be the bipartition of the vertex set of T. Let v be an arbitrary fixed vertex in T. Then for every positive integer k the tree $P_k \textcircled{v} T$ admits a super (a, d)-EAT labeling for all $d \in \{0, 1, 2, 3\}$.*

Proof Let T be an α-tree and v be an arbitrary fixed vertex in T. From Theorem 7.2.3 we get that for every positive integer k the tree $P_k \textcircled{v} T$ is also an α-tree. Moreover, if $\{A, B\}$ is a bipartition of the vertex set of T with $||A| - |B|| \le 1$ and $\{\mathbb{A}, \mathbb{B}\}$ is a bipartition of the vertex set of the α-tree $P_k \textcircled{v} T$, then $||\mathbb{A}| - |\mathbb{B}|| \le 1$. Thus, according to Theorem 7.1.2 we obtain that $P_k \textcircled{v} T$ admits a super (a, d)-EAT labeling for all $d \in \{0, 1, 2, 3\}$. □

Figure 7.9 gives a super $(39, 3)$-EAT labeling of $P_5 \textcircled{v} T''$ obtained from α-labeling of the graph in Fig. 7.8 by Theorem 7.3.2.

Denote by T_1 the graph $P_k \textcircled{v} T$. According to Theorem 7.2.2, if T is graceful and k is an even positive integer, then T_1 is an α-tree. Thus, by induction, we can see that the graph $T_{n+1} = P_{k_n} \textcircled{v_n} T_n$, where k_n is an even positive integer and v_n is an arbitrary fixed vertex in T_n, is also an α-graph according to Theorem 7.2.2. Moreover, from the construction of the α-tree T_{n+1} it follows that $|A_{n+1}| = |B_{n+1}|$, where $\{A_{n+1}, B_{n+1}\}$ is the bipartition of the vertex set of T_{n+1}. Thus, according to Theorem 7.3.1, the tree T_{n+1} admits a super (a, d)-EAT labeling for all $d \in \{0, 1, 2, 3\}$. The next theorem summarizes this result.

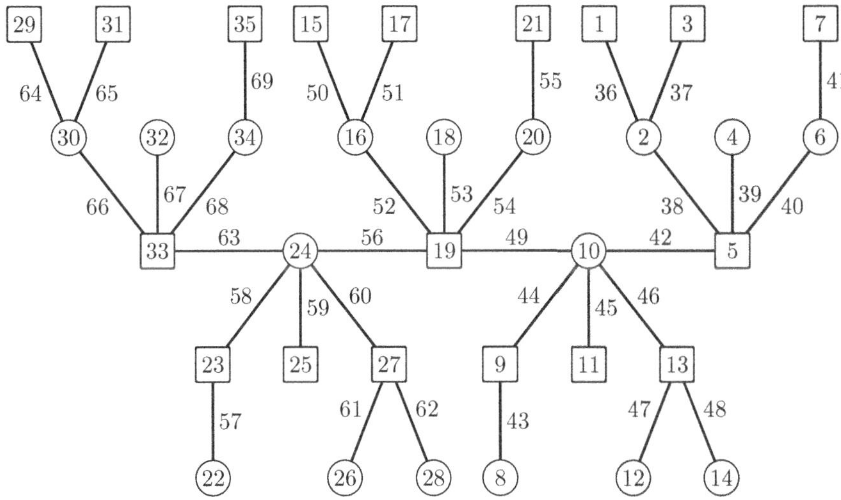

Fig. 7.9 Super (39, 3)-EAT labeling of $P_5 \textcircled{v} T''$

Theorem 7.3.3 ([46]) *Let k be an even positive integer. Let v be an arbitrary fixed vertex in tree T and let $T_1 = P_k \textcircled{v} T$. Let n be a positive integer and k_n be an even positive integer. Let v_n be an arbitrary fixed vertex in T_n and let $T_{n+1} = P_{k_n} \textcircled{v_n} T_n$. The graph T_{n+1} is super (a, d)-EAT for all $d \in \{0, 1, 2, 3\}$ if T is a graceful tree.*

Notice that for a graceful tree T, by using different fixed vertices of trees T_i, $i = 1, 2, \ldots, n$, we can find many different, up to isomorphism, α-trees T_{n+1} and their super (a, d)-EAT labelings.

Theorem 7.3.4 ([25]) *Every α-tree of size q produces a super (a, d)-EAT tree of size $2q$, for every $d \in \{0, 1, 2, 3\}$.*

Proof Let T be an α-tree of size q. It follows from Theorem 7.2.4 that a new tree $\mathcal{T}$ of size $2q$ is also the α-tree. Since the cardinalities of the bipartite sets of $\mathcal{T}$ differ by one, we have that $\mathcal{T}$ satisfies the conditions of Theorem 7.1.2 and therefore $\mathcal{T}$ admits labelings that are super (a, d)-EAT, for every $d \in \{0, 1, 2, 3\}$. □

7.4 Certain Classes of Super (a, d)-Edge-Antimagic Total Trees

Rosa [224] proved that caterpillars (trees whose removal of all end vertices produces a path) admit α-labeling. Figure 7.7 provides an example of α-labeling of caterpillar on seven vertices.

Bermond [58] conjectured that all lobsters (trees with the property that the removal of all end vertices produces a caterpillar) are graceful. Special classes of this conjecture are shown to be graceful. Ng in [201] describes graceful labelings for lobsters in which each vertex of the central path is attached to the centers of exactly two branches and in addition to this each of the vertices v_0 and v_m is attached to the center of a pendant branch. Wang et al. in [292] and Mishra and Panigrahi in [194] give graceful labelings to the lobsters having diameter at least five in which the degree of v_m is odd and the degree of the rest of the vertices in H are even. Chen, Lu, and Yeh in [81] give graceful labelings to some classes of lobsters in which the vertices of the central path are attached to the isomorphic copies of at most two different branches. Morgan [197] has proved that all lobsters with perfect matchings are graceful. In [195] and [196] graceful lobsters have the property that the degree of v_0 is even and the degrees of some (or all) vertices v_i, for $1 \leq i \leq m$, may be odd. Figure 7.10 exhibits graceful labeling of a lobster on nine vertices.

A *symmetric tree* ST is a rooted tree in which every level contains vertices of the same degree. In [60, 212] it is shown that symmetric trees are graceful. Figure 7.11 shows the symmetric tree with a graceful labeling. In [249] graceful symmetric trees are used for describing object-oriented software architecture.

A *banana tree* $(a_1K_{1,1}, \ldots, a_{t-1}K_{1,t-1}, a_tK_{1,t}, a_{t+1}K_{1,t+1}, \ldots, a_nK_{1,n})$ denotes the tree obtained by adding a vertex, the *apex*, to the union of a_i copies of the stars $K_{1,i}$, and joining the apex to a leaf of each star. Bhat-Nayak and Deshmukh

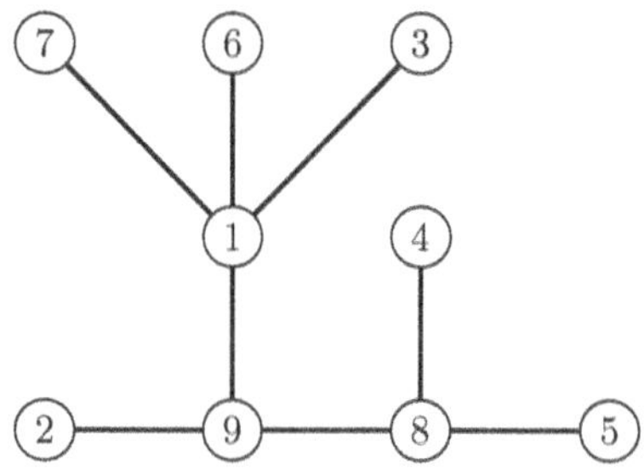

Fig. 7.10 Graceful labeling of a lobster

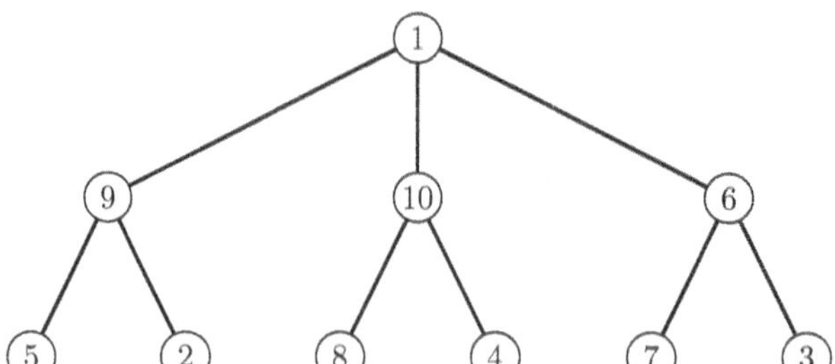

Fig. 7.11 Graceful labeling of a symmetric tree

[65] constructed three new families of graceful banana trees using an algorithmic labeling. They have shown that the following banana trees are graceful.

$BK = (K_{1,1}, \dots, K_{1,t-1}, (\beta + 1)K_{1,t}, K_{1,t+1}, \dots, K_{1,n})$, where $0 \le \beta < t$,
$BKK = (2K_{1,1}, \dots, 2K_{1,t-1}, (\beta+2)K_{1,t}, 2K_{1,t+1}, \dots, 2K_{1,n})$, where $0 \le \beta < t$, and
$BKKK = (3K_{1,1}, 3K_{1,2}, \dots, 3K_{1,n})$.

Murugan and Arumugan [199] additionally showed by construction that any banana tree BKR, where all stars have the same size, is graceful.

Regular *bamboo trees* are rooted trees consisting of the branches, the paths from the root to the leaves, of equal length, the leaves of which are identified with leaves of stars of equal size. These were shown to be graceful by Sekar in [232], see also [93] and [109]. *Olive trees* $\mathcal{T}(t)$ are rooted trees with t branches, the ith branch of which is a path of length i. Abhyankar and Bhat-Nayak [1] gave direct graceful labeling methods for olive trees.

By $\mathcal{F}$ we denote the family of graceful trees that contains caterpillars, symmetrical trees, lobsters from [81, 194, 195, 196, 197, 201, 292], olive trees, bamboo trees, and banana trees of type BK, BKK, $BKKK$, or BKR.

As the consequences of Theorems 7.3.1, 7.3.2, and 7.3.3, we have the following corollary.

Corollary 7.4.1 ([46]) *Let $T \in \mathcal{F}$ and v be an arbitrary fixed vertex in T.*

(i) For an even positive integer k, the tree $T_1 = P_k \circledcirc T$ admits a super (a, d)-EAT labeling for all $d \in \{0, 1, 2, 3\}$.
(ii) For every positive integer n, the tree $T_{n+1} = P_{k_n} \circledcirc T_n$, where k_n is a positive integer and v_n is an arbitrary fixed vertex in T_n, admits a super (a, d)-EAT labeling for all $d \in \{0, 1, 2, 3\}$.

7.5 Disjoint Union of α-Graphs

7.5.1 Arithmetic Sequences

This section contains the tools that allow us to determine the type of a sequence constructed by combining two different sequences. It will be useful later.

Lemma 7.5.1 ([35]) *Let $\mathcal{M}$ be an arithmetic sequence $\mathcal{M} = \{a + d(i - 1) : 1 \le i \le k + 1\}$, for the positive integers a, d and k, k even. Then there exists a permutation $\mathcal{P}(\mathcal{M})$ of the elements of $\mathcal{M}$ such that $\mathcal{M} + \mathcal{P}(\mathcal{M})$ is an arithmetic sequence with first term $2a + kd/2$ and a common difference d.*

Proof Suppose that $\mathcal{M} = \{p_i : p_i = a + d(i-1), 1 \le i \le k+1\}$ for k even and $a, d > 0$. Consider the permutation $\mathcal{P}(\mathcal{M}) = \{q_i : 1 \le i \le k+1\}$ of the elements of $\mathcal{M}$, where

$$q_i = \begin{cases} a + \frac{(k-i+1)d}{2} & \text{if } i \text{ is odd, } 1 \le i \le k+1 \\ a + \frac{(2k-i+2)d}{2} & \text{if } i \text{ is even, } 2 \le i \le k. \end{cases}$$

We claim that $\mathcal{M} + \mathcal{P}(\mathcal{M})$ is an arithmetic sequence. In fact,

$$p_i + q_i = \begin{cases} 2a + \frac{(k+i-1)d}{2} & \text{if } i \text{ is odd, } 1 \le i \le k+1 \\ 2a + \frac{(2k+i)d}{2} & \text{if } i \text{ is even, } 2 \le i \le k. \end{cases}$$

Thus $\mathcal{M} + \mathcal{P}(\mathcal{M})$ is the arithmetic sequence with the first term $2a + kd/2$ and the common difference d. □

Lemma 7.5.2 ([35]) *Let $\mathcal{N}$ be a sequence $\mathcal{N} = \{c+d(i-1) : 1 \le i \le (k+1)/2\} \cup \{c+di : (k+3)/2 \le i \le k+1\}$, for positive integers c, d and k, k odd. Then there exists a permutation of the elements of an arithmetic sequence $\mathcal{S} = \{r + d(i-1) : 1 \le i \le k+1\}$ such that $\mathcal{N} + \mathcal{P}(\mathcal{S})$ is an arithmetic sequence with the first term $c + r + (k+1)d/2$ and the common difference d.*

Proof Let $\mathcal{N} = \{n_i : n_i = c + d(i-1), 1 \le i \le (k+1)/2\} \cup \{n_i : n_i = c + di, (k+3)/2 \le i \le k+1\}$ be a sequence, for k odd and $c, d > 0$. Let $\mathcal{S} = \{r + d(i-1) : 1 \le i \le k+1\}$ be an arithmetic sequence. There are three cases to describe a required permutation $\mathcal{P}(\mathcal{S}) = \{h_i : 1 \le i \le k+1\}$.

Case A For $k \equiv 1 \pmod 6$, where $k \ge 7$, we construct

$$h_i = \begin{cases} r + (k-2i)d & \text{if } i \equiv 1 \pmod 3 \text{ and } 1 \le i < \frac{k-1}{2} \\ r + (k-2i)d & \text{if } i \equiv 2 \pmod 3 \text{ and } 2 \le i < \frac{k-1}{2} \\ r + (k+3-2i)d & \text{if } i \equiv 0 \pmod 3 \text{ and } 3 \le i \le \frac{k-1}{2} \\ r + kd & \text{if } i = \frac{k+1}{2} \\ r + (k-1)d & \text{if } i = \frac{k+3}{2} \\ r + (2k-2i)d & \text{if } i \equiv 0 \pmod 3 \text{ and } \frac{k+5}{2} \le i \le k-1 \\ r + (2k-2i)d & \text{if } i \equiv 1 \pmod 3 \text{ and } \frac{k+7}{2} \le i \le k \\ r + (2k+3-2i)d & \text{if } i \equiv 2 \pmod 3 \text{ and } \frac{k+9}{2} \le i \le k+1 \end{cases}$$

and for $k = 1$

$$h_i = \begin{cases} r + d & \text{if } i = 1 \\ r & \text{if } i = 2. \end{cases}$$

Case B For $k \equiv 5 \pmod 6$, where $k \geq 11$, we define

$$h_i = \begin{cases} r + (k-1)d & \text{if } i = 1 \\ r + (k-3)d & \text{if } i = 2 \\ r + (k-2i)d & \text{if } i \equiv 0 \pmod 3 \text{ and } 3 \leq i < \frac{k-1}{2} \\ r + (k-2i)d & \text{if } i \equiv 1 \pmod 3 \text{ and } 4 \leq i < \frac{k-1}{2} \\ r + (k+3-2i)d & \text{if } i \equiv 2 \pmod 3 \text{ and } 5 \leq i \leq \frac{k-1}{2} \\ r + kd & \text{if } i = \frac{k+1}{2} \\ r + (k-4)d & \text{if } i = \frac{k+3}{2} \\ r + (k-2)d & \text{if } i = \frac{k+5}{2} \\ r + (k-5)d & \text{if } i = \frac{k+7}{2} \\ r + (2k-2i)d & \text{if } i \equiv 1 \pmod 3 \text{ and } \frac{k+9}{2} \leq i \leq k-1 \\ r + (2k-2i)d & \text{if } i \equiv 2 \pmod 3 \text{ and } \frac{k+11}{2} \leq i \leq k \\ r + (2k+3-2i)d & \text{if } i \equiv 0 \pmod 3 \text{ and } \frac{k+13}{2} \leq i \leq k+1. \end{cases}$$

For $k = 5$ the permutation is

$$h_i = \begin{cases} r + 4d & \text{if } i = 1 \\ r + 2d & \text{if } i = 2 \\ r + 5d & \text{if } i = 3 \\ r + d & \text{if } i = 4 \\ r + 3d & \text{if } i = 5 \\ r & \text{if } i = 6. \end{cases}$$

Case C For $k \equiv 3 \pmod 6$, where $k \geq 9$, we define

$$h_i = \begin{cases} r + (k-1)d & \text{if } i = 1 \\ r + (k-2i)d & \text{if } i \equiv 2 \pmod 3 \text{ and } 2 \leq i < \frac{k-1}{2} \\ r + (k-2i)d & \text{if } i \equiv 0 \pmod 3 \text{ and } 3 \leq i < \frac{k-1}{2} \\ r + (k+3-2i)d & \text{if } i \equiv 1 \pmod 3 \text{ and } 4 \leq i \leq \frac{k-1}{2} \\ r + kd & \text{if } i = \frac{k+1}{2} \\ r + (2k-2i)d & \text{if } i \equiv 0 \pmod 3 \text{ and } \frac{k+3}{2} \leq i \leq k \\ r + (2k+3-2i)d & \text{if } i \equiv 1 \pmod 3 \text{ and } \frac{k+5}{2} \leq i \leq k+1 \\ r + (2k-2i)d & \text{if } i \equiv 2 \pmod 3 \text{ and } \frac{k+7}{2} \leq i \leq k-1. \end{cases}$$

For $k = 3$ we define the permutation in the following way

$$h_i = \begin{cases} r + 2d & \text{if } i = 1 \\ r + 3d & \text{if } i = 2 \\ r & \text{if } i = 3 \\ r + d & \text{if } i = 4. \end{cases}$$

There is no problem in seeing that, in all the considered cases, each integer h_i, $1 \le i \le k+1$, belongs to $\mathcal{S}$ and $\{n_i + h_i : 1 \le i \le k+1\} = \{c+r+(k+1)d/2, c+r+(k+3)d/2, \dots, c+r+(3k+1)d/2\}$. This produces the desired result. □

Some results are known for the super edge-antimagicness of forests. Namely, Ivančo and Lučkaničová [141] described some constructions of super EMT (super $(a, 0)$-EAT) labelings for $K_{1,m} \cup K_{1,n}$. Super (a, d)-EAT labelings for $P_n \cup P_{n+1}$, $nP_2 \cup P_n$ and $nP_2 \cup P_{n+2}$ have been described by Sudarsana et al. in [268], and EMT labelings for nP_3 can be found in [54].

Let G be a graph of order n and size $n-1$. We denote by mG a disjoint union of m copies of G. Our main goal in this section is to show that if G admits an α-labeling, then mG admits a super (a, d)-EAT labeling.

We start by basic counting to determine an upper bound for the difference d of a super (a, d)-EAT labeling. Let (p, q) graph be a super (a, d)-EAT. It is easy to see that the minimum possible edge-weight is at least $p + 4$ and the maximum possible edge-weight is no more than $3p + q - 1$. Thus $a + (q - 1)d \le 3p + q - 1$ and $d \le (2p + q - 5)/(q - 1)$. For $p = mn$, $q = m(n - 1)$ and $m \ge 1$, $n \ge 3$, we have that $d < 4$.

The next lemma presents a connection between α-labeling of G and $(a, 1)$-EAV labeling of mG.

Lemma 7.5.3 ([35]) *Let G be a graph of order n and size $n - 1$, $n \ge 3$. If G admits an α-labeling, and m is odd, $m \ge 1$, then mG admits an $(a, 1)$-EAV labeling.*

Proof Suppose that G is an α-graph. It is known, see [198] or [25], that if graph G of order n and size $n - 1$ admits an α-labeling, then G also admits an $(a, 1)$-EAV labeling. Hence, for $m = 1$ we have the desired result.

Figueroa-Centeno, Ichishima, and Muntaner-Batle [98] showed that a (p, q) graph H is super edge-magic if and only if there exists a bijective function $f : V(H) \to \{1, 2, \dots, p\}$ such that the set $\{f(u) + f(v) : uv \in E(H)\}$ consists of q consecutive integers. In our terminology this means that a (p, q) graph H is super EMT if and only if there exists a $(b - p - q, 1)$-EAV labeling of H. With respect to the previous result it follows that if a graph G of order n and size $n - 1$ admits an α-labeling, then G also admits a super edge-magic labeling.

It was proved by Figueroa-Centeno, Ichishima, and Muntaner-Batle in [100] that if H is a super edge-magic bipartite or tripartite graph, and m is odd, then mH is super edge-magic. Evidently, if G admits an α-labeling, and m is odd, then mG admits an $(a, 1)$-EAV labeling. □

Figure 7.12 gives a (24, 1)-EAV labeling of disjoint union of five copies of a caterpillar on nine vertices.

Lemma 7.5.4 ([35]) *Let G be a graph of order n and size $n-1$, $n \geq 3$. If G admits an α-labeling, and m is odd, $m \geq 1$, then mG admits a super $(a+2mn-m, 0)$-EAT labeling and a super $(a+mn+1, 2)$-EAT labeling.*

Proof In light of Lemma 7.5.3 we assume that f is an $(a, 1)$-EAV labeling of mG, where the set of the edge-weights forms the sequence $\{a, a+1, \ldots, a+mn-m-1\}$.

Case A The difference is $d = 0$. We extend the vertex labeling f into a labeling g such that

$$g(u) = f(u), \qquad \text{for every vertex } u \in V(mG)$$
$$g(uv) = 2mn - m + a - (g(u) + g(v)), \qquad \text{for every edge } uv \in E(mG).$$

Since

$$a \leq g(u) + g(v) \leq a + mn - m - 1,$$

we have that

$$mn + 1 \leq g(uv) \leq 2mn - m$$

and thus g is a total labeling. Every edge $uv \in E(mG)$ has the edge-weight $g(u) + g(uv) + g(v) = a + 2mn - m$. This implies that mG is super $(a+2mn-m, 0)$-EAT.

Case B The difference is $d = 2$. We consider a labeling h defined in the following way.

$$h(u) = f(u), \qquad \text{for every vertex } u \in V(mG)$$
$$h(uv) = mn + 1 - a + (h(u) + h(v)), \qquad \text{for every edge } uv \in E(mG).$$

Evidently, h is a total labeling and since

$$a \leq h(u) + h(v) \leq a + mn - m - 1$$

and

$$mn + 1 \leq h(uv) \leq 2mn - m$$

the set of the edge-weights is $\{a + mn + 1, a + mn + 3, \ldots, a + 3mn - 2m - 1\}$. Thus mG is super $(a + mn + 1, 2)$-EAT. □

Fig. 7.12 (24, 1)-EAV labeling of $5T$

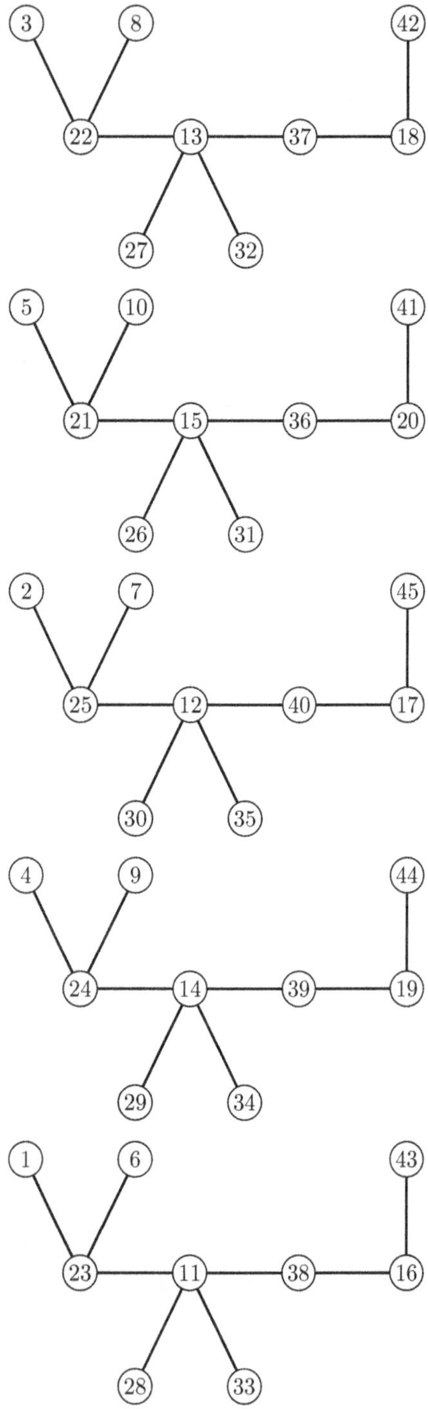

Lemma 7.5.5 ([35]) *Let G be a graph of order n and size $n-1$, $n \geq 4$ even. If G admits an α-labeling, then mG admits a super $(b, 1)$-EAT labeling for every $m \geq 1$.*

Proof Let us distinguish two cases.

Case A: m Odd As G is an α-graph of order n and size $n-1$, according to Lemma 7.5.3 there exists an $(a, 1)$-EAV labeling f of mG. Thus the set of the edge-weights gives the sequence $\mathcal{M} = \{a + (i-1) : 1 \leq i \leq k+1\}$, where $k = m(n-1) - 1$. Since n is even and if m odd, then k is even. With respect to Lemma 7.5.1, for $d = 1$, there exists a permutation $\mathcal{P}(\mathcal{M})$ of the elements of $\mathcal{M}$ such that $\mathcal{M} + (\mathcal{P}(\mathcal{M}) - a + mn + 1)$ is an arithmetic sequence with the first term $a + (m(3n-1)+1)/2$ and the common difference $d = 1$.

If $(\mathcal{P}(\mathcal{M}) - a + mn + 1)$ is an edge labeling of mG with the labels $mn + 1, mn+2, \dots, 2mn - m$, then $\mathcal{M} + (\mathcal{P}(\mathcal{M}) - a + mn + 1)$ determines the set of the edge-weights under the resulting total labeling. Hence, mG is super $(b, 1)$-EAT for $b = a + (m(3n-1)+1)/2$.

Case B: m Even Assume that f is an α-labeling of a graph G with n vertices and $n-1$ edges, and A, B are its bipartite sets. Without loss of generality, we may assume that the vertex labeled by the boundary value λ belongs to A. So, $f(u) < f(v)$ for any $u \in A$ and $v \in B$.

We denote by $V(mG) = \bigcup_{j=1}^{m}\{u_j, v_j : u_j \in A^j, v_j \in B^j\}$ the vertex set of a disjoint union of m copies of G, i.e., $\bigcup_{j=1}^{m}\{A^j \cup B^j\} = V(mG)$.

Consider the vertex labeling g of mG such that, for every $u_j \in A^j$, $1 \leq j \leq m$, we assign

$$g(u_j) = m(f(u) - 1) + j, \quad \text{if } u \in A,$$

and for every $v_j \in B^j$, $1 \leq j \leq m$, we assign

$$g(v_j) = \begin{cases} m(n + \lambda - f(v)) + \frac{m+1-j}{2} & \text{if } v \in B \text{ and } j \text{ is odd} \\ m(n + \lambda + 1 - f(v)) + \frac{2-j}{2} & \text{if } v \in B \text{ and } j \text{ is even.} \end{cases}$$

Since $1 \leq f(u) \leq \lambda$ and $\lambda + 1 \leq f(v) \leq n$, thus the function g assigns the labels $1, 2, 3, \dots, m\lambda - 1, m\lambda$ to all vertices $u_j \in A^j$, $1 \leq j \leq m$, and the labels $m\lambda + 1, m\lambda + 2, \dots, mn - 1, mn$ to all vertices $v_j \in B^j$, $1 \leq j \leq m$. Therefore g is an injective function from $\bigcup_{j=1}^{m}\{A^j \cup B^j\}$ into $\{1, 2, \dots, mn\}$.

If uv is an edge in G, $u \in A$, $v \in B$, then $u_j v_j$ is the edge in mG, where $u_j \in A^j$, $v_j \in B^j$, for $1 \leq j \leq m$. For the edge-weight of $u_j v_j$, we have

$$g(u_j) + g(v_j) = \begin{cases} m(n + \lambda - (f(v) - f(u))) + \frac{1+j-m}{2} & \text{if } j \text{ is odd} \\ m(n + \lambda - (f(v) - f(u))) + \frac{2+j}{2} & \text{if } j \text{ is even.} \end{cases}$$

We can see that, for each edge $uv \in E(G)$, the edge-weights of the corresponding edges in mG produce the sequence

$$\mathcal{N} = \{c + d(i-1) : 1 \le i \le \tfrac{k+1}{2}\} \cup \{c + di : \tfrac{k+3}{2} \le i \le k+1\}$$

for $c = m(n + \lambda - 1/2 - (f(v) - f(u))) + 1$, $d = 1$ and $k = m - 1$. For $f(v) - f(u) = l$, we have $n-1$ sequences $\mathcal{N}_l$, $1 \le l \le n-1$.

Now we define an arithmetic sequence $\mathcal{S}_l = \{r_l + d(i-1) : 1 \le i \le k+1\}$, for $d = 1$, $k = m-1$, and

$$r_l = \begin{cases} \frac{m}{2}(2n - 1 + l) + 1 & \text{if } l \text{ is odd} \\ \frac{m}{2}(3n - 2 + l) + 1 & \text{if } l \text{ is even.} \end{cases}$$

We can see that $\bigcup_{l=1}^{n-1} \mathcal{S}_l = \{mn+1, mn+2, \ldots, 2mn-m\}$. From Lemma 7.5.2, it follows that for each sequence $\mathcal{N}_l$, $1 \le l \le n-1$, there exists a permutation of the elements of the arithmetic sequence $\mathcal{S}_l$ such that $\mathcal{N}_l + \mathcal{P}(\mathcal{S}_l)$, $1 \le l \le n-1$, is an arithmetic sequence with the first term

$$\begin{cases} \frac{m}{2}(4n + 2\lambda - l - 1) + 2 & \text{if } l \text{ is odd} \\ \frac{m}{2}(5n + 2\lambda - l - 2) + 2 & \text{if } l \text{ is even} \end{cases}$$

and the common difference $d = 1$. It is a matter of routine checking to see that $\bigcup_{l=1}^{n-1}\{\mathcal{N}_l + \mathcal{P}(\mathcal{S}_l)\} = \{m(3n+2\lambda)/2 + 2, m(3n+2\lambda)/2 + 3, \ldots, m(5n + 2\lambda - 2)/2 + 1\}$.

If the arithmetic sequence $\bigcup_{l=1}^{n-1} \mathcal{S}_l$ is the set of the edge labels of mG, then $\bigcup_{l=1}^{n-1}\{\mathcal{N}_l + \mathcal{P}(\mathcal{S}_l)\}$ describes the set of the corresponding edge-weights of mG. This implies that mG has a super $(m(3n+2\lambda)/2 + 2, 1)$-EAT labeling. □

The next theorem follows from the three previous lemmas.

Theorem 7.5.1 ([35]) *Let G be an α-graph of order n and size $n-1$, $n \ge 3$. The graph mG is super (a, d)-EAT if one of the following conditions is satisfied.*

(i) $d \in \{0, 2\}$ and m is odd, $m \ge 1$.
(ii) $d = 1$ and n is even, $m \ge 1$.

The next result gives a connection between α-labelings and $(a, 2)$-EAV labelings.

Lemma 7.5.6 ([35]) *Let G be an α-graph of order n and size $n-1$ and let $\{A, B\}$ be the bipartition of its vertex set. If $||A| - |B|| \le 1$, then mG is $(m+2, 2)$-EAV, for every $m \ge 1$.*

Proof It is proved in [25] that if G is an α-graph of order n and size $n-1$ and $||A| - |B|| \le 1$, then G is $(3, 2)$-EAV. Hence the desired result holds for $m = 1$.

Let f be an α-labeling of graph G of order n and size $n-1$ and A, B be the bipartite sets of G. We may assume that $0 \le |A| - |B| \le 1$ and the vertex labeled by the boundary value λ belongs to A. In the case that the vertex labeled by the boundary value λ does not belong to A under the α-labeling f, then a new labeling

$$f^{\star}(v) = n + 1 - f(v), \quad \text{for } v \in V(G)$$

is an α-labeling as well and its boundary value $n - \lambda$ appears on a vertex of A.

Now we consider a vertex labeling g of mG such that, for every $u_j \in A^j$, $1 \leq j \leq m$, we define

$$g(u_j) = m(2f(u) - 2) + j, \quad \text{if } u \in A$$

and for every $v_j \in B^j$, $1 \leq j \leq m$, we define

$$g(v_j) = m(2n + 1 - 2f(v)) + j, \quad \text{if } v \in B.$$

Since $1 \leq f(u) \leq \lambda$ and $\lambda + 1 \leq f(v) \leq n$, it follows that the function g assigns the labels $\{1, 2, \dots, m\} \cup \{2m+1, 2m+2, \dots, 3m\} \cup \dots \cup \{m(2\lambda-4)+1, m(2\lambda-4)+2, \dots, m(2\lambda-3)\} \cup \{m(2\lambda-2)+1, m(2\lambda-2)+2, \dots, m(2\lambda-1)\}$ to all vertices $u_j \in A^j$, $1 \leq j \leq m$, and the labels $\{m+1, m+2, \dots, 2m\} \cup \{3m+1, 3m+2, \dots, 4m\} \cup \dots \cup \{m(2n-2\lambda-3)+1, m(2n-2\lambda-3)+2, \dots, m(2n-2\lambda-2)\} \cup \{m(2n-2\lambda-1)+1, m(2n-2\lambda-1)+2, \dots, m(2n-2\lambda)\}$ to all vertices $v_j \in B^j$, $1 \leq j \leq m$. If $0 \leq |A| - |B| \leq 1$, then $\lambda = \lceil n/2 \rceil$ and evidently g is an injective function with the labels $1, 2, \dots, mn$.

Moreover, if uv is an edge in G, $u \in A$, $v \in B$, then $u_j v_j$ is the edge in mG, where $u_j \in A^j$, $v_j \in B^j$, for $1 \leq j \leq m$. For the edge-weight of $u_j v_j$, $1 \leq j \leq m$, we have

$$g(u_j) + g(v_j) = m(2n - 1) + 2j - 2m(f(v) - f(u)).$$

Since f is an α-labeling, then $1 \leq f(v) - f(u) \leq n - 1$ for $uv \in E(G)$, and the edge-weights of mG form the arithmetic sequence $\{m+2, m+4, \dots, 2mn - m\}$. Thus g is an $(m+2, 2)$-EAV labeling of mG. □

Figure 7.13 illustrates a (6, 2)-EAV labeling of the disjoint union of 4 copies of caterpillar on 13 vertices.

Theorem 7.5.2 ([35]) *Let G be an α-graph of order n and size $n-1$ and let $\{A, B\}$ be a bipartition of the vertex set of G. If $||A| - |B|| \leq 1$, then mG is super (a, d)-EAT, for $d \in \{1, 3\}$ and every $m \geq 1$.*

Proof It follows from Lemma 7.5.6 that if a graph G satisfies the assumptions of the theorem, then mG is $(m+2, 2)$-EAV for every $m \geq 1$. Let g be an $(m+2, 2)$-EAV labeling of mG with the set of edge-weights $\{g(u) + g(v) : uv \in E(mG)\} = \{m+2, m+4, \dots, 2mn - m\}$.

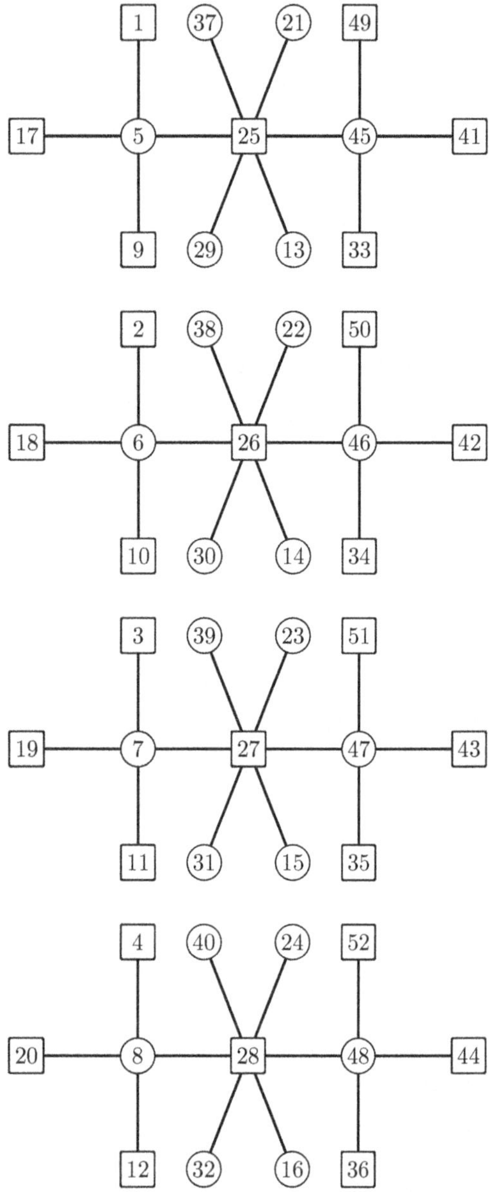

Fig. 7.13 (6, 2)-EAV labeling of $4T$

We extend the vertex labeling g into a total labeling h_1 and a total labeling h_2 by adding the edge labels from the set $\{mn+1, mn+2, \ldots, 2mn-m\}$, where

$$h_1(u) = h_2(u) = g(u), \qquad \text{for every vertex } u \in V(mG)$$

$$h_1(uv) = 2mn - m + 1 + \frac{m - (h_1(u) + h_1(v))}{2},$$

$$h_2(uv) = mn + \frac{(h_2(u) + h_2(v)) - m}{2}, \qquad \text{for every edge } uv \in E(mG).$$

It easily follows that if $\{h_1(u) + h_1(v) : uv \in E(mG)\} = \{m+2, m+4, \ldots, 2mn-m\}$, then the set of edge-weights is $\{h_1(u) + h_1(v) + h_1(uv) : uv \in E(mG)\} = \{2mn+2, 2mn+3, \ldots, 3mn-m+1\}$. The reader can also easily verify that $\{h_2(u) + h_2(v) + h_2(uv) : uv \in E(mG)\} = \{mn+m+3, mn+m+6, \ldots, 4mn-2m\}$. This implies the desired result. □

7.6 Disjoint Union of Caterpillars

In this section we study a super edge-antimagicness of forests in which every component is a caterpillar. Recall that the *caterpillar* is a graph derived from a path by hanging any number of leaves from the vertices of the path. Sugeng et al. in [273] described some constructions of the super (a, d)-EAT labelings of the caterpillars for $d \in \{0, 1, 2, 3\}$.

Let T be a caterpillar of order n and mT be the disjoint union of m copies of T. Rosa [224] showed that all caterpillars have an α-labeling. Therefore all results from the previous section hold for T and mT. Moreover, we complete one case when $d = 1$ and n odd.

Lemma 7.6.1 ([35]) *There is a super $(a, 1)$-EAT labeling for a caterpillar of order n, $n \geq 3$ odd.*

Proof We consider a caterpillar T of order n, $n \geq 3$ odd. Any caterpillar is bipartite. We denote by $\{A, B\}$ the bipartition of the vertex set of the caterpillar T, i.e., $V(T) = A(T) \cup B(T)$. We can draw the vertices of T in two rows, such that each row contains only the vertices from one partite set. Clearly, it is possible to make the drawing of T such that there are no edge crossings. Let $e_1^*, e_2^*, \ldots, e_{n-1}^*$ be the edges of T ordered from left to right. If one of the endpoints of the edge $e_{(n+1)/2}^*$ is of degree 1, then we denote it by v_1. If both endpoints of $e_{(n+1)/2}^*$ have degrees greater than 1, we denote by v_1 the vertex which is the common vertex of the edges $e_{(n+1)/2}^*$ and $e_{(n+3)/2}^*$. The next vertices ordered from v_1 to the right in the same partition we denote by $v_2, v_3, \ldots, v_t$. We continue in the same partition at the beginning and we denote the vertices ordered from left to v_1 by $v_{t+1}, v_{t+2}, \ldots, v_{t+s}$, that is, $v_{t+1}, v_{t+2}, \ldots, v_{t+s}, v_1, v_2, \ldots, v_t$ are ordered vertices in the first partition, say $A(T)$. Let $u_1, u_2, \ldots, u_{n-t-s}$ be the vertices in the second partition, say $B(T)$, ordered from left to right.

Consider the labeling $f : V(T) \to \{1, 2, \ldots, n\}$ defined by

$$f(v_l) = \begin{cases} l & \text{if } 1 \le l \le t \\ n - t - s + l & \text{if } t + 1 \le l \le t + s \end{cases}$$

$$f(u_l) = t + l \qquad \text{if } 1 \le l \le n - t - s.$$

Now, we redefine the edges of T such that

$$e_i = \begin{cases} e^*_{\frac{n+1}{2}-1+i} & \text{if } 1 \le i \le \frac{n-1}{2} \\ e^*_{i+1-\frac{n+1}{2}} & \text{if } \frac{n+1}{2} \le l \le n - 1. \end{cases}$$

We can see that the set of the edge-weights gives the sequence $\mathcal{N} = \{wt(e_i) : wt(e_i) = c + (i - 1), 1 \le i \le (k + 1)/2\} \cup \{wt(e_i) : wt(e_i) = c + i, (k + 3)/2 \le i \le k + 1\}$ for $k = n - 2$, where c is an edge-weight of the edge $e^*_{(n+1)/2} = e_1$. With respect to Lemma 7.5.2, for $d = 1$, there exists a permutation of the elements of the arithmetic sequence $\mathcal{S} = \{r + d(i - 1) : 1 \le i \le k + 1\}$ for $d = 1$, $k = n - 2, r = n + 1$, such that $\mathcal{N} + \mathcal{P}(\mathcal{S})$ is an arithmetic sequence with the first term $c + (3n + 1)/2$ and the common difference $d = 1$. If $\mathcal{S}$ is a set of edge labels of T, then $\mathcal{N} + \mathcal{P}(\mathcal{S})$ describes the set of the corresponding edge-weights of T. Thus T admits a super $(c + (3n + 1)/2, 1)$-EAT labeling. □

Figure 7.14 illustrates super (37, 1)-EAT labeling of a caterpillar of odd order described in the proof of Lemma 7.6.1.

Let us remark that the previous lemma was proved in [273] by a different construction. We described only one convenient vertex labeling f which will be useful in the next theorem.

Theorem 7.6.1 ([35]) *Let T be a caterpillar of order n, $n \ge 3$ odd. If T admits a super $(a, 1)$-EAT labeling, then mT also admits a super $(b, 1)$-EAT labeling, for every $m \ge 2$.*

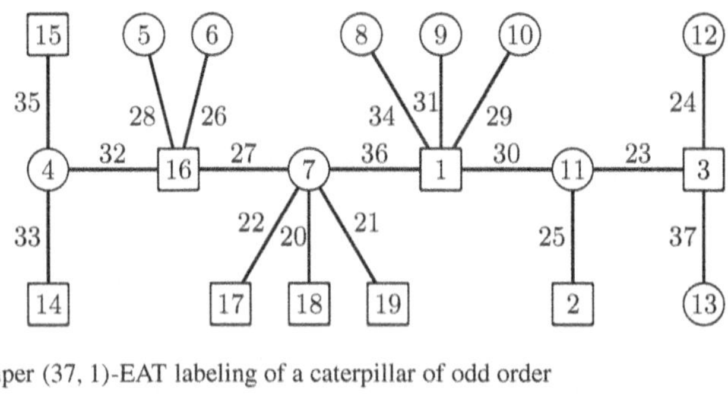

Fig. 7.14 Super (37, 1)-EAT labeling of a caterpillar of odd order

Proof Assume that a caterpillar T of order n, $n \geq 3$ odd, with vertices and edges denoted as in Lemma 7.6.1, admits a super $(a, 1)$-EAT labeling. We denote by $V(mT) = \bigcup_{j=1}^{m}\{A^j(T) \cup B^j(T)\}$ the vertex set of the disjoint union of m copies of the caterpillar T where $A^j(T) = \{v_l^j : 1 \leq l \leq t+s\}$, $B^j(T) = \{u_l^j : 1 \leq l \leq n-t-s\}$, $1 \leq j \leq m$. Let $E(mT) = \bigcup_{j=1}^{m}\{e_i^j : 1 \leq i \leq n-1\}$ be the edge set of mT. Evidently every edge e_i^j has one endpoint in $A^j(T)$ and other one in $B^j(T)$.

Let us distinguish two cases.

Case A: m odd We extend the vertex labeling f from Lemma 7.6.1 to a labeling g_1 such that for every $1 \leq l \leq t+s$ we put

$$g_1(v_l^j) = \begin{cases} m(f(v_l)-1) + \frac{m+3}{2} - j & \text{if } 1 \leq j \leq \frac{m+1}{2} \\ m(f(v_l)-1) + \frac{3m+3}{2} - j & \text{if } \frac{m+3}{2} \leq j \leq m, \end{cases}$$

and for every $1 \leq l \leq n-t-s$ we set

$$g_1(u_l^j) = \begin{cases} m(f(u_l)-1) + 2j - 1 & \text{if } 1 \leq j \leq \frac{m+1}{2} \\ m(f(u_l)-1) + 2j - m - 1 & \text{if } \frac{m+3}{2} \leq j \leq m. \end{cases}$$

It is a routine procedure to verify that if

$$f(v_l) \in \{1, 2, \ldots, t\} \cup \{n-s+1, n-s+2, \ldots, n\}$$

and

$$f(u_l) \in \{t+1, t+2, \ldots, n-s\},$$

then the vertex labeling g_1 is a bijective function from $V(mT)$ onto the set $\{1, 2, \ldots, mn\}$. Moreover, for the edge-weights we have

$$wt_{g_1}(e_i^j) = m\,wt_f(e_i) + \frac{1-3m}{2} + j, \quad \text{for } 1 \leq i \leq n-1 \text{ and } 1 \leq j \leq m.$$

It follows from Lemma 7.6.1 that

$$wt_f(e_i) = \begin{cases} c + (i-1) & \text{if } 1 \leq i \leq \frac{n-1}{2} \\ c + i & \text{if } \frac{n+1}{2} \leq i \leq n-1, \end{cases}$$

thus the edge-weights of the corresponding edges in each copy of mT produce a sequence $\mathcal{N}_j = \{wt_{g_1}(e_i^j) : wt_{g_1}(e_i^j) = c_j + m(i-1), 1 \leq i \leq (k+1)/2\} \cup \{wt_{g_1}(e_i^j) : wt_{g_1}(e_i^j) = c_j + mi, (k+3)/2 \leq i \leq k+1\}$ for $c_j = mc + (1-3m)/2 + j$, $k = n-2$ and $1 \leq j \leq m$.

According to Lemma 7.5.2, it follows that for each sequence $\mathcal{N}_j$, $1 \le j \le m$, there exists a permutation of the elements of the arithmetic sequence $\mathcal{S}_j = \{r_j + m(i-1) : 1 \le i \le k+1\}$ for $k = n-2$ and

$$r_j = \begin{cases} \frac{m(2n+1)-j}{2} + 1 & \text{if } j \text{ is odd} \\ mn + m + \frac{2-j}{2} & \text{if } j \text{ is even,} \end{cases}$$

such that $\mathcal{N}_j + \mathcal{P}(\mathcal{S}_j)$, $1 \le j \le m$, is an arithmetic sequence with the first term

$$a_j = \begin{cases} \frac{m(2c+2n-2)+3+j}{2} + \frac{(n-1)m}{2} & \text{if } j \text{ is odd} \\ \frac{m(2c+2n-1)+3+j}{2} + \frac{(n-1)m}{2} & \text{if } j \text{ is even} \end{cases}$$

and the common difference m.

If $\bigcup_{j=1}^{m} \mathcal{S}_j$ is an edge labeling of mT with the labels $mn+1, mn+2, \dots, 2mn-m$, then $\bigcup_{j=1}^{m}\{\mathcal{N}_j + \mathcal{P}(\mathcal{S}_j)\} = \bigcup_{j=1}^{m}\{a_j + m(i-1) : 1 \le i \le n-2\} = \{m(c+(3n-3)/2)+2, m(c+(3n-3)/2)+3, \dots, m(c+(5n-5)/2)+1\}$ is the set of the edge-weights and we arrive at the desired result.

Case B: m even We extend the vertex labeling f to a labeling g_2 in the following way, where for every $1 \le l \le t+s$,

$$g_2(v_l^j) = \begin{cases} m(f(v_l)-1) + \frac{m+2}{2} - j & \text{if } 1 \le j \le \frac{m}{2} \\ m(f(v_l)-1) + \frac{3m+2}{2} - j & \text{if } \frac{m+2}{2} \le j \le m \end{cases}$$

and, for every $1 \le l \le n-t-s$,

$$g_2(u_l^j) = \begin{cases} m(f(u_l)-1) + 2j - 1 & \text{if } 1 \le j \le \frac{m}{2} \\ m(f(u_l)-1) + 2j - m & \text{if } \frac{m+2}{2} \le j \le m. \end{cases}$$

Again it is not difficult to verify that if

$$f(v_l) \in \{1, 2, \dots, t\} \cup \{n-s+1, n-s+2, \dots, n\}$$

and

$$f(u_l) \in \{t+1, t+2, \dots, n-s\},$$

then the vertex labeling $g_2 : V(mT) \to \{1, 2, \dots, mn\}$ is a bijective function.

For the edge-weights we have

$$wt_{g_2}(e_i^j) = \begin{cases} mw_f(e_i) - \frac{3m}{2} + j & \text{if } 1 \le j \le \frac{m}{2} \\ mw_f(e_i) - \frac{3m}{2} + j + 1 & \text{if } \frac{m}{2} + 1 \le j \le m. \end{cases}$$

Now, we define the arithmetic sequences $\mathcal{S}_j = \{r_j + m(i-1) : 1 \le i \le k+1\}$ for $k = n-2$, $1 \le j \le m$, where

$$\left.\begin{array}{l} \text{for } k' = m-1 \equiv 5 \pmod 6,\ k' \ge 5 \\ \text{for } k' = m-1 \equiv 1 \pmod 6,\ k' \ge 1 \\ \text{for } k' = m-1 \equiv 3 \pmod 6,\ k' \ge 3 \end{array}\right\} r_j = mn + 1 - r + h_j.$$

We are using the labeling h from the proof of Lemma 7.5.2 for $d = 1$ and for every $k' = m-1$.

We will use a similar argument as in Case A that the edge-weights of the corresponding edges in each copy of mT produce a sequence $\mathcal{N}_j = \{wt_{g_2}(e_i^j) : wt_{g_2}(e_i^j) = c_j + m(i-1), 1 \le i \le (k+1)/2\} \cup \{wt_{g_2}(e_i^j) : wt_{g_2}(e_i^j) = c_j + mi, (k+3)/2 \le i \le k+1\}$, for $k = n-2$, and

$$c_j = \begin{cases} \frac{m}{2}(2c-3) + j & \text{if } 1 \le j \le \frac{m}{2} \\ \frac{m}{2}(2c-3) + j + 1 & \text{if } \frac{m}{2} + 1 \le j \le m. \end{cases}$$

With respect to Lemma 7.5.2, for each sequence $\mathcal{N}_j$, $1 \le j \le m$, there exists a permutation of the elements of the arithmetic sequence $\mathcal{S}_j = \{r_j + m(i-1) : 1 \le i \le k+1\}$, $1 \le j \le m$, such that $\mathcal{N}_j + \mathcal{P}(\mathcal{S}_j)$, $1 \le j \le m$, is an arithmetic sequence with the first term $c_j + r_j + (k+1)m/2$ and the common difference m.

If $\bigcup_{j=1}^{m} \mathcal{S}_j = \{mn+1, mn+2, \dots, 2mn-m\}$ is a set of edge labels of mT, then $\bigcup_{j=1}^{m}\{\mathcal{N}_j + \mathcal{P}(\mathcal{S}_j)\} = \{m(c+(3n-3)/2)+2, m(c+(3n-3)/2)+3, \dots, m(c+(5n-5)/2)+1\}$ determines the set of the edge-weights of mT and the resulting total labeling is super $(b, 1)$-EAT. □

In Theorem 7.5.2 it is proved that if G is an α-graph of order n and size $n-1$ and $||A| - |B|| \le 1$, where $\{A, B\}$ is the bipartition of the vertex set of G, then mG is super $(a, 3)$-EAT, for every $m \ge 1$. Paper [25] gives a super $(13, 3)$-EAT labeling of a caterpillar which does not satisfy the restriction for the cardinalities of bipartite sets A and B because in this case $|A| = 2$ and $|B| = 2n-1$. What can we say about a super $(a, 3)$-EAT labeling of mG in the case when a graph G of order n and size $n-1$ does not satisfy the restriction for the cardinalities of bipartite sets A and B? At this time we have no answer to this question. Bača, Lascsáková, and Semaničová propose the following open problem for further investigation.

Open Problem 7.6.1 ([35]) Let T be a caterpillar of order n and $||A| - |B|| > 1$, where $\{A, B\}$ is the bipartition of its vertex set. For the graph mT determine if there is a super $(a, 3)$-EAT labeling.

Chapter 8
Conclusion

Following the introduction, the first three chapters of this monograph were devoted to magic graphs. Characterizations were given for magic graphs and regular magic graphs. Necessary and sufficient conditions for supermagic graphs were reviewed. Labeled constructions were presented for the complete graph minus an edge as well as for the complete graph minus two nonadjacent edges. Chapter 2 concluded with the construction of supermagic graphs based on graph factorization.

Chapter 3 introduced vertex-magic total labelings and presented new results and constructions for regular graphs including cycles, complete graphs, and generalized Petersen graphs as well as for non-regular graphs such as complete bipartite graphs, complete multipartite graphs, and wheel-related graphs.

A similar approach was taken in Chap. 4 which was devoted to edge-magic total labelings. Constructions were given for several connected graphs such as fans, friendship graphs, ladders, generalized prisms, paths, and path-like trees. Also considered were labeling constructions for families of disjoint isomorphic structures such as stars, paths, and path-like trees.

The main concern in this monograph was paid to the antimagic total labelings. This section followed the structured format of the magic graph sections by concentrating separately on vertex-antimagic and edge-antimagic labelings. Results on super vertex-antimagic total labelings were presented for cycles, paths, generalized Petersen graphs, trees, and unicyclic graphs.

In the following chapter, constructions were given for maximal edge-antimagic vertex labeled graphs with difference $d = 1$ and $d = 2$ as well as for super edge-antimagic total vertex labeled regular graphs with difference $d = 1$. Super edge-antimagic total labelings were also presented for circulant graphs, toroidal polyhexes, and certain families of disconnected graphs.

Chapter 7 concentrated on the connection between α-labelings and edge-antimagic labelings, resulting in a method for generating large classes of edge-antimagic trees from smaller graceful trees.

M. Bača et al., *Magic and Antimagic Graphs*, Developments in Mathematics 60,
https://doi.org/10.1007/978-3-030-24582-5_8

This work concludes with a summary of the intriguing and challenging conjectures and open problems peppered throughout the text.

8.1 Open Problems

Open Problem 2.3.1 ([152]) Find the smallest magic index of a magic graph.

Open Problem 2.8.1 ([231]) Decide whether the Möbius ladder M_{2m+1} is supermagic for some m, $m \neq 2$.

Open Problem 2.9.1 ([124]) Is the graph $K_{n,n}-$(1-factor) supermagic when $n \equiv 3 \pmod 4$?

Open Problem 3.1.1 ([189]) Find a VMT labeling for the antiprism A_n, for all odd $n \geq 3$.

Open Problem 3.3.1 ([79]) Do all graphs satisfying Theorem 3.3.5 have a VMT labeling?

Open Problem 4.3.1 ([257]) For wheel W_n, $n \equiv 2 \pmod 8$, determine if there is an EMT labeling.

Open Problem 4.3.2 ([176]) For the generalized prism, $C_m \square P_n$, $n > 2$ and m even, determine if there is a super EMT labeling.

Open Problem 4.5.1 ([39]) Let $G \cong (2m)P_n$, $n \neq 2$, $m \geq 1$. Is G a strong super EMT?

Open Problem 4.5.2 ([39]) Let $G \cong (2m)P_n$, $n \neq 2$, $m \geq 1$. How many nonisomorphic strong super EMT labelings does G admit?

Open Problem 4.5.3 ([39]) Let $G \cong \bigcup_{j=1}^{2m} T_j$ be a disjoint union of an even number of path-like trees, all of them of the same order, and such that $T_j \neq P_2$ for $j = 1, 2, \ldots, 2m$. Is G a super EMT graph?

Open Problem 5.2.1 ([272]) For the complete graph K_n and complete bipartite graph $K_{n,n}$, determine if there is an (a, d)-VAT labeling for every feasible value of $d > 1$.

Open Problem 5.4.1 Find an $(a, 4)$-VAT labeling of cycle C_n and path P_n for n even, $n \geq 4$.

Open Problem 5.4.2 For the cycles C_n and the paths P_n, determine if there is an $(a, 5)$-VAT labeling.

Open Problem 5.5.1 For the generalized Petersen graph $P(n, m)$, find (if there is) a construction of a super (a, d)-VAT labeling.

(i) For n even, $n \geq 4$, $3 \leq m \leq n/2 - 1$, and $d \in \{3, 4\}$.
(ii) For n odd, $n \geq 3$, $2 \leq m < n/2$, and $d \in \{0, 2, 3, 4\}$.

Open Problem 5.6.1 ([12]) For the graphs mP_3 and mP_4, determine if there is a super (a, d)-VAT labeling, for every $m \geq 2$ and $d \in \{1, 2\}$.

Open Problem 5.6.2 ([12]) For the graph mP_n, $n \geq 5$ and $m > 1$, determine if there is a super (a, d)-VAT labeling for the feasible values of the difference d.

Open Problem 6.3.1 ([37]) For the friendship graph f_n, determine if there is a super $(a, 0)$-EAT or a super $(a, 2)$-EAT labeling, for $n > 7$.

Open Problem 6.3.2 ([38]) Determine the complexity of deciding if a given tree of maximum degree 4 is a path-like tree.

Open Problem 6.3.3 ([273]) For the caterpillar $S_{n_1,n_2,\ldots,n_r}$, determine if there is a super $(a, 3)$-EAT labeling, for r odd and $N_2 = N_1 + 1$.

Open Problem 6.3.4 ([273]) For the caterpillar $S_{n_1,n_2,\ldots,n_r}$, determine feasible pairs (N_1, N_2), $N_1 \neq N_2$ and $|N_1 - N_2| \neq 1$, which make a super $(a, 3)$-EAT labeling impossible.

Open Problem 6.3.5 ([26]) For the circulant graph $C_n(a_1, n/2)$, for $n \equiv 2 \mod 4$, $n \geq 6$, if a_1 is even and $\gcd(a_1, n/2) > 1$ or if a_1 is odd, determine whether there exists an $(a, 1)$-EAV labeling.

Open Problem 6.3.6 ([85]) For mP_n, $m \geq 2$ even, $n \geq 4$, determine if there is a super (a, d)-EAT labeling, with $d \in \{0, 2\}$.

Open Problem 6.3.7 ([83]) For the graph $K_{1,m} \cup K_{1,n}$, $m \geq n \geq 2$, if m is not a multiple of $n + 1$ determine whether there is a super $(a, 2)$-EAT labeling.

Open Problem 6.3.8 ([83]) For the graph $K_{1,m} \cup K_{1,n}$, $m + n$ even and $m \neq n$, determine if there is a super $(a, 1)$-EAT labeling.

Open Problem 6.3.9 ([83]) For the graph $K_{1,m} \cup K_{1,n}$, if $m \neq n$, determine if there is a super $(a, 3)$-EAT labeling.

Open Problem 6.3.10 ([30]) For the graph $mS_{t_1,t_2,\ldots,t_n}$, for $t_1 = t_2 = \cdots = t_n = t$, determine if there is a super $(a, 3)$-EAT labeling, for every $m \geq 2$, $n \geq 2$ and $t \neq 2$.

Open Problem 6.3.11 ([30]) Find, if possible, some structural characteristics of a graph $mS_{t_1,t_2,\ldots,t_n}$ which make a super (a, d)-EAT labeling impossible.

Open Problem 6.3.12 ([29]) For $mK_{3,3}$, $m \geq 3$ odd, determine if there is a super $((33m + 3)/2, 0)$-EAT labeling.

Open Problem 6.3.13 ([84]) For the graph $mK_{s[n]}$, mn odd, $m \geq 3$, $n \geq 1$ and $s \equiv 2, 3 \pmod 4$, $s \geq 6$, determine if there is a super $(2mns + (mn^2s(s-1) + 6)/4, 0)$-EAT labeling.

Open Problem 6.3.14 ([40]) For a forest $F \cong \bigcup_{j=1}^{m} T_j$, $m \geq 2$ even, determine if there is a super (a, d)-EAT labeling with $d \in \{0, 2\}$.

Open Problem 7.6.1 ([35]) Let T be a caterpillar of order n and $||A| - |B|| > 1$, where $\{A, B\}$ is the bipartition of its vertex set. For the graph mT determine if there is a super $(a, 3)$-EAT labeling.

8.2 Conjectures

Conjecture 1.1 ([125]) Every connected graph other than K_2 is antimagic.

Conjecture 1.2 ([125]) Every tree other than K_2 is antimagic.

Conjecture 2.8.1 ([137]) Let G be an r-regular bipartite graph of order $2n$. If $r > n/2$, then G is supermagic except for $n \equiv 0 \pmod 2$ and $d \equiv 1 \pmod 2$.

Conjecture 2.10.1 ([262]) Every regular magic graph is prime-magic.

Conjecture 2.10.2 ([31]) If $n \geq 5$, then the minimum value of index σ which can be assigned to the prime-magic graph $K_{n,n}$ is $\sigma(S_n)$.

Conjecture 3.1.1 ([179]) For each $n \geq 5$ there is a VMT labeling of K_n, for every feasible value of k.

Conjecture 3.1.2 ([189]) There is a VMT labeling for the plane graph R_n, for every $n \geq 3$.

Conjecture 3.1.3 ([178]) All regular graphs other than K_2 and $2K_3$ possess VMT labelings.

Conjecture 3.2.1 ([49]) Super VMT (p, q) graph such that $2q < \sqrt{10p^2 - 6p + 1}$ and minimum degree is 2 exists for all integer values of the magic constant $k = 2q + q(q+1)/p + (p+1)/2$.

Conjecture 3.4.1 ([117]) If k is an odd integer, $k > 1$, then kK_4 admits a super VMT labeling.

Conjecture 4.3.1 ([96]) W_n is EMT if $n \not\equiv 3 \pmod 4$.

Conjecture 4.1.1 ([164, 221]) Every tree is EMT.

Conjecture 4.1.2 ([96]) Every tree is super EMT.

Conjecture 4.2.1 ([27]) There is a super EMT labeling for the generalized Petersen graph $P(n, m)$, for every n odd, $n \geq 9$, and $4 \leq m \leq (n-3)/2$.

Conjecture 4.3.2 ([270]) The ladder $L_n \cong P_n \square P_2$ is super EMT if n is even.

Conjecture 4.6.1 ([119]) All trees are harmonious.

Conjecture 5.1.1 ([32]) If n is odd, $n \geq 7$, then the prism $C_n \square P_2$ is $((n+7)/2, 4)$-VAE.

Conjecture 5.1.2 ([188]) For $n \geq 4$, the antiprism A_n has a $(2n + 5, 6)$-VAE labeling.

Conjecture 5.1.3 ([205]) For n odd, $n \geq 3$, $K_{n,n+2}$ is $((n + 1)(n^2 - 1)/2, n + 1)$-VAE.

Conjecture 5.5.1 ([203]) There is an $(a, 1)$-VAT labeling of generalized Petersen graph $P(n, m)$ for n odd and $1 \leq m < n/2$.

Conjecture 6.3.1 ([270]) If m is even, $m \geq 4$, $n \geq 3$, and $d \in \{0, 2\}$, then the generalized prism $C_m \Box P_n$ has a super (a, d)-EAT labeling.

Conjecture 6.3.2 ([27]) There is a super (a, d)-EAT labeling for the generalized Petersen graph $P(n, m)$, for every n odd, $n \geq 9, d \in \{0, 2\}$, and $3 \leq m \leq (n - 3)/2$.

Conjecture 6.3.3 ([30]) There is a super (a, d)-EAT labeling of the graph $mS_{t_1,t_2,\ldots,t_n}$, for $t_1 = t_2 = \cdots = t_n = t \geq 1$, $d \in \{0, 1, 2\}$ and for every $m \geq 2$ and $n \geq 2$.

Conjecture 6.3.4 ([29]) There is a super $((15m + 5)/2, 2)$-EAT labeling for $mK_{3,3}$, for all m odd.

Glossary of Abbreviations

EAT Edge-antimagic total
EAV Edge-antimagic vertex
EMT Edge-magic total
VAE Vertex-antimagic edge
VAT Vertex-antimagic total
VMT Vertex-magic total

M. Bača et al., *Magic and Antimagic Graphs*, Developments in Mathematics 60,
https://doi.org/10.1007/978-3-030-24582-5

References

1. V.J. Abhyankar and V.N. Bhat-Nayak, Easiest graceful labeling of olive trees, *Bull. Bombay Math. Coll.* **14** (2000), 16–25.
2. A.A. Abiyev, *Sayılı siirli Karelerin Doğal şifresi*, Enderun Ofset Matbaacılık, Ankara, 1996.
3. A.A. Abiyev, A. Baykasoğlu, T. Dereli, İ.H. Filiz and A. Abiyev, Investigation of center mass by magic squares and its possible engineering applications, *Robot. Auton. Syst.* **49** (2004), 219–226.
4. J. Abrham and A. Kotzig, Exponential lower bounds for the number of graceful numbering of snakes, *Congr. Numer.* **72** (1990), 163–174.
5. B.D. Acharya and S.M. Hegde, Strongly indexable graphs, *Discrete Math.* **93** (1991), 275–299.
6. A. Ahmad, K. Ali, M. Bača, P. Kovář and A. Semaničová-Feňovčíková, Vertex-antimagic labelings of regular graphs, *Acta Math. Sin. (Engl. Ser.)* **28**(9) (2012), 1865–1874.
7. A. Ahmad, K. Ali and E.T. Baskoro, On super edge-magic total labelings of a forest of banana trees, *Utilitas Math.* **83** (2010), 323–332.
8. A. Ahmad, A.Q. Baig and M. Imran, On super edge-magicness of graphs, *Utilitas Math.* **89** (2012), 373–380.
9. A. Ahmad, S.C. López, F.A. Muntaner-Batle and M. Rius-Font, Enumerating super edge-magic labelings for the union of nonisomorphic graphs, *Bull. Austral. Math. Soc.* **84**(2) (2011), 310–321.
10. R.E.L. Aldred and B.D. McKay, Graceful and harmonious labellings of trees, *Bull. Inst. Combin. Appl.* **23** (1998), 69–72.
11. R.E.L. Aldred, J. Širáň and M. Širáň, A note on the number of graceful labelings of paths, *Discrete Math.* **261** (2003), 27–30.
12. G. Ali, M. Bača and F. Bashir, On super vertex-antimagic total labelings of disjoint union of paths, *AKCE Int. J. Graphs Combin.* **6**(1) (2009), 11–20.
13. G. Ali, M. Bača, Y. Lin and A. Semaničová-Feňovčíková, Super vertex-antimagic labelings of disconnected graphs, *Discrete Math.* **309** (2009), 6048–6054.
14. K. Ali, M. Hussain and A. Razzaq, Super edge-magic total labeling of a tree, *Utilitas Math.* **91** (2013), 355–364.
15. N. Alon, G. Kaplan, A. Lev, Y. Roditty and R. Yuster, Dense graphs are antimagic, *J. Graph Theory* **47**(4) (2004), 297–309.
16. B. Alspach, D. Dyer and D.L. Kreher, On isomorphic factorizations of circulant graphs, *J. Combin. Des.* **14**(5) (2006), 406–414.
17. W.S. Andrews, *Magic Squares and Cubes*, Dover, New York, 1960.

M. Bača et al., *Magic and Antimagic Graphs*, Developments in Mathematics 60,
https://doi.org/10.1007/978-3-030-24582-5

18. A. Armstrong and D. McQuillan, Vertex-magic total labelings of even complete graphs, *Discrete Math.* **311** (2011), 676–683.
19. S. Arumugam, M. Miller, O. Phanalasy and J. Ryan, Antimagic labeling of generalized pyramid graphs, *Acta Math. Sin. (Engl. Ser.)* **30**(2) (2014), 283–290.
20. S. Arumugam and M. Nalliah, Super (a, d)-edge antimagic total labelings of friendship graphs, *Austral. J. Combin.* **53** (2012), 237–243.
21. M. Bača, On certain properties of magic graphs, *Utilitas Math.* **37** (1990), 259–264.
22. M. Bača, Antimagic labelings of antiprisms, *J. Combin. Math. Combin. Comput.* **35** (2000), 217–224.
23. M. Bača, Consecutive-magic labeling of generalized Petersen graphs, *Utilitas Math.* **58** (2000), 237–241.
24. M. Bača and C. Barrientos, On super edge-antimagic total labelings of mK_n, *Discrete Math.* **308** (2008), 5032–5037.
25. M. Bača and C. Barrientos, Graceful and edge-antimagic labelings, *Ars Combin.* **96** (2010), 505–513.
26. M. Bača, Y. Bashir, M.F. Nadeem and A. Shabbir, On super edge-antimagicness of circulant graphs, *Graphs and Combin.* **31** (2015), 2019–2028.
27. M. Bača, E.T. Baskoro, R. Simanjuntak and K.A. Sugeng, Super edge-antimagic labelings of the generalized Petersen graph $P(n, (n-1)/2)$, *Utilitas Math.* **70** (2006), 119–127.
28. M. Bača, F. Bertault, J.A. MacDougall, M. Miller, R. Simanjuntak and Slamin, Vertex-antimagic total labelings of graphs, *Discuss. Math. Graph Theory* **23** (2003), 67–83.
29. M. Bača and L. Brankovic, Edge-antimagicness for a class of disconnected graphs, *Ars Combin.* **97**A (2010), 145–152.
30. M. Bača, Dafik, M. Miller and J. Ryan, Edge-antimagic total labeling of disjoint union of caterpillars, *J. Combin. Math. Combin. Comput.* **65** (2008), 61–70.
31. M. Bača and I. Holländer, Prime-magic labelings of $K_{n,n}$, *J. Franklin Inst.* **327** (1990), 923–926.
32. M. Bača and I. Holländer, On (a, d)-antimagic prisms, *Ars Combin.* **48** (1998), 297–306.
33. M. Bača, I. Holländer and K.W. Lih, Two classes of super-magic quartic graphs, *J. Combin. Math. Combin. Comput.* **23** (1997), 113–120.
34. M. Bača, P. Kovář, A. Semaničová-Feňovčíková and M.K. Shafiq, On super $(a, 1)$-edge-antimagic total labelings of regular graphs, *Discrete Math.* **310** (2010), 1408–1412.
35. M. Bača, M. Lascsáková and A. Semaničová, On connection between α-labelings and edge-antimagic labelings of disconnected graphs, *Ars Combin.* **106** (2012), 321–336.
36. M. Bača, Y. Lin, M. Miller and R. Simanjuntak, New constructions of magic and antimagic graph labelings, *Utilitas Math.* **60** (2001), 229–239.
37. M. Bača, Y. Lin, M. Miller and M.Z. Youssef, Edge-antimagic graphs, *Discrete Math.* **307** (2007), 1232–1244.
38. M. Bača, Y. Lin and F.A. Muntaner-Batle, Super edge-antimagic labelings of the path-like trees, *Utilitas Math.* **73** (2007), 117–128.
39. M. Bača, Y. Lin and F.A. Muntaner-Batle, Strong labelings of linear forests, *Acta Math. Sin. (Engl. Ser.)* **25**(12) (2009), 1951–1964.
40. M. Bača, Y. Lin and F.A. Muntaner-Batle, Edge-antimagic labelings of forests, *Utilitas Math.* **81** (2010), 31–40.
41. M. Bača, Y. Lin and A. Semaničová-Feňovčiková, Note on super antimagicness of disconnected graphs, *AKCE Int. J. Graphs Combin.* **6**(1) (2009), 47–55.
42. M. Bača and M. Miller, *Super Edge-Antimagic Graphs: A Wealth of Problems and Some Solutions*, Brown Walker Press, Boca Raton, Florida, 2008.
43. M. Bača, M. Miller, O. Phanalasy and A. Semaničová-Feňovčíková, Constructions of antimagic labelings for some families of regular graphs, *J. Algorithm. Comput.* **44** (2013), 1–7.
44. M. Bača, M. Miller and Slamin, Vertex-magic total labelings of generalized Petersen graphs, *Intern. J. Comput. Math.* **79** (2002), 1259–1263.

45. M. Bača, F.A. Muntaner-Batle, A. Semaničová-Feňovčíková and M.K. Shafiq, On super $(a, 2)$-edge-antimagic total labeling of disconnected graphs, *Ars Combin.* **113** (2014), 129–137.
46. M. Bača, A. Semaničová-Feňovčíková and M.K. Shafiq, A method to generate large classes of edge-antimagic trees, *Utilitas Math.* **86** (2011), 33–43.
47. M. Bača and A. Shabbir, Total labelings of toroidal polyhexes, *Sci. Int.* **24**(3) (2012), 239–241.
48. M. Bača and M.Z. Youssef, Further results on antimagic graph labelings, *Austral. J. Combin.* **38** (2007), 163–172.
49. C. Balbuena, E. Barker, K.C. Das, Y. Lin, M. Miller, J. Ryan, Slamin, K. Sugeng and M. Tkáč, On the degrees of a strongly vertex-magic graph, *Discrete Math.* **306** (2006), 539–551.
50. C. Barrientos, Graceful labelings of cyclic snakes, *Ars Combin.* **60** (2001), 85–96.
51. C. Barrientos, Graceful labelings of chain and corona graphs, *Bull. Inst. Combin. Appl.* **34** (2002), 17–26.
52. C. Barrientos, Difference Vertex Labelings, *Ph.D. Thesis*, Universitat Politécnica de Catalunya, Barcelona, 2004.
53. C. Barrientos, The gracefulness of unions of cycles and complete bipartite graphs, *J. Combin. Math. Combin. Comput.* **52** (2005), 69–78.
54. E.T. Baskoro and A.A.G. Ngurah, On super edge-magic total labeling of nP_3, *Bull. ICA* **37** (2003), 82–87.
55. A.F. Beardon, The maximum degree in a vertex-magic graph, *Austral. J. Combin.* **30** (2004), 113–116.
56. O. Berkman, M. Parnas and Y. Roditty, All cycles are edge-magic, *Ars Combin.* **59** (2001), 145–151.
57. E.R. Berlekamp, J.H. Conway and R.K. Guy, *Winning Ways*, Academic Press, New York, 1982.
58. J.C. Bermond, Graceful graphs, radio antennas and French windmills, *Graph T. Combin.*, Pitman, London (1979), 18–37.
59. J.C. Bermond, F. Comellas and D.F. Hsu, Distributed loop computer networks: A survey, *J. Parallel Distrib. Comput.* **24** (1995), 2–10.
60. J.C. Bermond and D. Sotteau, Graph decompositions and G-design, Proc. 5th British Combin. Conf., 1975, *Congr. Numer.* **15** (1976), 53–72.
61. Ľ. Bezegová, Balanced degree-magic complements of bipartite graphs, *Discrete Math.* **313**(19) (2013), 1918–1923.
62. Ľ. Bezegová and J. Ivančo, An extension of regular supermagic graphs, *Discrete Math.* **310** (2010), 3571–3578.
63. Ľ. Bezegová and J. Ivančo, On conservative and supermagic graphs, *Discrete Math.* **311** (2011), 2428–2436.
64. Ľ. Bezegová and J. Ivančo, A characterization of complete tripartite degree-magic graphs, *Discuss. Math. Graph Theory* **32**(2) (2012), 243–253.
65. V. Bhat-Nayak and U. Deshmukh, New families of graceful banana trees, *Proc. Indian Acad. Math. Sci.* **106** (1996), 201–216.
66. G.S. Bloom and S.W. Golomb, Applications of numbered undirected graphs, *Proc. IEEE* **65** (1977), 562–570.
67. G.S. Bloom and S.W. Golomb, Numbered complete graphs, unusual rules, and assorted applications, *In: Theory and Applications of Graphs, Lecture Notes in Math.* **642** (1978), 53–65.
68. R. Bodendiek and G. Walther, *Aritmethisch antimagische Graphen*, In: K. Wagner and R. Bodendiek, eds., Graphentheorie III, BI-Wiss. Verl., Mannheim, 1993.
69. R. Bodendiek and G. Walther, On number theoretical methods in graph labelings, *Res. Exp. Math.* **21** (1995), 3–25.
70. R. Bodendiek and G. Walther, (a, d)-antimagic parachutes, *Ars Combin.* **42** (1996), 129–149.
71. R. Bodendiek and G. Walther, (a, d)-antimagic parachutes II, *Ars Combin.* **46** (1997), 33–63.

72. R. Bodendiek and G. Walther, On arithmetic antimagic edge labelings of graphs, *Mitt. Math. Ges. Hamburg* **17** (1998), 85–99.
73. F. Boesch and R. Tindell, Circulants and their connectivities, *J. Graph Theory* **8** (1984), 487–499.
74. J. Bondy and M. Kouider, Hamilton cycles in regular 2-connected graphs, *J. Combin. Theory, Ser. B* **44** (1988), 177–186.
75. M. Burzio and G. Ferrarese, The subdivision graph of a graceful tree is a graceful tree, *Discrete Math.* **181** (1998), 275–281.
76. I. Cahit, Cordial graphs: a weaker version of graceful and harmonious graphs, *Ars Combin.* **23** (1987), 201–208.
77. I. Cahit, On cordial and 3-equitable labellings of graphs, *Utilitas Math.* **37** (1990), 189–198.
78. G.G. Cash, Simple means of computing the Kekulé structure count for toroidal polyhex fullerenes, *J. Chem. Inf. Comput. Sci.* **38** (1998), 58–61.
79. R. Cattell, Vertex magic total labellings of complete multipartite graphs, *J. Combin. Math. Combin. Comput.* **55** (2005), 187–197.
80. C.M. Cavalier, Graceful Labelings, *Ph.D. Thesis*, University of South Carolina, Columbia, 2009.
81. W.C. Chen, H.I. Lu and Y.N. Yeh, Operation of interlaced trees and graceful trees, *Southeast Asian Bulletin of Math.* **4** (1997), 337–348.
82. J. Cohen, P. Fraigniaud and C. Gavoille, Recognizing Knödel graphs, *Discrete Math.* **250** (2002), 41–62.
83. Dafik, M. Miller, J. Ryan and M. Bača, Antimagic labeling of the union of two stars, *Austral. J. Combin.* **42** (2008), 35–44.
84. Dafik, M. Miller, J. Ryan and M. Bača, On antimagic labelings of disjoint union of complete s-partite graphs, *J. Combin. Math. Combin. Comput.* **65** (2008), 41–49.
85. Dafik, M. Miller, J. Ryan and M. Bača, On super (a, d)-edge antimagic total labeling of disconnected graphs, *Discrete Math.* **309** (2009), 4909–4915.
86. Dafik, M. Miller, J. Ryan and M. Bača, Super edge-antimagic total labelings of $mK_{n,n,n}$, *Ars Combin.* **101** (2011), 97–107.
87. U. Derings and B. Hünten, Magic graphs - a new characterization, *Report No. 83265-OR, Universität Bonn*, 1983.
88. M. Deza, P.W. Fowler, A. Rassat and K.M. Rogers, Fullerenes as tilings of surfaces, *J. Chem. Inf. Comput. Sci.* **40** (2000), 550–558.
89. M. Doob, Generalisations of magic graphs, *J. Combin. Theory, Ser. B* **17** (1974), 205–217.
90. M. Doob, Characterizations of regular magic graphs, *J. Combin. Theory, Ser. B* **25** (1978), 94–104.
91. S. Drajnová, J. Ivančo and A. Semaničová, Numbers of edges in supermagic graphs, *J. Graph Theory* **52** (2006), 15–26.
92. A. Dürer, *Melencolia I* – engraving, 1514, http://en.wikipedia.org/wiki/Melencolia I.
93. M. Edwards and L. Howard, A survey of graceful trees, *Atlantic Electronic J. of Math.* **1** (2006), 5–30.
94. A. Elumalai and G. Sethuraman, Gracefulness of union of cycle with parallel chords and complete bipartite graphs or paths, *J. Discrete Math. Sci. Cryptogr.* **12** (2009), 245–252.
95. A. Elumalai and G. Sethuraman, Gracefulness of a cycle with parallel chords and parallel P_k-chords of different lenghts, *Ars Combin.* **104** (2012), 143–148.
96. H. Enomoto, A.S. Lladó, T. Nakamigawa and G. Ringel, Super edge-magic graphs, *SUT J. Math.* **34** (1998), 105–109.
97. G. Fertin and A. Raspaud, A survey on Knödel graphs, *Discrete Appl. Math.* **137** (2004), 173–195.
98. R.M. Figueroa-Centeno, R. Ichishima and F.A. Muntaner-Batle, The place of super edge-magic labelings among other classes of labelings, *Discrete Math.* **231** (2001), 153–168.
99. R.M. Figueroa-Centeno, R. Ichishima and F.A. Muntaner-Batle, On super edge-magic graphs, *Ars Combin.* **64** (2002), 81–95.

100. R.M. Figueroa-Centeno, R. Ichishima and F.A. Muntaner-Batle, On edge-magic labelings of certain disjoint unions of graphs, *Austral. J. Combin.* **32** (2005), 225–242.
101. R.M. Figueroa-Centeno, R. Ichishima and F.A. Muntaner-Batle, A magical approach to some labeling conjectures, *Discuss. Math. Graph Theory* **31**(1) (2011), 79–113.
102. R.M. Figueroa-Centeno, R. Ichishima, F.A. Muntaner-Batle and M. Rius-Font, Labeling generating matrices, *J. Combin. Math. Combin. Comput.* **67** (2008), 189–216.
103. P. Fraigniaud and J.G. Peters, Minimum linear gossip graphs and maximal linear (Δ, k)-gossip graphs, *Networks* **38**(3) (2001), 150–162.
104. D. Fronček, P. Kovář and T. Kovářová, Vertex magic total labeling of products of cycles, *Australas. J. Combin.* **33** (2005), 169–181.
105. R. Frucht and F. Harary, On the corona of two graphs, *Aequationes Math.* **4** (1970), 322–325.
106. Y. Fukuchi, A recursive theorem for super edge-magic labelings of trees, *SUT J. Math.* **36** (2000), 279–285.
107. Y. Fukuchi, Edge-magic labelings of generalized Petersen graph $P(n, 2)$, *Ars Combin.* **59** (2001), 253–257.
108. Y. Fukuchi, Edge-magic labelings of wheel graphs, *Tokyo J. Math.* **24** (2001), 153–167.
109. J.A. Gallian, A Dynamic Survey of Graph Labeling, *Electron. J. Combin.*, 2017, #DS6.
110. M. Gardner, Mathematical games: The graceful graphs of Solomon Golomb, or how to number a graph parsimoniously, *Scientific American* **226**(3) (1972), 108–112; **226**(4) (1972), 104; **226**(6) (1972), 118.
111. A. Gaudí, *Sagrada Família*, http://www.gaudiallgaudi.com/AA012b.htm#La_facana_de_la_Passio
112. R.D. Godbold and P.J. Slater, All cycles are edge-magic, *Bull. Inst. Combin. Appl.* **22** (1998), 93–97.
113. J.W. von Goethe, *Faust*, http://en.wikipedia.org/wiki/Goethe's_Faust
114. S.W. Golomb, How to number a graph, Graph Theory and Computing, *R.C. Read ed., Academic Press*, New York (1972), 23–37.
115. J. Gómez, Solution of the conjecture: If $n \equiv 0 \pmod 4$, $n > 4$, then K_n has a super vertex-magic total labeling, *Discrete Math.* **307** (2007), 2525–2534.
116. J. Gómez, Two new methods to obtain super vertex-magic total labelings of graphs, *Discrete Math.* **308** (2008), 3361–3372.
117. J. Gómez and P. Kovář, On super vertex-magic total labeling of the disjoint union of k copies of K_n, *Ars Combin.* **113** (2014), 175–192.
118. T. Grace, On sequential labelings of graphs, *J. Graph Theory* **7** (1983), 195–201.
119. R.L. Graham and N.J.A. Sloane, On additive bases and harmonious graphs, *SIAM J. Alg. Discrete Math.* **1** (1980), 382–404.
120. I.D. Gray, Vertex-magic total labelings of regular graphs, *SIAM J. Discrete Math.* **21** (2007), 170–177.
121. I.D. Gray, J.A. MacDougall, J.P. McSorley and W.D. Wallis, Vertex-magic labelings of trees and forests, *Discrete Math.* **261** (2003), 285–298.
122. I.D. Gray, J.A. MacDougall, R.J. Simpson and W.D. Wallis, Vertex-magic total labelings of complete bipartite graphs, *Ars Combin.* **69** (2003), 117–127.
123. I.D. Gray, J.A. MacDougall and W.D. Wallis, On vertex-magic labeling of complete graphs, *Bull. Inst. Combin. Appl.* **38** (2003), 42–44.
124. N. Hartsfield and G. Ringel, Supermagic and antimagic graphs, *J. Recreat. Math.* **21**(2) (1989), 107–115.
125. N. Hartsfield and G. Ringel, *Pearls in Graph Theory: A Comprehensive Introduction*, Academic Press, Boston – San Diego – New York – London, 1990.
126. S.M. Hegde, On indexable graphs, *J. Combin. Inform. System Sci.* **17** (1992), 316–331.
127. C. Heuberger, On planarity and colorability of circulant graphs, *Discrete Math.* **268** (2003), 153–169.
128. Y.S. Ho and S.M. Lee, Some initial results on the supermagicness of regular complete k-partite graphs, *J. Combin. Math. Combin. Comput.* **39** (2001), 3–17.

129. P. Hrnčiar and A. Haviar, All trees of diameter five are graceful, *Discrete Math.* **233** (2001), 133–150.
130. C. Huang, A. Kotzig, and A. Rosa, Further results on tree labellings, *Utilitas Math.* **21** (1982), 31–48.
131. M. Hussain, K. Ali, M.T. Rahim and E.T. Baskoro, On (a, d)-vertex-antimagic total labeling of Harary graphs, *Utilitas Math.* **83** (2010), 73–80.
132. M. Hussain, E.T. Baskoro and K. Ali, On super antimagic total labeling of harary graph, *Ars Combin.* **104** (2012), 225–233.
133. M. Hussain, E.T. Baskoro and Slamin, On super edge-magic total labeling of banana trees, *Utilitas Math.* **79** (2009), 243–251.
134. R. Ichishima, S.C. López, F.A. Muntaner-Batle and M. Rius-Font, The power of digraph products applied to labelings, *Discrete Math.* **312**(2) (2012), 221–228.
135. R. Ichishima, F.A. Muntaner-Batle and A. Oshima, Enlarging the classes of super edge-magic 2-regular graphs, *AKCE Int. J. Graphs Combin.* **10**(2) (2013), 129–146.
136. J. Ivančo, On supermagic regular graphs, *Math. Bohemica* **125** (2000), 99–114.
137. J. Ivančo, Magic and supermagic dense bipartite graphs, *Discuss. Math. Graph Theory* **27** (2007), 583–591.
138. J. Ivančo, A construction of supermagic graphs, *AKCE Int. J. Graphs Combin.* **6**(1) (2009), 91–102.
139. J. Ivančo, P. Kovář and A. Semaničová, On the existence of regular supermagic graphs, *J. Combin. Math. Combin. Comput.* **71** (2009), 49–64.
140. J. Ivančo, Z. Lastivková and A. Semaničová, On magic and supermagic line graphs, *Math. Bohemica* **129** (2004), 33–42.
141. J. Ivančo and I. Lučkaničová, On edge-magic disconnected graphs, *SUT J. Math.* **38**(2) (2002), 175–184.
142. J. Ivančo and T. Polláková, Supermagic graphs having a saturated vertex, *Discuss. Math. Graph Theory* **34** (2014), 75–84.
143. J. Ivančo and A. Semaničová, Some constructions of supermagic graphs using antimagic graphs, *SUT J. Math.* **42**(2) (2006), 177–186.
144. J. Ivančo and A. Semaničová, Some constructions of supermagic non-regular graphs, *Australas. J. Combin.* **38** (2007), 127–139.
145. M. Javaid, A.A. Bhatti, M. Hussain and K. Ali, Super edge-magic total labeling on forest of extended w-trees, *Utilitas Math.* **91** (2013), 155–162.
146. M. Javaid, M. Hussain, K. Ali and K.H. Dar, Super edge-magic total labeling on w-trees, *Utilitas Math.* **86** (2011), 183–191.
147. M. Javaid, M. Hussain, K. Ali and H. Shaker, On super edge-magic total labeling on subdivision of trees, *Utilitas Math.* **89** (2012), 169–177.
148. J. Jeba Jesintha and G. Sethuraman, A new class of graceful rooted trees, *J. Discrete Math. Sci. Cryptogr.* **11** (2008), 421–435.
149. J. Jeba Jesintha and G. Sethuraman, Generation of graceful trees, *Internat. J. Algor. Comput. Math.* **2** (2009), 33–38.
150. J. Jeba Jesintha and G. Sethuraman, All arbitrarily fixed generalized banana trees are graceful, *Math. Comput. Sci.* **5** (2011), 51–62.
151. R.H. Jeurissen, *Magic graphs, a characterization*, Mathematisch Instituut Universiteit Toernooiveld, 6525 ED Nijmegen (1982), The Netherlands.
152. R.H. Jeurissen, Magic graphs, a characterization, *Europ. J. Combin.* **9** (1988), 363–368.
153. S. Jezný and M. Trenkler, Characterization of magic graphs, *Czechoslovak Math. J.* **33** (1983), 435–438.
154. Jirimutu and J. Wang, On (a, d)-antimagic labelings of generalized Petersen graphs $P(n, 2)$, *Ars Combin.* **90** (2009), 161–174.
155. E.C. Kirby, R.B. Mallion and P. Pollak, Toridal polyhexes, *J. Chem. Soc. Faraday Trans.* **89**(12) (1993), 1945–1953.
156. E.C. Kirby and P. Pollak, How to enumerate the connectional isomers of a toridal polyhex fullerene, *J. Chem. Inf. Comput. Sci.* **38** (1998), 66–70.

157. D.J. Klein, Elemental benzenoids, *J. Chem. Inf. Comput. Sci.* **34** (1994), 453–459.
158. D.J. Klein and H. Zhu, Resonance in elemental benzenoids, *Discrete Appl. Math.* **67** (1996), 157–173.
159. W. Knödel, New gossips and telephones, *Discrete Math.* **13**(1) (1975), 95.
160. K.M. Koh, D.G. Rogers and T. Tan, On graceful trees, *Nanta Math.* **10** (1977), 207–211.
161. K.M. Koh, D.G. Rogers and T. Tan, Two theorems on graceful trees, *Discrete Math.* **25** (1979), 141–148.
162. K.M. Koh, D.G. Rogers and T. Tan, Product of graceful trees, *Discrete Math.* **31** (1980), 279–292.
163. A. Kotzig and A. Rosa, Magic valuations of finite graphs, *Canad. Math. Bull.* **13** (1970), 451–461.
164. A. Kotzig and A. Rosa, Magic valuations of complete graphs, *Publ. CRM* **175** (1972).
165. P. Kovář, Unified approach to magic labeling of copies of regular graphs, *Congr. Numer.* **168** (2004), 197–206.
166. P. Kovář, Vertex magic total labeling of products of regular VMT graphs and regular supermagic graphs, *J. Comb. Math. Comb. Comput.* **54** (2005), 21–31.
167. P. Kovář, Magic labelings of regular graphs, *AKCE Int. J. Graphs Combin.* **4** (2007), 261–275.
168. H.K. Krishnappa, K. Kopthapalli and V.C. Venkaiah, Vertex magic total of complete graphs, *AKCE Int. J. Graphs Combin.* **6** (2009), 143–154.
169. M.J. Lee, On super $(a, 1)$-edge-antimagic total labelings of grids and crowns, *Ars Combin.* **104** (2012), 97–105.
170. M.J. Lee, On super $(a, 1)$-edge-antimagic total labelings of Cartesian product graphs, *J. Discrete Math. Sciences and Cryptography* **16**(2–3) (2013), 117–124.
171. M.J. Lee, C. Lin and W.H. Tsai, On antimagic labeling for power of cycles, *Ars Combin.* **98** (2011), 161–165.
172. S.M. Lee and Q.X. Shan, All trees with at most 17 vertices are super edge-magic, *16th MCCCC Conference*, Carbondale, University Southern Illinois, 2002.
173. M.J. Lee, W.H. Tsai and C. Lee, On super $(a, 1)$-edge-antimagic total labelings of subdivision of stars, *Utilitas Math.* **88** (2012), 355–365.
174. K.W. Lih, Bao Qi-Shou and his polyhedral Hun Yuan Tu, Philosophy and Conceptual History of Science in Taiwan, *Kluwer Academic Publishers*, 1993, 209–220.
175. Y. Lin and M. Miller, Vertex-magic total labelings of complete graphs, *Bull. Inst. Combin. Appl.* **33** (2001), 68–76.
176. S.C. López and F.A. Muntaner-Batle, *Graceful, Harmonious and Magic Type Labelings: Relations and Techniques*, Springer, 2017.
177. S.C. López, F.A. Muntaner-Batle and M. Rius-Font, On super edge-magic decomposable graphs, *Indian J. Pure Appl. Math.* **43** (2012), 455–473.
178. J.A. MacDougall, *Vertex-magic labeling of regular graphs*, Lecture, DIMACS Connect Institute, July, 2002.
179. J.A. MacDougall, M. Miller, Slamin and W.D. Wallis, Vertex-magic total labelings of graphs, *Utilitas Math.* **61** (2002), 3–21.
180. J.A. MacDougall, M. Miller and K.A. Sugeng, Super vertex-magic total labelings of graphs, *Proc. of AWOCA* (2004), 119–122.
181. J.A. MacDougall, M. Miller and W.D. Wallis, Vertex-magic total labelings of wheels and related graphs, *Utilitas Math.* **62** (2002), 175–183.
182. A.M. Marr and W.D. Wallis, *Magic Graphs*, Birkhäuser, New York, 2013.
183. D. McQuillan, Vertex-magic cubic graphs, *J. Combin. Math. Combin. Comput.* **48** (2004), 103–106.
184. D. McQuillan, A technique for constructing magic labelings of 2-regular graphs, *J. Combin. Math. Combin. Comput.* **75** (2010), 129–135.
185. D. McQuillan and J. McQuillan, Magic labelings of triangles, *Discrete Math.* **309** (2009), 2755–2762.
186. D. McQuillan and K. Smith, Vertex-magic total labelings of odd complete graphs, *Discrete Math.* **305** (2005), 240–249.

187. M. Miller and M. Bača, Antimagic valuations of generalized Petersen graphs, *Austral. J. Combin.* **22** (2000), 135–139.
188. M. Miller, M. Bača and Y. Lin, On two conjectures concerning (a, d)-antimagic labelings of antiprisms, *J. Combin. Math. Combin. Comput.* **37** (2001), 251–254.
189. M. Miller, M. Bača and J.A. MacDougall, Vertex-magic total labeling of generalized Petersen graphs and convex polytopes, *J. Combin. Math. Combin. Comput.* **59** (2006), 89–99.
190. M. Miller, J.A. MacDougall, Slamin and W.D. Wallis, Problems in total graph labelings, *Proc. of AWOCA*, Perth, 1999, 19–25.
191. M. Miller, O. Phanalasy and J. Ryan, All graphs have antimagic total labelings, *Electron. Notes Discrete Math.* **38** (2011), 645–650.
192. M. Miller, O. Phanalasy, J. Ryan and L. Rylands, Antimagicness of some families of generalized graphs, *Austral. J. Combin.* **53** (2012), 179–190.
193. M. Miller, O. Phanalasy, J. Ryan and L. Rylands, Sparse graphs with vertex antimagic edge labelings, *AKCE Int. J. Graphs Comb.* **10**(2) (2013), 193–198.
194. D. Mishra and P. Panigrahi, Graceful lobsters obtained by component moving of diameter four trees, *Ars Combin.* **81** (2006), 129–147.
195. D. Mishra and P. Panigrahi, Some graceful lobsters with both odd and even degree vertices on the central path, *Utilitas Math.* **74** (2007), 155–177.
196. D. Mishra and P. Panigrahi, Some graceful lobsters with all three types of branches incident on the vertices of the central path, *Comput. Math. with Appl.* **56** (2008), 1382–1394.
197. D. Morgan, All lobsters with perfect matchings are graceful, *Technical Report, University of Alberta*, **TR05-01**, 2005.
198. F.A. Muntaner-Batle, Magic Graphs, *Ph.D. Thesis*, Universitat Politécnica de Catalunya, Barcelona, 2001.
199. M. Murugan and G. Arumugan, Are banana trees graceful?, *Math. Ed. (Siwan)* **35** (2001), 18–20.
200. M. Nalliah and S. Arumugam, Super (a, d)-edge-antimagic total labelings of generalized friendship graphs, *J. Combin. Math. Combin. Comput.* **84** (2013), 81–90.
201. H.K. Ng, Gracefulness of a class of lobsters, *Notices AMS* **7** (1986), abstract no. 825-05-294.
202. A.A.G. Ngurah and E.T. Baskoro, On magic and antimagic total labeling of generalized Petersen graph, *Utilitas Math.* **63** (2003), 97–107.
203. A.A.G. Ngurah, E.T. Baskoro and R. Simanjuntak, On antimagic total labelings of generalized Petersen graph, *J. Combin. Math. Combin. Comput.* **55** (2005), 570–70.
204. A.A.G. Ngurah, E.T. Baskoro and R. Simanjuntak, On new families of (super) edge-magic graphs, *Utilitas Math.* **74** (2007), 111–120.
205. T. Nicholas, S. Somasundaram and V. Vilfred, On (a, d)-antimagic special trees, unicyclic graphs and complete bipartite graphs, *Ars Combin.* **70** (2004), 207–220.
206. A. Parestu, D.R. Silaban and K.A. Sugeng, Pelabelan simpul-ajaib total dari gabungan graf matahari, *Prosising Seminar Nasional Matematika Universitas Parahyangan Bandung* **3** (2008), 407–414.
207. A. Parestu, D.R. Silaban and K.A. Sugeng, Vertex-antimagic total labeling of the union of suns, *J. Combin. Math. Combin. Comput.* **71** (2009), 179–188.
208. O. Phanalasy, M. Miller, C.S. Iliopoulos, S.P. Pissis and E. Vaezpour, Construction of antimagic labeling for the Cartesian product of regular graphs, *Math. Comput. Sci.* **5**(1) (2011), 81–87.
209. O. Phanalasy, M. Miller, L. Rylands and P. Lieby, On a relationship between completely separating systems and antimagic labeling of regular graphs, *Lecture Notes in Comput. Sci.* **6460** (2011), 238–241.
210. N.C.K. Phillips, R.S. Rees and W.D. Wallis, Edge-magic total labelings of wheels, *Bull. Inst. Combin. Appl.* **31** (2001), 21–30.
211. K. Pinn and C. Wieczerkowski, Number of magic squares from parallel tempering Monte Carlo, *Int. J. Mod. Phy. C (IJMPC)* **9** (1998), 541–546.
212. S. Poljak and M. Sûra, An algorithm for graceful labeling of a class of symmetrical trees, *Ars Combin.* **14** (1982), 57–66.

213. P.R.L. Pushpam and A. Saibulla, On super (a, d)-edge antimagic total labeling of certain families of graphs, *Discuss. Math. Graph Theory* **32**(3) (2012), 535–543.
214. P.R.L. Pushpam and A. Saibulla, Super (a, d)-edge antimagic total labeling of some classes of graphs, *SUT J. Math.* **48**(1) (2012), 1–12.
215. M.T. Rahim, K. Ali and I. Javaid, On antimagic total labeling of some families of graphs, *Ars Combin.* **95** (2010), 225–234.
216. M.T. Rahim and Slamin, Most wheel related graphs are not vertex magic, *Utilitas Math.* **77** (2008), 193–199.
217. M.T. Rahim and Slamin, Vertex-magic total labeling of the union of suns, *Ars Combin.* **103** (2012), 305–310.
218. M.T. Rahim, I. Tomescu and Slamin, On vertex-magic total labeling of some wheel related graphs, *Utilitas Math.* **73** (2007), 97–104.
219. S. Rahmawati, K.A. Sugeng and D.R. Silaban, Construction of $(a, 2)$-edge antimagic vertex graph using adjacency matrix, *Austral. J. Combin.* **56** (2013), 257–272.
220. S. Rahmawati, K.A. Sugeng, D.R. Silaban, M. Miller and M. Bača, Construction of new larger (a, d)-edge antimagic vertex graphs by using adjacency matrices, *Austral. J. Combin.* **56** (2013), 257–272.
221. G. Ringel and A.S. Lladó, Another tree conjecture, *Bull. Inst. Combin. Appl.* **18** (1996), 83–85.
222. R. Robinson and N. Wormald, Almost all regular graphs are hamiltonian, *Random Struct. Algor.* **5** (1994), 363–374.
223. Y. Roditty and T. Bachar, A note on edge-magic cycles, *Bull. Inst. Combin. Appl.* **29** (2000), 94–96.
224. A. Rosa, On certain valuations of the vertices of a graph, *Theory of Graphs, Internat. Symposium, Rome, July 1966, Gordon and Breach, N.Y. and Dunod Paris* (1967), 349–355.
225. A. Rosa and J. Širáň, Bipartite labelings of trees and the gracesize, *J. Graph Theory* **19** (1995), 201–215.
226. L. Rylands, O. Phanalasy, J. Ryan and M. Miller, Construction for antimagic generalized web graphs, *AKCE Int. J. Graphs Comb.* **8**(2) (2011), 141–149.
227. A.N.M. Salman, A.A.G. Ngurah and N. Izzati, On (super) edge-magic total labelings of a subdivision of a star S_n, *Utilitas Math.* **81** (2010), 275–284.
228. Ľ. Šándorová and M. Trenkler, On characterization of magic graphs, *Colloquia Math. Societatis J. Bolyai* **52** Combinatorics, North-Holland, Amsterdam, 1988, 447–452.
229. T.G. Schmalz, W.A. Seitz, D.J. Klein and G.E. Hite, Elemental carbon cages, *J. Am. Chem. Soc.* **110** (1988), 1113–1127.
230. J. Sedláček, Problem 27, *In: Theory and Its Appl., Proc. Symp. Smolenice*, (1963), 163–164.
231. J. Sedláček, On magic graphs, *Math. Slovaca* **26** (1976), 329–335.
232. C. Sekar, Studies in Graph Theory, *Ph.D. Thesis*, Madurai Kamaraj University, Madurai, 2002.
233. A. Semaničová, Graph Labellings, *Ph.D. Thesis*, P.J. Šafárik University in Košice, Košice, 2006.
234. A. Semaničová, On magic and supermagic circulant graphs, *Discrete Math.* **306** (2006), 2263–2269.
235. A. Semaničová, Magic graphs having a saturated vertex, *Tatra Mt. Math. Publ.* **36** (2007), 121–128.
236. G. Sethuraman and R. Dhavamani, Graceful numbering of an edge-gluing of shell graphs, *Discrete Math.* **218** (2000), 283–287.
237. G. Sethuraman and A. Elumalai, On graceful graphs: Pendant edge extensions of a family of complete bipartite and complete tripartite graphs, *Indian J. Pure Appl. Math.* **32** (2001), 1283–1296.
238. G. Sethuraman and A. Elumalai, Gracefulness of a cycle with parallel P_k-chords, *Australas. J. Combin.* **32** (2005), 205–211.
239. G. Sethuraman and A. Elumalai, Packing of any set of graphs into a graceful, harmonious, elegant graph, *Ars Combin.* **76** (2005), 297–301.

240. G. Sethuraman and J. Jesintha, A new class of graceful lobsters, *J. Combin. Math. Combin. Comput.* **67** (2008), 99–109.
241. G. Sethuraman and J. Jesintha, A new class of graceful rooted trees, *J. Disc. Math. Sci. Crypt.* **11** (2008), 421–435.
242. G. Sethuraman and J. Jesintha, Gracefulness of a family of rooted trees, *Far East J. Appl. Math.* **30** (2008), 143–159.
243. G. Sethuraman and J. Jesintha, All banana trees are graceful, *Advances Appl. Disc. Math.* **4** (2009), 53–64.
244. G. Sethuraman and S.P.M. Kishore, On graceful graphs: Union of n copies of edge deleted subgraphs of K_4, *Indian J. Pure Appl. Math.* **30** (1999), 801–808.
245. G. Sethuraman and K. Sankar, On graceful and cordial labeling of shell graphs, *Ars Combin.* **108** (2013), 515–532.
246. G. Sethuraman and P. Selvaraju, Gracefulness of arbitrary supersubdivisions of graphs, *Indian J. Pure Appl. Math.* **32** (2001), 1059–1064.
247. G. Sethuraman and P. Selvaraju, On graceful graphs: one vertex unions of nonisomorphic complete bipartite graphs, *Indian J. Pure Appl. Math.* **32** (2001), 975–980.
248. G. Sethuraman and S. Venkatesh, Decomposition of complete graphs and complete bipartite graphs into α-labelled trees, *Ars Combin.* **93** (2009), 371–385.
249. H.H. Shen, C. Liu, M.M. Shen and W.M. Zheng, An algorithm for describing object-oriented software architecture using graph, *Proc. of Technology of Object-Oriented Languages and Systems*, TOOLS 31 (1999), 225–231.
250. W.C. Shiu, P.C.B. Lam and H.L. Cheng, Supermagic labeling of an s-duplicate of $K_{n,n}$, *Proc. of the Thirty-first Southeastern International Conference on Combinatorics, Graph Theory and Computing* **146** (2000), 119–124.
251. W.C. Shiu, P.C.B. Lam and S.M. Lee, On construction of supermagic graphs, *J. Combin. Math. Combin. Comput.* **42** (2002), 147–160.
252. W. Sierpiński, *Teoria Liczb*, Państwowe Wydawnictwo Naukowe, Warszawa, 1959.
253. D.R. Silaban, A. Parestu, B.N. Herawati, K.A. Sugeng and Slamin, Vertex-magic total labelings of unions of generalized Petersen graphs and union of special circulant graphs, *J. Combin. Math. Combin. Comput.* **71** (2009), 201–207.
254. D.R. Silaban and K.A. Sugeng, Edge antimagic total labeling on paths and unicycles, *J. Combin. Math. Combin. Comput.* **65** (2008), 127–132.
255. R. Simanjuntak and M. Miller, Survey of (a, d)-antimagic graph labelings, *MIHMI* **6** (2000), 179–184.
256. R. Simanjuntak, M. Miller and F. Bertault, Two new (a, d)-antimagic graph labelings, *Proc. of the Eleventh Australasian Workshop on Combinatorial Algorithms* (2000), 179–189.
257. Slamin, M. Bača, Y. Lin, M. Miller and R. Simanjuntak, Edge-magic total labelings of wheels, fans and friendship graphs, *Bull. Inst. Combin. Appl.* **35** (2002), 89–98.
258. Slamin and M. Miller, On two conjectures concerning vertex-magic total labelings of generalized Petersen graphs, *Bull. Inst. Combin. Appl.* **32** (2001), 9–16.
259. Slamin, A.C. Prihandoko, T.B. Setiawan, V. Rosita and B. Shaleh, Vertex-magic total labelings of disconnected graphs, *J. Prime Resaerch in Math.* **2** (2006), 147–156.
260. H. Snevily, New families of graphs that have α-labelings, *Discrete Math.* **170** (1997), 185–194.
261. R.A. Stanton and C.R. Zarnke, Labelling of balanced trees, *Proc. of the Fourth Southeastern Conference on Combinatorics, Graph Theory and Computing* (Boca Raton), 1973, 479–495.
262. B.M. Stewart, Magic graphs, *Canad. J. Math.* **18** (1966), 1031–1059.
263. B.M. Stewart, Supermagic complete graphs, *Canad. J. Math.* **19** (1967), 427–438.
264. I.W. Sudarsana, E.T. Baskoro, D. Ismaimuza and H. Assiyatun, Creating new super edge-magic total labelings from old ones, *J. Combin. Math. Combin. Comput.* **55** (2005), 83–90.
265. I.W. Sudarsana, E.T. Baskoro, S. Uttunggadewa and D. Ismaimuza, An expansion technique on super edge-magic total graphs, *Ars Combin.* **91** (2009), 231–241.

266. I.W. Sudarsana, E.T. Baskoro, S. Uttunggadewa and D. Ismaimuza, Expansion techniques on the super edge antimagic total graphs, *J. Combin. Math. Combin. Comput.* **71** (2009), 189–199.
267. I.W. Sudarsana, A. Hendra, Adiwijaya and D.Y. Setyawan, On super edge anti magic total labeling for t-joint copies of wheel, *Far East J. Math. Sciences* **69**(2) (2012), 275–283.
268. I.W. Sudarsana, D. Ismaimuza, E.T. Baskoro and H. Assiyatun, On super (a, d)-edge-antimagic total labeling of disconnected graphs, *J. Combin. Math. Combin. Comput.* **55** (2005), 149–158.
269. K.A. Sugeng and M. Miller, Relationship between adjacency matrices and super (a, d)-edge-antimagic total labeling of graphs, *J. Combin. Math. Combin. Comput.* **55** (2005), 71–82.
270. K.A. Sugeng, M. Miller and M. Bača, Super edge-antimagic total labelings, *Utilitas Math.* **71** (2006), 131–141.
271. K.A. Sugeng, M. Miller and M. Bača, Super antimagic total labeling of graphs, *Utilitas Math.* **76** (2008), 161–171.
272. K.A. Sugeng, M. Miller, Y. Lin and M. Bača, Super (a, d)-vertex-antimagic total labelings, *J. Combin. Math. Combin. Comput.* **55** (2005), 91–102.
273. K.A. Sugeng, M. Miller, Slamin and M. Bača, (a, d)-edge-antimagic total labelings of caterpillars, *Lecture Notes in Comput. Sci.* **3330** (2005), 169–180.
274. K.A. Sugeng and D.R. Silaban, Super (a, d)-vertex antimagic total labeling on a disjoint union of regular graphs, *J. Combin. Math. Combin. Comput.* **71** (2009), 217–225.
275. V. Swaminathan and P. Jeyanthi, Super vertex-magic labeling, *Indian J. Pure Appl. Math.* **34** (2003), 935–939.
276. M. Tezer and I. Cahit, A note on (a, d)-vertex antimagic total labeling of paths and cycles, *Utilitas Math.* **68** (2005), 217–221.
277. M. Trenkler, Some results on magic graphs, Graphs and Others Combinatorial Topics, *Proc. of the Third Czechoslovak Symposium on Graph Theory, Teubner-texte zur Mathematik - Band* **59**, Taubner Verlagsgellschaft, Leipzig, (1983), 328–332.
278. M. Trenkler, A construction of magic cubes, *The Math. Gazette* **84** (2000), 36–41.
279. M. Trenkler, Numbers of vertices and edges of magic graph, *Ars Combin.* **53** (2000), 93–96.
280. M. Trenkler, Magic p-dimensional cubes, *Acta Arith.* **96** (2001), 361–364.
281. M. Trenkler, Super-magic complete n-partite hypergraphs, *Graphs and Combin.* **17** (2001), 117–175.
282. M. Trenkler and V. Vetchý, Magic powers of graphs, *Math. Bohemica* **122** (1997), 121–124.
283. W. Trump, How many magic squares are there? http://www.trump.de/magic-squares/howmany.html.
284. M. Tsuchiya and K. Yukomura, Some families of edge-magic graphs, *Proc. of the Eight International Conference on Combinatorics, Graph Theory and Algorithms*, Kalamazoo, Michigan, **2** (1999), 817–822.
285. H. Ullah, G. Ali, M. Ali and A. Semaničová-Feňovčíková, On super (a, d)-edge-antimagic total labeling of special types of crown graphs, *J. Appl. Math.* **2013** (2013), no. 896815.
286. L. Valdés, Edge-magic K_p, Paper delivered at *Thirty-Second South-Eastern International Conference on Combinatorics, Graph Theory and Computing*, (Baton Rouge, 2001).
287. W.D. Wallis, *Magic Graphs*, Birkhäuser, Boston - Basel - Berlin, 2001.
288. W.D. Wallis, Vertex magic labelings of multiple graphs, *Congr. Numer.* **152** (2001), 81–83.
289. W.D. Wallis, E.T. Baskoro, M. Miller and Slamin, Edge-magic total labelings, *Austral. J. Combin.* **22** (2000), 177–190.
290. T.M. Wang, Toroidal grids are antimagic, *In: Computing and Combinatorics, Lecture Notes in Comput. Sci.* **3595** (2005), 671–679.
291. T.M. Wang and C.C. Hsiao, On antimagic labelling for graph products, *Discrete Math.* **308** (2008), 3624–3633.
292. J.G. Wang, D.J. Jin, X.G. Lu and D. Zhang, The gracefulness of a class of lobster trees, *Math. Comput. Model.* **20** (1994), 105–110.
293. M.E. Watkins, A theorem on Tait colorings with an application to the generalized Petersen graphs, *J. Combin. Theory* **6** (1969), 152–164.

294. X. Xirong, Y. Yuansheng, X. Yue, K.M.M. Haque and S. Lixin, Super edge-magic labelings of generalized Petersen graphs $P(n, 3)$, *Ars Combin.* **85** (2007), 19–31.
295. X. Xirong, Y. Yuansheng, X. Yue and L. Huijun, On (a, d)-antimagic labelings of generalized Petersen graphs $P(n, 3)$, *Ars Combin.* **86** (2008), 23–31.
296. X.R. Xu, J.M. Xu, M. Lu, Z. Baosheng and C. Nan, On (a, d)-antimagic labelings of generalized Petersen graphs, *Ars Combin.* **90** (2009), 411–421.
297. D. Ye, Z. Qi and H. Zhang, On k-resonant fullerene graphs, *SIAM J. Discrete Math.* **23**(2) (2009), 1023–1044.
298. D. Ye and H. Zhang, 2-extendability of toroidal polyhexes and Klein-bottle polyhexes, *Discrete Appl. Math.* **157**(2) (2009), 292–299.
299. D. Ye and H. Zhang, Extremal fullerene graphs with the maximum Clar number, *Discrete Appl. Math.* **157**(14) (2009), 3152–3173.
300. X. Yue, Y. Yuansheng, Mominul and W. Liping, Super vertex-magic total labelings of $W_{3,n}$, *Ars Combin.* **86** (2008), 121–128.

Index

M. Bača et al., *Magic and Antimagic Graphs*, Developments in Mathematics 60,
https://doi.org/10.1007/978-3-030-24582-5

The manufacturer's authorised representative in the EU is Springer Nature Customer Service Centre GmbH, Europaplatz 3, 69115 Heidelberg, Germany. If you have any concerns regarding our products, please contact ProductSafety@springernature.com

Printed and bound by CPI Group (UK) Ltd, Croydon, CR0 4YY
23/07/2026
02175013-0005